ARCHITECTURAL THINKING IN A CLIMATE EMERGENCY

AF833783

What is the role of architecture – and the architect – in the Anthropocene?

It's an immutable fact: human activity is driving the climate and biodiversity crises that now threaten all life. The damage we inflict on the planet undermines basic human rights, displaces millions, and intensifies structural racism, sexism and segregation, with the greatest burden falling on the most vulnerable people and ecosystems.

Architects must act.

The design and construction of most buildings and urban environments today are rooted in an economic model that pursues infinite growth. Our profession is embedded in a paradigm that favours individual gain over collective benefit. We are rewarded for overlooking the interdependence between people and the natural world. Architecture contributes directly to environmental degradation and social injustice.

Architects agree that the profession must change. But we lack the tools and knowledge to undertake the transformations that are urgently needed. This book aims to address that need through 15 chapters illustrating how we can act collectively to make a difference.

Architectural Thinking in a Climate Emergency brings together writers, researchers, educators, students and practitioners working at the forefront of this transformation. Contributors come from fourteen countries across Africa, Asia, Europe, North and South America and Oceania. Many are leading voices in their fields; others are emerging thinkers introducing fresh perspectives from both academia and practice.

Above all, they each affirm the architect's responsibility to help transition toward carbon-neutral, sustainable practices that advance social and environmental justice.

Sofie Pelsmakers (she/her) is Professor of Sustainable Housing Design at Tampere University, Finland. She is a chartered architect (UK), author of several books and has dedicated the past two decades to teaching, designing and researching sustainable architecture. She brings experience from Denmark (the Aarhus School of Architecture) and the UK (Sheffield School of Architecture, The Bartlett, University of East London). Her collaborative work connects research, teaching and practice for the training of the architects of the future.

James Benedict Brown is an Associate Professor of Architecture at Umeå University in Sweden. He teaches architectural history and theory and researches the relationship between architectural education, practice, values and materials.

ARCHITECTURAL THINKING IN A CLIMATE EMERGENCY

Edited by
Sofie Pelsmakers and James Benedict Brown

Routledge
Taylor & Francis Group
LONDON AND NEW YORK

Cover image: Essi Nisonen

First published 2026
by Routledge
4 Park Square, Milton Park, Abingdon, Oxon OX14 4RN

and by Routledge
605 Third Avenue, New York, NY 10158

Routledge is an imprint of the Taylor & Francis Group, an informa business

© 2026 selection and editorial matter, Sofie Pelsmakers and James Benedict Brown; individual chapters, the contributors

The right of James Benedict Brown and Sofie Pelsmakers to be identified as the authors of the editorial material, and of the authors for their individual chapters, has been asserted in accordance with sections 77 and 78 of the Copyright, Designs and Patents Act 1988.

The Open Access version of this book, available at www.taylorfrancis.com, has been made available under a Creative Commons Attribution-Non Commercial-No Derivatives (CC-BY-NC-ND) 4.0 International license.

Any third party material in this book is not included in the OA Creative Commons license, unless indicated otherwise in a credit line to the material. Please direct any permissions enquiries to the original rightsholder.

Trademark notice: Product or corporate names may be trademarks or registered trademarks, and are used only for identification and explanation without intent to infringe.

British Library Cataloguing-in-Publication Data
A catalogue record for this book is available from the British Library

Library of Congress Cataloging-in-Publication Data
Names: Brown, James Benedict editor | Pelsmakers, Sofie editor
Title: Architectural thinking in a climate emergency / edited by James Benedict Brown and Sofie Pelsmakers.
Description: Abingdon, Oxon ; New York, NY : Routledge, 2026. | Includes bibliographical references and index.
Identifiers: LCCN 2025018781 (print) | LCCN 2025018782 (ebook) | ISBN 9781032277523 hardback | ISBN 9781032277486 paperback | ISBN 9781003293903 ebook
Subjects: LCSH: Architecture—Philosophy | Architecture—Environmental aspects | Architecture and climate
Classification: LCC NA2500 .A7116 2026 (print) | LCC NA2500 (ebook) | DDC 720.1/03—dc23/eng/20250619
LC record available at https://lccn.loc.gov/2025018781
LC ebook record available at https://lccn.loc.gov/2025018782

ISBN: 9781032277523 (hbk)
ISBN: 9781032277486 (pbk)
ISBN: 9781003293903 (ebk)

DOI: 10.4324/9781003293903

Typeset in Sabon LT Pro
by codeMantra

This book is dedicated to future generations of human and non-human users of architecture and the built environment and its students, educators, researchers, policy-makers and practitioners. You are part of creating a better more equitable and fair world – both through your work and your actions as a citizen. Don't ever stop trying.

CONTENTS

ACKNOWLEDGEMENTS

This book has been in the making for several years. It was developed through discussions as co-editors, starting in 2019. At that time, Stephen Parnell was instrumental in shaping the embryonic ideas that form the background of this book. We cannot thank him enough for those early prompts and conversations.

Thank you to all the chapter authors for their patience during the book's review, editorial and production process. Further credit goes to each contributor who has studied their field of research and developed their work over years, sometimes decades. We thank them for sharing snippets of their wisdom in their chapters, without which this book would not exist.

During the book's writing and editing, our societies faced much adversity, such as a global pandemic, increased global warming extremes, world conflicts, trade wars between neighbouring communities and countries, democratic backsliding and increasingly polarising political language from elected (and unelected) representatives that has excluded and divided rather than uniting people against the threat of the climate emergency.

Press and academic freedoms, along with minority and gender rights, are under attack in our democratic societies. In a world that sometimes verges on the surreal, we thank all contributors for providing hope and inspiring each of us to understand how we can, through architectural thinking, help create a better, more sustainable world. Whether authors are located in or have heritage from Africa, Asia, Europe, North or South America, or Oceania, what unites us is the belief that we should care and stand up for those without a voice (whether human or non-human), as exemplified in various ways in each research chapter.

Special thanks are due to Essi Nisonen for the book cover design and Tampere University for funding the Open-Access e-book publication, which makes this book more inclusive and accessible globally, which is in line with its core values and message. At Routledge, thanks to Hannah Studd and Nick Craggs for their patience and trust in our project.

As two academics who landed in Scandinavia around the same time, we wanted to work together for some time. We are thankful that this book project was a great opportunity to do so and exchange ideas, perspectives and knowledge. This happened almost entirely online, but in later stages, also on the sofas and around the dinner tables of our respective homes in Finland and Sweden.

James would like to thank his colleagues at Umeå School of Architecture, especially those in the Architecture and Wood research group. To Amanda: thank you for your love, support and willingness to critique and improve everything that I do.

Sofie would like to thank her Tampere University colleagues and close peers, some of whom also became friends and collaborators around the world. My knowledge builds on those before me and on the invigorating conversations I have the privilege to be part of every day. It is all of you and the 'ASUTUT Society' and (past and present) students that continue to influence and shape and advance my thinking, which has also influenced the shaping and content of this book. All of you remind me every single day what an honour it is to work together, and how we collectively can make change. Finally, thank you to JH, for simply being amazing and for being in my life.

ABOUT THE AUTHORS

Mercia Abbott is a research investigator at NUWAO (Nature-based Urban design for Wellbeing and Adaptation in Oceania) and Wellington City Council, Aotearoa New Zealand.

Elina Alatalo is an architect and doctoral researcher in Environmental Policy of Tampere University, Finland. Her recent research has concentrated on new proactive forms of urban activism, getting vacant spaces back into use and developing sustainable urban neighbourhoods. She also teaches landscape architecture at Aalto University and coordinates interaction in the CO-CARBON research project that looks at urban nature. Elina is a co-founder of Insurgent Spatial Practices collective that explores the valuable knowledge that alternative cultures develop.

Felicity Atekpe is an architect, Associate Professor and Director of Practice and Design of Practice at UCL's Bartlett School of Architecture, London. Her academic interests encompass innovative pedagogies addressing the role of education, ethics and alternative qualification routes. Her research explores interdisciplinary approaches to equitable urban landscapes examining how architects can adapt their professional practices conceptually, ethically and practically. She also writes about domesticity, belonging, language, anthropology and The Everyday. She provides academic design practice leadership for programmes across The Bartlett, emphasising professionalism, ethics, competency and architecture's social purpose. Atekpe's work contributes to bridging the gap between theory and practice, preparing students for the complex and dynamic nature of the architectural profession and its related disciplines beyond the conventional notions of practice. Her approach contributes to educating architects who are well-equipped to address global challenges and create more inclusive and sustainable built environments.

Anika van Aswegen is a research associate in the Department of Architecture at the University of Pretoria, South Africa. She is a former lecturer and postgraduate coordinator for the Interior Architecture programmes at the same institution. Her research expertise lies at the intersection of transformative learning in the spatial design studio, empathy in design and arts-based research methods. Her research and design values are grounded in bottom-up approaches with reciprocal collaboration, co-creation and shared authorship. Her research has shown that transformative learning is possible as a slow process of discovery of self and the awareness of and a respect for multiple perspectives, resulting in deep epistemological shifts in students. She currently lives in a small town in South Africa's dry country, the Karoo. Here she works as a glass artist and explores creative art and writing projects for children and the youth, outside the conventional frameworks of academia. She plans to expand this into projects with local collaborators and co-creators.

Dalia Milián Bernal is a doctoral researcher, teacher and project manager at the School of Architecture at Tampere University in Tampere, Finland. Her work lies at the intersection of architecture, urban planning and critical urban studies. Her research examines how individuals transform their living environments through grassroots and radical practices and seeks in these processes clues for the construction of a more just, caring and sustainable world.

James Benedict Brown is an Associate Professor of Architecture at Umeå University in Sweden. He teaches architectural history and theory and researches the relationship between architectural education, practice, values and materials.

Carin Combrinck is an Associate Professor and Head of the Department of Architecture in the Faculty of Engineering, Built Environment and Information Technology at the University of Pretoria, where she also serves as the director of the Unit for Urban Citizenship. She is responsible for several courses that collectively contribute to the teaching of Urban Citizenship within the UP Department of Architecture. Her work has been published in peer-reviewed journals and scholarly books and she has contributed to international conferences related to her teaching and research. Through her collaboration with Chalmers University of Technology in Gothenburg, Sweden, she has contributed towards the development of the ukuDoba methodology, aimed at fine-grained geo-spatial data capture in informal settlements, and co-ordinated the collaboration between the Urban Citizen Studios and Reality Studio. Through this collaboration, significant shifts in the design studio could be achieved towards the role of architects in empowering communities in marginal urban settings. Carin has been acknowledged for her work in communities by way of the UP Community Engagement Award in 2019 as well as the UP Laureate Award for Teaching and Learning in 2023.

Elizabeth Donovan is associate professor at Aarhus School of Architecture in Aarhus, Denmark. She is a researcher and educator with a focus on how to embed a holistic approach to sustainable architecture in both learning and practice.

Ng'eno Chelang'at Faith is a registered Kenyan architect and educator who is currently a lecturer at the University of Sheffield in the United Kingdom. She leads the Msc. Sustainable Architecture Studies Post Graduate Taught Program and teaches Environment and Technology. Through the lens of decoloniality, her research challenges Eurocentric sustainable design approaches and instead centres on local knowledge, cultural practices, and the specific needs of communities particularly in the Global South.

Lauren Ferguson is a postdoctoral scholar at Harvard TH Chan School of Public Health. She holds a PhD in Environmental Engineering from University College London (UCL). Her work focuses on quantifying population exposure to indoor environmental hazards, using building physics and statistical models.

Susan Galavan is a lecturer in Architecture at the Atlantic Technological University, Sligo, Ireland. She is a fully qualified architect with over 14 years in practice in Ireland, Italy and Germany. Her research draws on this experience to focus on nineteenth-century urbanism in Europe, with a special interest in housing. She holds a first-class Masters' degree from University College Dublin and a PhD in architectural history from Trinity College Dublin. Her book *Dublin's Bourgeois Homes: building the Victorian suburbs 1850–1901* (2017), was the first in-depth analysis of Dublin's Victorian housing. Between 2017 and 2022, she held academic positions in Belgium: she was a Marie Sklodowska-Curie Postdoctoral Fellow at the KULeuven/UAntwerp and lectured at the University of Liège. She also worked with *e2Arc-Architects for Future Cities* on a collaborative European-funded project to re-nature six pilot cities across Europe. She is currently interested in the links between sustainability and historic built heritage, and the role of the architectural humanities in facing the challenges of the climate crisis.

Jenny Hughes is an Associate Director and environmental sustainability specialist at Nature Positive, UK, with over 11 years of experience advising businesses and organisations on sustainability, nature and climate change. She has experience in delivering environmental studies for large-scale nature restoration projects in the UK and is currently working with businesses to support them to understand their relationship with nature and manage impacts, risks and opportunities. She is experienced in developing sustainability frameworks, conducting double materiality assessments and developing corporate nature strategies.

Jan Hugo is a registered architect and senior lecturer in the Architecture Department at the University of Pretoria. He completed his PhD focusing on the climate change adaptation potential of building-integrated agriculture. His current research focus is on the role of the built environment in developing climate change resilient cities, specifically focused on heat health solutions. Jan Hugo currently coordinates the Professional Masters in Architecture programme at UP and participates in the Regenerative and Resilient cities research group.

Salla Jokela is University Lecturer at the Faculty of Social Sciences at Tampere University. She holds a PhD in human geography from the University of Helsinki. Her research focuses on social inclusion effects of waterfront developments, city branding, visual culture and identity politics, urban discourses, and social impact of tourism in cities.

Rebecca Kiddle is a Kāhui (Advisory Board) member at NUWAO (Nature-based Urban design for Wellbeing and Adaptation in Oceania) and Director of Te Manawahoukura,

Te Wānanga o Aotearoa, Aotearoa New Zealand. She has a background in the intersection between Indigenous knowledge and urbanism.

David Kroll is a Senior Lecturer in Architecture at the University of Adelaide in Australia. An architect, educator and researcher, he is dedicated to promoting architecture that enhances people's lives. His research interests include housing, sustainable design and built heritage, drawing from his extensive background in architecture, applied design research and architectural history and theory. Before joining the University of Adelaide in 2020, he held academic positions at the University of Sydney, University of South Australia, Anglia Ruskin University, University College London, University of Kent, and the University of East London. David's professional experience includes working with CF Møller Architects in London on projects such as the Athletes' Village and the Darwin Centre Two at the Natural History Museum. He earned his PhD from the University of London, focusing on London housing history, a thesis that was shortlisted for the RIBA President's Medal for Research in 2015 and led to the book *Mobilising Housing Histories: Learning from London's Past*, co-edited with Peter Guillery (UCL). He has published widely and continues to influence the field through his teaching, research and commitment to sustainable architectural practices.

Mikko Kyrönviita is a researcher at the Faculty of Social Sciences, and doctoral researcher in Environmental Policy at the Faculty of Management and Business, Tampere University, Finland. His research interests are in urban activism, everyday politics, skateboarding and DIY urbanism. He is co-founder of the Insurgent Spatial Practices research collective that aims to envision sustainable urban futures through more political, democratic and creative spatial practices.

Markus Laine is a University Lecturer at the Faculty of Management and Business at Tampere University. He holds a PhD in Administrative Sciences from Tampere University. His research focuses on sustainable urban development and planning, policy studies, persuasive (visual) storytelling, urban discourses, and the social impact of urban development processes.

Sanna Lehtinen is a Senior University Lecturer at Aalto University's School of Arts, Design and Architecture in Espoo, Finland. Her research focuses on philosophical and applied environmental and urban aesthetics. Sanna's other professional interests include currently also everyday aesthetics, philosophy of technology and the study of contemporary forms of environmental and urban art. She has a special interest in finding out how philosophical and applied aesthetics can support human wellbeing, environmental sustainability, and the development of different design practices. Sanna has published over 30 peer-reviewed articles and book chapters and is the Co-Founder and Editor-in-Chief of *Philosophy of the City Journal* as well as the Assistant Editor of *Contemporary Aesthetics Journal*.

Antti Lehto is Assistant Professor at Aalto University, where he holds the chair of Housing Design. His current research focuses on circular construction, mainly through constructed pilot projects. In addition to research and teaching in academia, he's currently designing several high-profile projects in Finland as a founding partner at Helsinki-based architectural design practice INARO.

Pieta McAleer-Harding is a research investigator at NUWAO (Nature-based Urban design for Wellbeing and Adaptation in Oceania).

Cara Mulholland is a Lecturer in Project Management in Manchester Metropolitan University, Manchester, UK. She is a social value specialist who promotes new, sustainable approaches to project management in academic, industrial and policy-based settings. She believes in a holistic approach to achieving the best overall value from complex projects based on her experience on megaprojects, nationally significant infrastructure, digitalisation of established methods, and trialling approaches to strategic change management in multinational organisations. Her research interests lie in the intersection of multidisciplinary topics that explore triple bottom-line sustainability in technical projects and organisations.

Mirjami Myllymäki is an architect currently based in Helsinki, Finland. She completed her master's degree in sustainable architecture at Tampere University (Tampere, Finland) with a design thesis (featured in Chapter 4), that explores the principles of user-oriented design, circular economy and empathy towards all living beings.

Essi Nisonen is an architect, doctoral researcher and the responsible teacher in sustainable housing design at Tampere University (Tampere, Finland), working in the ASUTUT Sustainable Housing Design Research Group. She has worked in research projects related both to sustainable housing design and sustainable architecture education. Her dissertation addresses the professional values and cultures transmitted in architecture education and their connection to architects' ability to work within planetary boundaries and to promote eco-social justice. She is particularly interested in exploring, defining and imagining paths towards resilient, restorative and democratic cultures of learning and practicing architecture in the climate emergency.

Maibritt Pedersen Zari is an Associate Professor at Huri Te Ao Hoahoanga School of Future Environments, Te Wānanga Aronui O Tāmaki Makau Rau Auckland University of Technology, Aotearoa New Zealand, and is the lead of NUWAO (Nature-based Urban design for Wellbeing and Adaptation in Oceania), and the Regenerative Ecologies Design Lab (REDLab). Her research into regenerative design redefines sustainable architecture and urban design through emulating ecosystems, working with ecologies and nature, and integrating complex social factors into architectural and urban design. Her current work focus is on Moananui Oceania. Pedersen Zari is the co-author/editor of *Ecologies Design: Transforming Architecture, Landscape, and Urbanism* (2020), author of *Regenerative Urban Design and Ecosystem Biomimicry* (2018), and co-author of *Materials for a Healthy, Ecological and Sustainable Built Environment* (2017).

Sofie Pelsmakers (she/her) is Professor of sustainable housing design at Tampere University, Finland. She is a chartered architect (UK), author of several books and has dedicated the past two decades to teaching, designing and researching sustainable architecture. She brings experience from Denmark (the Aarhus School of Architecture) and the UK (Sheffield School of Architecture, The Bartlett, University of East London). Her collaborative work connects research, teaching and practice for the training of the architects of the future.

Chrisna du Plessis is former Chair of the School for the Built Environment and now Research Professor in the Department of Architecture at the University of Pretoria. Her interest is in exploring ways of thinking, doing and being that will lead towards a regenerative future. She has taught internationally at postgraduate level at a number of leading universities and teaches a foundational course in ecosystemic thinking for spatial designers at the University of Pretoria. Her research concentrates on developing new ways of thinking about the problems of the global holocrisis to expand our understanding of concepts such as sustainability, resilience and regenerative development, how they are related, and how they influence the way we think about, and the approaches we use, to create cities, neighbourhoods and buildings. She co-authored the award-winning *Designing for Hope: Pathways to Regenerative Sustainability* (2014) that outlined the principles of regenerative thinking and how these influence design practice.

Jenni Poutanen works as the Head of Degree Programme and University Lecturer at the Tampere University, Finland. Her research concerns user-centred architecture and the effects of societal changes on learning and knowledge work environments, leisure services and innovation centres. She is particularly interested in the transformation of existing facilities, sharing of spaces, adaptability, effectiveness and spatial interventions. Currently, she is leading a multi-disciplinary research project that examines knowledge work environments supporting well-being and productivity. Additionally, she has been involved in campus development at Tampere University. Prior to joining academia, she worked as a designer for several years.

Flora Samuel is a Professor of Architecture and Head of Department at the University of Cambridge. Her strength is in connecting ideas and people from different backgrounds and fields – the scholarship of integration and application, making connections across disciplines and sectors, contextualising specialisms in the larger context and making sense of them to non-specialists. Her current work focuses on inclusion through planning policy and processes evidenced through digital map making with communities and others. Flora is a Trustee of the Quality of Life Foundation and on the Board of the Taskforce for Planning.

Dayle Shand is a lecturer at the University of Pretoria, in the Department of Architecture. She is a landscape architect by training, but teaches across the disciplines of architecture, landscape architecture and interior architecture, in the Earth Studies module, and runs various design studios. She supervises students in the design and research streams. Dayle is also professionally registered with the South African Council for the Landscape Architectural Profession (SACLAP) and gained practical experience working for well-known South African landscape architectural firms on various types of public open space projects, including local community parks. Dayle completed her PhD thesis (2023) on the topic of 'Nature-based park making', wherein she considered environmental justice related to local community parks in South Africa. Her research activities have expanded to place-based landscape approaches, environmental justice in South African cities, designed ecologies and local community knowledge informants for design.

Veronica Soebarto is a Professor at The University of Adelaide, School of Architecture and Civil Engineering in Adelaide, South Australia, Australia. She has been teaching architecture students since 1988, first at University of Indonesia (1988–1990), then at the University of Adelaide (1998–now). She obtained her Master of Architecture in 1992 from Texas A&M University and PhD from the same university in 1996. In addition to teaching and research, she currently leads the Green Urban Futures initiative of the Environment Institute and the Sustainable Futures theme of the Adelaide-Nottingham Alliance.

Jonathon Taylor is an Associate Professor in Urban Physics at the Faculty of the Built Environment, Tampere University, Finland. His research interests are in healthy and energy-efficient housing, combining building physics and Geographic Information Systems to evaluate population-level exposure to various indoor environmental health hazards. He currently studies indoor air quality, disease transmission risks, indoor heat and cold exposures, and moisture in the built environment; climate change and housing adaptation; and informal housing and household air pollution in the global south.

Veera Turku is a doctoral candidate in The Department of Geography, Planning and Environment at Concordia University in Montréal, Canada. Her research focuses on innovative post-collaborative planning practices, emphasizing citizen agency in socio-ecological transitions and the strategic reuse of vacant and abandoned spaces. In her PhD research, Turku examines the redevelopment of former industrial sites in Montréal, analysing the pivotal roles of civil society in these transformation processes and how various stakeholders shape redevelopment efforts towards sustainable development. Additionally, she is also part of a research group called *Observatoire des milieux de vie urbains* at The Université du Québec à Montréal (UQAM).

Renata Tyszczuk is Professor of Architecture and Climate, and Chair in Architectural Humanities at the School of Architecture, University of Sheffield, UK. She is an academic and artist whose work explores questions concerning global environmental change, provisionality, climate futures and scenario making. She has been awarded a British Academy Mid-Career Fellowship (2013–2014) and a Leverhulme Trust Major Research Fellowship (2019–2022) for research in these areas. Her book *Provisional Cities: Cautionary Tales for the Anthropocene* (2018) explored the condition of the Anthropocene through critical storytelling. Renata convenes Culture and Climate Change, a framework for a series of projects and initiatives that support collaborative and creative responses to climate change.

Laura Uimonen is a University Instructor and postdoctoral researcher at Tampere University in Tampere Finland. Cities as part of nature is her main value with a strong effort plan for co-living in shared environments. Ecological and cultural values of a place are vital for the radical change needed. Every site plays a part in the valuable living ecosystem connected to the neighbourhood, part of town, the whole city, the entire region and beyond to seas and oceans, up to air and deep down into the soil. Her research focuses on biodiversity and urban design in research projects Biodiversity interventions on wellbeing (BIWE) and Enabling multispecies transitions (MUST). In urban planning and design and landscape planning teaching she addresses the importance of a reflective architect in multidisciplinary co-development for a truly sustainable way of life. Uimonen has experience with perspectives on critical environmental challenges from field biology, urban ecology, environmental education and urban planning and design. Uimonen is an architect, a Doctor of Arts and has an education in biological nature inventories. She has researched railway stations and railway parks as public spaces and public art processes in urban planning.

Maria Vidali (she/they) is an architect and she is currently teaching urban design at the CYA/DIKEMES, a study abroad programme in Athens, Greece. From 2020 to 2023, she had been teaching architecture at the University of Thessaly school of architecture and landscape architecture at Ioannina University school of architecture. Her research work: Village and Land, The outlying chapels of the island of Tinos was published as a book in Greece, in 2009 by Futura Publications & Tinos Municipal Enterprise. Her interests expand on various topics based on phenomenology and hermeneutics. Her articles and published work exist in several foreign and Greek scientific journals mainly on the subject of architecture, vernacular architecture and the farming landscape, narrative, metaphor and fiction as a methodology/tool in architectural education.

1

SETTING THE SCENE

Architectural thinking in a time of climate emergency

James Benedict Brown, Sofie Pelsmakers

A statement of the problem

We pollute the water we drink, the air we breathe, and the land we inhabit. We extract finite resources to create cities and spaces that require vast amounts of fossil fuels to operate. We exploit the environment and its non-human inhabitants, encroaching on their natural habitats (IPCC, 2018, 2022). The Anthropocene is now undeniable: the climate and biodiversity crises are direct consequences of our human activities, including those within our field, and our actions threaten everyone's way of life and basic human rights by damaging the very planet that sustains us (IPCC, 2018, 2022). Furthermore, these exploitative practices reinforce structural and systemic racism, sexism, segregation, and discrimination (Musterd, 2020; Pedraza, 2019).

Almost every building we create contributes to this pollution and exploitation of the environment. Buildings are part of an economic and societal model that promotes infinite growth and rewards individual gains over collective societal benefits, ignoring their mutual interdependence. Buildings are the most significant contributor to global heating in most European countries (Watts, 2018). The construction and use of buildings are responsible for around 36% of the European Union's CO_2 emissions (European Commission, 2019). It is now indisputable that we must stop burning fossil fuels and transition to a carbon-neutral society to prevent irreversible climatic changes. Many architects and architecture academics now agree that architecture can no longer continue to contribute to these problems. Systemic and transformative change is needed (see, for example, Architects Declare and the Architects Climate Action Network, ACAN). This includes reframing the meaning and purpose of architecture and the identity and value systems of architects (see e.g., Nisonen, 2022; Till, 2020). Despite the best efforts of many, we seem to lack a collective will (see e.g., Pelsmakers & Newman, 2021; Till, 2020). Change is slow and difficult to foster when day-to-day business continues to profit from a harmful status quo.

Witness how, in recent years, certain global architecture practices 'un-declared' their previous support for Architects Declare's co-written commitment to tackle the planetary

DOI: 10.4324/9781003293903-1
This chapter has been made available under a CC-BY-NC-ND 4.0 license.

emergency (Ing, 2020a, 2020b). The critique of new airports and the impact of flying proved to be the breaking point for the engagement of certain well-known architecture practices, highlighting that we must not address our associated organisational practices in creating architecture. But this is clearly not enough; we must also address the immense carbon footprint of building energy use and that of quotidian materials like bricks, concrete, and steel and the damaging effects of cultural and political practices that valorise new buildings over renovation and refurbishment or not building at all. The architecture we create is also typically without consideration of how a diversity of all species (human and non-human) might inhabit these spaces, leading to significant implications for societal and ecological justice.

We must overcome any false dichotomy: we must address all of these harmful practices by radically changing our interaction with the world and others around us. What does this mean for how architects interact with the world? How, as architects, can we influence this change? What can architects contribute to the necessarily radical societal shift? What systemic change must architects undertake to contribute to this societal transformation? Because 'if architects are unable to make changes within the system, they need to change the system' (Pelsmakers & Newman, 2021).

In pursuit of answers to these questions, this book was conceived: to critically engage with architects and other built environment actors on how to consider these questions, the opportunities and pitfalls of our current values and architectural thinking, and collect possible strategies and ways forward for change.

Defining architectural thinking

In *The Thinking Hand* Juhani Pallasmaa writes that 'architecture requires a special category of thinking, an embodied thinking through the very medium of architecture itself' (Pallasmaa, 2009, p. 148). Pallasmaa argues that architectural thinking is always *embodied* – where thoughts are shaped and explored through tangible, physical, hands-on making, exploration, and reflection. Thinking-by-doing and designing-by-making are common buzzwords used to describe curricula and even the pedagogical approach of some architecture schools.

If there is general agreement that drawing and making are two important and embodied methods of architectural thinking, what thought processes are going on as our hands move the pencil or shape the model? While design thinking is, on the whole, a divergent task (i.e., developing in different directions) it is often made up of many convergent design tasks (i.e., bringing things together) (Lawson, 2006). This means that the designer must be capable of flipping between and working concurrently in two modes of thought that have often been dismissed as oppositional (Deutsch, 2020). This iterative and multi-modal lateral thinking capacity is both essential to and characteristic of architectural thinking (see also Cross, 2011), and leads to critical modes of thinking that go far beyond simple problem-solving (Deutsch, 2020).

Architectural thinking is a variant of design thinking (Florida, 2002) and has transitioned from just design to organisational change and management, emphasising user experience and involvement. Yet, 'for all the claims about being "user-centred," [descriptions of design thinking] still emphasize the designer as the main agent within design' (Kimbell, 2011, p. 289). Design thinking is considered to be about '*how things ought to be*', in contrast to scientific thinking, which focuses on '*how things are*' (Simon, 1996, p. 114). Architectural thinking, though – by definition – distinguishes itself from design thinking in its attachment to the much older discipline of architecture and all the historical, cultural, and professional baggage it brings with it. It is reinforced from within the inter-subjective community of the architectural profession, and it typically reinforces 'conformity with theoretical prescriptions of what constitutes "proper" architecture and "good" design' (Rowe, 1987). In other words, the 'how things ought to be' is fundamentally tainted by inter-subjective normative theoretical positions, manifestos, movements, or tendencies regarding the question 'what constitutes proper architecture?' (ibid., p. 115). These inter-subjective 'guiding principles' have an 'impact on the design process [that] may be quite considerable' (Lawson, 2006, p. 160), and they have generally ignored the climate emergency or broader impacts of our profession and our own working methods (e.g., Till, 2020; Pelsmakers & Newman, 2021). This is reflected by the architectural profession embracing an outwardly efficient but inwardly exploitative and extractive mode of thinking about design, and in undertaking design. For example, the adoption of the competitive model of the *charrette* engenders a competitive and exploitative attitude to practice, one in which we compete for work in a free-market economy that reduces financial outlay for clients and which allows 'competitions [to become] an opportunity for exploitation by unscrupulous clients and a major source of unethical behaviour on the part of some architects' (Fisher, 2000, p. 136).

The dominant culture in which architectural thinking operates is competitive rather than collaborative. Construction has become so complex that what was once the domain of the architect is now the joint responsibility of a multitude of consultants and professions, a process identified as a loss in 'directional agency' (Fry, 2007, p. 7). There persists a perverse worship of architecture always being 'building': building something new, and the architect as the sole author of the enduring architectural masterpiece. While many architects lament a loss of respect and control over the process, this tendency also hinders true collaboration and a humble acknowledgement of our own professional limitations. Truly sustainable architecture is incompatible with this outdated and isolationist view of architectural practice. Architects cannot solve societal, ecological, and climate problems alone, and in fact these attitudes perpetuate the exploitative status quo.

A collaborative effort that challenges these attitudes, which are already entrenched from architectural education through to practice, would emphasise learning from specialist knowledge, but also localised expertise from city planning, landscaping, ecology, building performance, energy use, service engineering, and users and community-based experts, among many others. Architects often approach interdisciplinary collaboration in a peculiarly reductive way: protective of their own mastery without extending the courtesy to others' expertise. This attitude is hugely damaging to our reputation (claiming to be an expert in areas where we are not) and hinders our ability to create truly sustainable buildings.

The problems with architectural thinking

Design thinking is characterised by the principle of user-centred design (Norman, 1988), a design philosophy that emphasises empathy for the (human) user throughout the design process. A more sustaining design philosophy would also put the natural environment and non-human users at the centre of designers' focus from the very beginning of their education. This can be framed as a challenge for a 'redirective practice' whose 'ambition [...] is to bring about the ontological designing of an other habitus (cum ethos) that can advance, rather than undermine, sustainment' (Fry, 2007, p. 8).

Architectural thinking requires a space and a scale beyond other design disciplines, intrinsically entangling humans, nature, and the artificial within an environment – from 1:1 to 1:50,000 and more. Yet at present, despite the potential of architectural thinking to create a world of 'how things out to be', we have collectively failed to use architectural thinking to address the climate and biodiversity crises. In fact, architects are entirely complicit in ignoring these issues, and 'Architecture, as presently taught and practised, is a pervasive form of climate change denial' (Moe, 2020). Architects' design decisions impact not only the immediate environment, but also people and habitats thousands of miles away, and for decades to come. Such decisions might involve, for example, materials harvested or mined and shipped halfway across the world to achieve an idealised aesthetic or the amount of embodied carbon expended in the building's fabric.

With the scientific evidence and understanding of the climate crisis now irrefutable (IPCC, 2018, 2022), the fallibility of this kind of architectural thinking is exposed. Consider the power with which attention has been turned towards the climate crisis by the Swedish high school student Greta Thunberg, who has called out adults for their inaction. Her exposure of political and professional inaction inverts our inherited belief in the moral and professional superiority of adults over young people. Thunberg is neither an expert nor a professional but a global citizen driven not by self-interest and unafraid to name the elephant in the room. Architects, who frequently leave the responsibility of being global citizens at the studio door, could learn a great deal from Thunberg.

If architects are to make a positive contribution to addressing the climate crisis, they must redress not only their abandonment of global citizenship, but also begin to recognise that architectural thinking is neither perfect nor the exclusive domain of architects. While we need to become more user-centric and include the perspectives of the people we design for, simultaneously we must also elevate our relationship with the natural world and other species, as part of a radical inclusivity approach (e.g., Nawratek et al., 2015). It is about acknowledging that humans too are a constituent part of the environment and its ecology, i.e., a trans-species perspective (Martin, 2020). A broader, more inclusive definition of architectural thinking would neither be the exclusive domain of architects, nor of an architecture produced by predominantly Western, white, male architects. We need to decolonise the means – and the meaning – of architectural thinking: '[i]f the architect is not social, then it's [*sic*] not an architect' (Kundoo, 2020), a recurring theme in several book chapters. We need to be citizens acting with and for other citizens and the non-human with care, empathy, and solidarity.

On thinking and action

Despite the efforts of many, architectural education and media still broadly focus on a non-diverse architectural mode of production, perpetuating the same kind of thinking and practices that caused the current problems to begin with. The climate emergency demands that we re-imagine what architecture is and what it means to be an architect in a world in crisis; this is the key focus in Part 1 of the book. What will this architecture be, if built at all, and how will our design processes and practices need to shift? (the focus of Part 3). And what does this mean for our educational values, systems, and attitudes? (the focus of Part 2). We need to overcome outmoded models of professionalised knowledge, models which we, as teachers, recognise are deeply ingrained in architectural education. Architectural thinking has been theorised defensively and largely as a problem-oriented process. Much of what has been written about the way that architects think is written in a way that is flattering to the discipline (Stevens, 2002). An example is that of *The Reflective Practitioner* (Schön, 1983), where a teacher guides a student through a design problem, and has since served to simplify a complex theory of design education and practice. This has largely been absorbed uncritically into the culture of the profession, and the master-pupil hierarchy has remained as a way of teaching. And while the studio unit system has brought tremendous diversity to schools of architecture, it still engenders silos of architectural thinking in which students seek the approval of just one or two tutors and their carefully invited critics. As a result, a year spent in one unit is often a year spent acquiring the singular architectural worldview of those few tutors. Meanwhile, the weight of professional accreditation limits the ability of educators to write sustained collaborative and interdisciplinary courses. (Several chapters in Part 2 of this book offer ideas and examples to illustrate how it can be done differently.)

Why this extensive consideration of the faults of our systems of education? We recognise the need to bridge the different cultures of architectural education and practice. At their best, schools of architecture provide students with opportunities for wide-ranging explorations of the many qualitative aspects of designing spaces and places. And yet when opportunities to look beyond our own discipline present themselves, students often take no more than a superficial dip into the ideas of other disciplines – just long enough to acquire buzzwords about rhizomes or mycelium or topics related to sustainability which, without deeper understanding, can be used quickly and uncritically in the service of speculative design proposals. Later, in practice, early career architects often struggle with the cultural shift to an environment in which sustainability is engaged with not speculatively but often through technical standards. We have observed how architects also tend to reinvent the wheel without deeper learning or involvement of experts from the traditions, expertise or knowledge of other fields – a habit they picked up in architecture education.

The severity and urgency of the climate and biodiversity emergency now render thinking alone insufficient. Now, more than ever, we need action. There are different ways of practising architecture, but it requires architects to reframe their standing in the world, revoking traditional hierarchies and traditional ways of practising and teaching architecture (Awan, Schneider, & Till, 2011).

In the quest to reduce carbon emissions, a regulatory approach to sustainability has unintentionally but perversely limited the entire ambition to the reduction of calculable carbon emissions, a goal that is generally achieved with technical solutions. This regulatory focus on carbon emissions alone creates a sustainability paradox that can deliver objectively carbon-neutral buildings that are not very sustainable when considered from a more holistic perspective. For example, the qualitative aspects of the life inside a carbon-neutral building are typically ignored, as are the impacts of our activities and interventions on the natural world and other species.

Equally, we risk dismissing issues of material and energy scarcity simply as concerns of the increasingly emollient category of sustainability instead of giving them the urgency they really demand (Goodbun et al., 2012). The climate emergency requires an expanded kind of architectural thinking that is quantitative in understanding and qualitative in proposition. Since people's behaviour and use of a building's spaces affect energy use, ignoring the inter-relationship between people, energy use, and carbon emissions means that the carbon savings seen on paper are unlikely to be achieved in reality – the so-called 'performance gap' (Kimpian et al., 2021). In addition, placing higher value on the quantitative at the expense of the qualitative means that we inherently de-prioritise non-technical solutions and softer values such as high-quality spaces, biodiversification, health and well-being etc. In the words of one architect, we must highlight the importance of both qualitative and quantitative aspects: 'We take it for granted that we are going to be designing delightful spaces [...] but [if] you have kids sitting in a beautiful classroom, [and] they are overheating or the air quality is bad, it is not very clever really' (Pelsmakers & Hilberth, 2019). We all want our own children to attend the school that is described here, yet in the quest to reduce carbon emissions we risk creating joyless spaces in which we merely survive rather than thrive, or when only focusing on beautiful spaces (as often architects tend to do), they may not perform optimally and affect well-being. As Lance Hosey (2012) writes, 'Imagine a day when we've perfectly solved the challenges of energy, resources, and emissions, and everything we do and make is clean, harmless, and infinitely renewable. Is that enough?' This is why many – perhaps most – so-called sustainable buildings do not deserve the title. There is a risk that low-carbon buildings will not only fail to meet their high technical standards when user factors are ignored but also offer little else when disregarding our connection to the environment and other species.

Architects' role as change-makers

What can architects do to renew architectural thinking about the holistic nature of environmental problems? There are precedents. Feilden Clegg Bradley with Bioregional and developer Crest Nicholson used the ten 'one planet living' principles in One Brighton as a design framework to look beyond quantifiable parameters (Thorpe, 2015). More recently, besides leading the way in thinking about design for disassembly and circular design and construction methodologies, Danish architect Anders Lendager, with NREP, Årstiderne Arkitekter, MOE, and Arup have been creating a holistic approach based on the UN's seventeen Sustainable Development Goals (SDGs) to support not only sustainable buildings, but sustainable lifestyles in the UN17 Village. Here, the whole building lifecycle, community building, and biodiversity principles are placed at the centre, including a restorative landscape that enhances habitats for rare local wildlife and increased biodiversity. This is also at the heart of the three principles of the Living Building Challenge: regenerative buildings that connect occupants to light, air, food, nature, and community, self-sufficient buildings that remain within the resource limits of their site, and buildings that create a positive impact on the human and natural systems that interact with them (Living Building Challenge, 2020). This approach to design can create restorative interventions that not only minimise and neutralise impacts but create a positive impact.

Two British practices, Architype and Feilden Clegg Bradley, have demonstrated how to develop and sustain a commercially successful architectural practice that addresses climatic and energy concerns while also making a positive impact on the communities that use the buildings and the surrounding environment. They have separately developed software packages which can provide an estimate of the embodied carbon impact of a building. To recall our earlier observations about the importance of collaborative practices at the earliest stages of the design process, there are also ways in which architects can facilitate learning from a broad pool of expertise. We suggest, as one example, Kieran Timberlake Architects, where collaborative practices have been built into the design process with the appointment of a dedicated Head of Research who can connect in-house building and material scientists, landscape architects and ecologists to create and test sustainable design solutions that could never be accomplished without them.

Pioneers of material and ecological sustainability also tend to advocate for social sustainability. Helen & Hard in Norway have demonstrated how to develop a form of 'relational' architectural practice defined by an explicitly holistic approach that can shape the culture and methodologies of a commercial office (Stangeland & Hensel, 2019). Al Borde in Ecuador and Communal in Mexico are speculating about how architects and practices can 'do things with less', turning material and economic scarcity into an aesthetic and socially empowering asset. All around us are precedents of well-established practices that regard social and productive sustainability as two sides of the same coin. Cullinan Studio was established as a cooperative, and today it is a thriving employee-owned practice organised around principles of ecology, sustainability, and social responsiveness. A similar employee-run trust owns Collective Architecture in Glasgow, taking collective decisions about practice direction and management. Here, on one of our doorsteps in northern Sweden, Nordmark & Nordmark have made headlines by paying all employees – regardless of age or experience – the same salary (Ternby, 2023). NonCONFORM is a decentralised office spread across different cities in Austria and Germany that uses collaborative *ideenwerkstatt* workshops to create future change scenarios with local citizens and actors. Berlin-based raumlabor, a collective of architects, uses research-based design and collaboration with local stakeholders to create experimental responses to complex urban circumstances (Kolkwitz & Luotonen, 2021). Anupama Kundoo, who we have already quoted in this chapter, sees architects needing to be connected to their surrounding society and uses local materials and craftsmanship (Kundoo, 2020). Related to the origin of raw materials and energy, there is also much to be learned from the creativity of architectural thinking in critical contexts, where conflict, poverty or material scarcity demand lateral solutions (e.g., Zimmermann & Rodríguez Izquierdo, 2021). Architect Diébédo Francis Kéré has designed not only buildings but also participatory processes that engage entire communities in the construction of buildings made from locally available materials in different countries in Africa. Kéré's projects are testament to architectural design being generated from the context, rooted in locality and local traditions and culture and people. Where architects are working to overcome the cultural and material legacy of colonialism, architects such as Mariam Kamara of Atelier Masomi in Niger work with the environment to create 'design that sustains people'. This includes buildings such as the Niamey Cultural Center that address a concern not only for the climate but also concern for human rights and local culture. These projects are all testament to the transformative power of changing empathetic value systems, but individually and collectively we must do more.

An inherent characteristic of sustainable architecture is that one solution does not fit all. Each context is unique, and our interventions must respond to them with place-based approaches. Architects balance many duties and competing interests, and with increased regulation and decreased budgets comes increased pressure. If the architect is not one of the most important advocates for the overall enhancement of our environment and people's well-being, then what is their role?

The climate emergency demands a more dynamic mode of architectural thinking that does not solely depend on a library of solutions that have worked before. A broader definition of architectural thinking that includes lateral, critical, multimodal, divergent, and convergent thinking, would be well placed to respond to quantitatively measurable evidence while creating and advocating for qualitative responses. Humans are just a small part of the environment upon which we depend and of which we are simply temporary custodians. We must consider the environment, including non-humans, and (g)lobal people, community and the public as our client first and foremost as part of our responsibility to others we share the planet with.

A renewed architectural thinking

Never before has the scientific community issued such stark warnings about the urgency of addressing the climate crisis (IPCC 2018, 2022). The architectural profession has dragged its heels. But there are an increasing number of voices calling for change. Architects do not design and build architecture alone; they do not hold all the knowledge. Nor do they hold a monopoly on architectural thinking. To contribute to a systemic shift towards a carbon-neutral society, architectural thinking needs to undergo an equivalent systemic shift. We simply do not have any more time or resources to continue to think, work, and educate as we currently do.

First and foremost, we must act with the environment and (non-)human users as our clients. Otherwise our actions will lead to spaces that are not inclusive and that cause exploitation, climate, social, and spatial injustice elsewhere. By acknowledging the limitations of the architectural thinking of architects, and acknowledging and engaging the design thinking and architectural thinking of others, we can create truly collaborative processes towards sustainable environments. Our global citizenship does not pause when we enrol in architecture school or enter the architectural profession. If anything, our responsibility is amplified. Our architecture must not be permitted to transgress other (human and non-human) rights. Ecopedagogy principles provide a useful starting point not only for achieving transformative education, but also transformative architectural practice and research. Ecopedagogy places inclusion, care for the natural environment, and social and collective transformation central. It is a dynamic co-creative process that is about an attitude to life-long learning and continuously shifts in order to meet today's challenges.

Secondly, high standards in all aspects of sustainability are needed. While a focus on carbon emissions is crucial, it cannot be at the expense of other less easily quantified aspects (see Pelsmakers et al., 2022). It is also paramount that design-thinking, concerned with the user, also elevates the environment and the non-human user as part of architectural thinking. It means that the architect de-centres themselves and re-centres others and the environment instead. Only in doing so will we ensure our built environment's sustainable transformation towards carbon neutrality while avoiding the sustainability paradox mentioned earlier.

A small number of architects are already exploiting such an attitude to create local transformative change in some aspects, and some examples were highlighted earlier. However, we must do more, because demonstrating relative sustainability is insufficient: 'doing less harm is not the same as doing no harm' (Pelsmakers & Newman, 2021, p. 145).

If architects are to address the climate crisis, it must involve all actors from the earliest stages, and success should be measured by conversations that replace 'no because' with 'yes-and…'. Our aim for this book has been to bring together different expert voices to do exactly that: to share our expertise and learn from each other to enable us to change the status quo based on the best available evidence and knowledge available at this moment in time. The key is to now build on this to collectively advance and create change with the urgency that is required.

Bibliography

Awan, N., Schneider, T., & Till, J. 2011. *Spatial Agency: Other Ways of Doing Architecture.* Routledge. See also www.spatialagency.net/

Cross, N., 2011. *Design Thinking: Understanding How Designers Think and Work.* Berg.

Deutsch, R., 2020. *Think Like an Architect: How to Develop Critical, Creative and Collaborative Problem-Solving Skills.* RIBA Publishing.

European Commission, 2019. New rules for greener and smarter buildings will increase quality of life for all Europeans. [Online] 15 April. Available: https://ec.europa.eu/info/news/new-rules-greener-and-smarter-buildings-will-increase-quality-life-all-europeans-2019-apr-15_en (Accessed 17 February 2025).

Fisher, T., 2000. *In the Scheme of Things: Alternative Thinking on the Practice of Architecture.* University of Minnesota Press.

Florida, R., 2002. *The Rise of the Creative Class: And How it's Transforming Work, Leisure, Community and Everyday Life.* Basic Books.

Fry, T., 2007. 'Redirective practice: an elaboration.' *Design Philosophy Papers*, 5.1. DOI:10.2752/144871307X13966292017072

Goodbun, J., Till, J., & Iossifova, D., 2012. *Scarcity: Architecture in an Age of Depleting Resources.* Wiley.

Harriss, H., Hyde, R., & Marcaccio, R., 2021. *Architects After Architecture: Alternative Pathways for Practice.* Routledge.

Hosey, L., 2012. *The Shape of Green: Aesthetics, Ecology, and Design.* Island Press.

Ing, W., 2020a. 'Foster + Partners withdraws from Architects Declare.' *Architects Journal* [Online], 2 December. Available: www.architectsjournal.co.uk/news/foster-partners-withdraws-from-architects-declare (Accessed 17 February 2025).

Ing, W., 2020b. 'Zaha Hadid Architects quits Architects Declare and blasts its leadership.' *Architects Journal* [Online], 3 December. Available: www.architectsjournal.co.uk/news/zaha-hadid-architects-quits-architects-declare (Accessed 17 February 2025).

IPCC [Masson-Delmotte, V., P. Zhai, H.-O. Pörtner, D. Roberts, J. Skea, P.R. Shukla, A. Pirani, W. Moufouma-Okia, C. Péan, R. Pidcock, S. Connors, J.B.R. Matthews, Y. Chen, X. Zhou, M.I. Gomis, E. Lonnoy, T. Maycock, M. Tignor, & T. Waterfield (eds.)], 2018. 'Summary for Policymakers' in *Global Warming of 1.5°C. An IPCC Special Report on the impacts of global warming of 1.5°C above pre-industrial levels and related global greenhouse gas emission pathways, in the context of strengthening the global response to the threat of climate change, sustainable development, and efforts to eradicate poverty.* Available: www.ipcc.ch/sr15/chapter/spm (Accessed 17 February 2025).

IPCC [H.-O. Pörtner, D.C. Roberts, M. Tignor, E.S. Poloczanska, K. Mintenbeck, A. Alegría, M. Craig, S. Langsdorf, S. Löschke, V. Möller, A. Okem, & B. Rama (eds.)], 2022. *Climate*

Change 2022: Impacts, Adaptation and Vulnerability. Contribution of Working Group II to the Sixth Assessment Report of the Intergovernmental Panel on Climate Change. Cambridge University Press. DOI:10.1017/9781009325844.
Kimbell, L., 2011 'Rethinking design thinking: Part I', *Design and Culture*, 3.3, 285–306. DOI: 10.2752/175470811X13071166525216.
Kimpian, J., Hartman.,H., & Pelsmakers, S., 2021. *Energy People Buildings: Making Sustainable Architecture Work*. RIBA Publishing.
Kolkwitz, M., & Luotonen, E., 2021. 'Contesting the architect's role through radical participatory design: a discussion with Al Borde and raumlabor' in Pelsmakers, S., & Newman, N. *Design Studio Vol. 1: Everything Needs to Change: Architecture and the Climate Emergency*. Routledge.
Kundoo, A., 2020. Interview: For Whom Are We Building. YouTube [Online]. Available: www.youtube.com/watch?v=OTGBD5bQ13s&ab_channel=LouisianaChannel (Accessed 17 February 2025).
Lawson, B., 2006. *How Designers Think: The Design Process Demystified* (4th edn). Elsevier.
Living Building Challenge, 2020. *Living Building Challenge Basics*. [Online] Available: https://living-future.org/lbc/basics (Accessed 17 February 2025).
Martin, C., 2020. Trans-species health: an aesthetic responsibility. *Foreground* [Online], 9 April. Available: www.foreground.com.au/agriculture-environment/trans-species-health/ (Accessed 17 February 2025).
Moe, K., 2020. Climate Change, Architecture Change. YouTube [Online], 18 February. Available: www.youtube.com/watch?v=iomFhd3Mmj8 (Accessed 17 February 2025).
Musterd., S. (ed.), 2020. *Handbook of Urban Segregation*. Edward Elgar Publishing.
Pedraza, I. N., 2019. *The Black Spatial Imaginary in Urban Design Practice: Lessons for Creating Black-affirming Public Spaces*. Master of Urban Planning thesis. University of Washington. Available: https://digital.lib.washington.edu/server/api/core/bitstreams/436d9733-4aaa-4c97-9376-bf3c4e727641/content (Accessed 17 February 2025).
Nawratek, K., Horton-Howe, A., Huggins, B., & Oxley, M. 2015. *Radical Inclusivity: Architecture and Urbanism*. Dpr-Barcelona.
Nisonen, E., 2022. *Architecture As Sustainability – Educating The Transformative Practitioner*. Master's Thesis, Faculty of the Built Environment, Tampere University. Available: https://trepo.tuni.fi/bitstream/handle/10024/139040/NisonenEssi.pdf (Accessed 17 February 2025).
Norman, D., 1988. *The Design of Everyday Things*. Basic Books.
Pallasmaa, J., 2009. *The Thinking Hand: existential and embodied wisdom in architecture*. John Wiley.
Pelsmakers, S. & Hilberth, T., 2019. 'Feedback in architecture: what does it mean, and why do it?' in *Proceedings of the Sustainable Architecture Conference* (S-Arch), Havana, Cuba.
Pelsmakers, S. & Newman, N., 2021. *RIBA Design Studio Volume 1. Everything Needs to Change: Architecture and the Climate Emergency*. RIBA Publishing.
Pelsmakers, S., Donovan, E. Hoggard, A., & Kozminska. U., 2022. *Designing for the Climate Emergency – a Guide for Architecture Students*. RIBA Publishing.
Rowe, P., 1987. *Design Thinking*. MIT Press.
Schön, D.A., 1983. *The Reflective Practitioner: How Professionals Think In Action*. Basic Books.
Simon, H., 1996. *The Sciences of the Artificial* (3rd edn). MIT Press.
Stangeland, S., & Hensel, M.U., 2019. 'The research and relational designs of Helen & Hard' in Hensel, M.U. & Nilsson, F. (eds.) *The Changing Shape of Architecture: Further Cases of Integrating Research and Design in Practice*. Routledge, 69–91.
Stevens, G., 2002. *The Favored Circle: The Social Foundations Of Architectural Distinction*. MIT Press.
Ternby, L., 2023. 'Chefen som införde samma lön för alla' ['The boss who introduced equal pay for everyone']. *Dagens Nyheter*, 30 May. [Online]. Available at: www.di.se/nyheter/chefen-som-inforde-samma-lon-for-alla/ (Accessed 17 February 2025).
Thorpe, D., 2015. *The 'One Planet' Life – A Blueprint for Low Impact Development*. Routledge.

Till, J., 2020. 'Architecture after Architecture' in Harriss, H., Hyde, R., & Maracaccio, R. (eds.), *Architects after Architecture – Alternative Pathways for Practice*. Routledge, 29–37.

Watts, J. (2018), Global warming should be called global heating, says key scientist. *The Guardian* [Online], 13 December. Available: www.theguardian.com/environment/2018/dec/13/global-heating-more-accurate-to-describe-risks-to-planet-says-key-scientist (Accessed 17 February 2025).

Zimmermann, A., & Rodríguez Izquierdo, F., 2021. 'Adaptation out of necessity' in Pelsmakers, S., & Newman, N. (eds.) *Design Studio Volume 1. Everything Needs To Change: Architecture And The Climate Emergency*. RIBA Publishing, 130–137.

PART 1

Architectural philosophies to address the climate emergency

Injustice, circularity, radical inclusivity, spatial insurgence

Part 1 of the book consists of four interrelated chapters that set out the impacts of current architectural thinking and practice on people, climate, resource use and biodiversity. These chapters highlight how these approaches lead to injustices and exclusion, but also how alternative thinking and practices can help mitigate these impacts.

While the climate emergency is driven by carbon emissions and technical solutions exist for reducing these (e.g., renewable energy technologies, design and zero carbon buildings, Chapter 2), this needs to be balanced with other important considerations, for example: reuse and retrofit of buildings (Chapters 2 and 3); occupant behaviour and real performance of buildings (Chapter 2); ethical issues (e.g., unequal exposure and ability to respond to climate and environmental hazards, Chapter 2); unequal division and extraction of resources (Chapters 3 and 5); unequal access to decision-making and spaces (Chapters 4 and 5) and considering the rights of non-humans and nature (Chapter 4). These ethical issues turn into injustices where, for example, people and communities of lower socio-economic groups have different exposures to hazards with a different vulnerability to the effects of these hazards and typically did the least to contribute to them. This extends to non-humans and nature too, and to the unequal division of resources and exposure and vulnerability to hazards between countries.

Chapter 2, *Architects and Climate Justice* by Jonathon Taylor, Salla Jokela, Markus Laine and Lauren Ferguson sets out the impact that the built environment has on the natural world, from energy consumption and associated carbon emissions, to waste production and resource use and also the impact on human health and well-being associated with these impacts and a changing climate that we are ill prepared to adapt to. They explain that this is closely intertwined with equity issues, where different socio-economic groups are exposed to different health inequalities. Injustices may increase as more vulnerable people are more exposed to the health impacts of pollution and the threats of a changing climate, particularly pronounced in housing given many people spend so much time at home.

Taylor et al. highlight the link between the built environment and unequal health outcomes between socio-economic groups (e.g., inequalities in adverse thermal, flood

DOI: 10.4324/9781003293903-2
This chapter has been made available under a CC-BY-NC-ND 4.0 license.

and/or pollution exposures), which is linked to the design and performance of buildings. The authors illustrate how architects' design decisions (from design, building and occupancy) can shape the impacts of the building and its performance, and can also perpetuate inequalities. The link to architects' and planners' work is clear: if architects build just three typical buildings in their entire career (most will design a multitude of this), they will 'generate 162 average American lifestyles worth of carbon emissions' (Hurst, 2021 in Taylor et al., 2025). They also argue that the impacts of individual buildings have a much broader impact on society than the building itself in terms of health and well-being and future resource use. And this is why the architect's role goes beyond environmental sustainability (e.g., mitigating climate change through efficient design and retrofit of the housing stock; reducing waste from buildings (also argued in Chapter 3)). Instead, the architect's role is also critical for climate and social justice to avoid future inequalities and to provide co-benefits for population health. They make three clear suggestions for architectural thinking to be broadened to include (1) design for future climates and hazards; (2) design for occupants and (3) design for the collective good.

Some of this shift includes energy-efficient transformations and reuse of buildings, instead of focusing on building new; design for disassembly becoming the norm, and extending involvement of architects and the design team beyond the day that users move in (i.e., evaluation of buildings once in use to ensure they are working as intended).

While Chapter 2 maps a vast amount of evidence regarding our unsustainable architectural thinking and practices and the role of architects to improve quality, performance and justice issues, Chapter 3, *The new aesthetics of circular architecture*, by Antti Lehto and Sanna Lehtinen highlights that there are significant gaps still in thinking around circularity and what it means for architectural conceptual frameworks, theory and practice. Based on prior key theories, Lehto and Lehtinen synthesise seven design strategies for circular construction, which is typically equated with design for disassembly and reuse of materials, but the seven strategies go beyond this: (1) refusing and reducing; (2) design for disassembly; (3) reuse of materials; (4) adaptability and flexibility; (5) durable design and repairability; (6) building for attachment and trust; and (7) generality of materials.

These seven strategies are not just design actions in themselves but about new mindsets that extend beyond simple actions that are driven from a critical examination of our role as architects in the extraction and exploitation of resources as a systemic issue. As discussed in Chapter 2, it requires architectural thinking to shift from prioritising the new (i.e., new materials, building anew), and from an attitude of operating in a world with infinite resources to a world of finite resources (scarcity). This was also referred to by Taylor et al. as 'sufficient design', which also includes a social dimension with focus on human behaviour and emotions.

To achieve this shift in practice, Lehto and Lehtinen argue for four necessary changes, focusing on new mindset and value chains needed to make circularity happen, including accepting scarcity; accepting a changing role and new aesthetics, for which new design processes and business models and innovations are needed to enable increased complexity and scope (e.g. deeper understanding of building components and how to use reused materials in innovative ways, ensuring they are also detachable and accessible). As the other chapters also highlight, design processes and the architect's role are also a social activity where collaboration with others is needed. Flexibility in architectural thinking is also necessary as highlighted by Taylor et al. in relation to future climate changes.

This also applies when working with reused materials and the uncertainty of availability as well as fluidity in function and use of spaces and perception of the aesthetics of working with aged and ageing materials and imperfection, as also discussed and illustrated with a design case study by Mirjami Myllymäki, Sofie Pelsmakers and Laura Uimonen in Chapter 4.

Lehto and Lehtinen argue that while circularity is seen as a technical solution to the climate emergency and resource scarcity, discussion around the aesthetics of circularity still matters because architecture can and should both act towards the climate emergency but also create a vision for a future world. They highlight that while beauty is crucial, and helps people look after buildings and creates long term sustainable architecture, it is one part of the circularity discussion, and it must not be at the expense of climate emergency action. Equally, the aesthetics of circular architecture can accelerate sustainable change and acceptance: what circular architecture looks like may be unconventional, but it expresses a new aesthetic and that its architecture was co-created in different ways with other stakeholders. The design case study of Myllymäki et al. also highlights that while the architectural expression is different, it simply reflects a more sustainable and empathetic architectural thinking.

In Chapter 4, Myllymäki et al. set out a philosophy of radical inclusivity, which means having empathy to human and non-human actors. The chapter highlights that while humans have been exploiting and extracting resources to shape the built environment (thereby affecting the climate and natural environment), it has been a minority of people who have been exploiting the majority of people and nature for the accumulation of wealth, creating injustices (some of which were discussed in Chapter 2). Wealth and profit-seeking mechanisms – at the heart of the Capitalocene – mean that people are omitted from design decision-making and rights of non-humans and nature are erased and only seen as to be exploited in the quest for ever more profit. This is exacerbated by urbanisation, which removes access to nature or connection with it – a separation which also leaves its exploitation unchallenged.

Myllymaki et al. further argue that empathy, solidarity and communality with humans, non-humans and nature should be part of architectural thinking's priority and they introduce the concepts of radical inclusivity (i.e., including humans and non-humans and enabling them to use space in various ways, after Nawratek and Nawratek (2016)); biophilia (i.e., emotional connection to nature that can be visual or non-visual (e.g., air-flow, scents and sounds); and topophilia (i.e. fondness of place that includes also non-living natural elements (e.g., stone)), all shown to impact health and well-being positively. As in Chapter 2, justice issues are also brought to the foreground, highlighting that everyone, regardless of background or age or income, has the right to connection with nature. Similarly to Chapter 3 and exposure to a new circular aesthetic, the authors argue that connection to nature can accelerate sustainable change and protect nature from destructive actions, leading to 'actions towards more sustainable living, rejecting exploitation and the systems that support it' (p. 72).

Architecture has been at the heart of spatialisation of capital (profit-making) and privilege and has legitimised the activity of those in power and led to activities that have prioritised the interests of the few, which is at the heart of the societal and planetary crises. It is also the reason why architecture is built around obsolescence (i.e. not designed to be permanent) instead of providing adaptable, diverse approaches (see Chapter 3).

In this world, nature only exists to serve humans, exacerbated further by the design and planning processes that separate management and technical solutions from participatory design, making it hard for designers to know who they design for. How it can be done differently is illustrated with a speculative design project and a call for change to our values from the extractive and exploitative profit-driven world of the Capitalocene towards an ecologies of repair, i.e., restorative practices that bring radical inclusivity and biophilia together with social and ecological values for collective thinking and changing. This emphasises architecture as a social activity (as also highlighted in Chapters 2 and 3) and that it is undertaken by a whole community instead of a single (or few) master-architect(s) making design decisions, though there is still a lot of resistance to this collective model in architectural thinking (as described in Part 2, Chapter 7). Similarly to new circular design aesthetics as discussed in Chapter 3, a new aesthetic arises from these new architectural design and thinking processes when their focus is on principles of radical inclusivity, i.e., going away from designing for the few by the few – as has typically been the case throughout history – and ensuring architecture is co-designed by the many that benefits the many, including non-humans and nature.

How to challenge this current prevailing narrative of top-down processes and profit-driven decision-making in reality is further highlighted in Chapter 5 by Dalia Milián Bernal, Mikko Kyrönviita, Elina Alatalo and Veera Turku by proposing the production of architecture in different ways through a philosophy of insurgent spatial practices. They argue and illustrate how – while we need the top-down processes for cities and governments to act – that we also need bottom-up processes and practices that occur outside and alongside profit-driven urbanisation processes that are not typically profit driven, i.e., insurgent spatial practices. Doing so will help shift architectural thinking and production to be more sustainable and democratic with new and diverse socio-spatial configurations and creation of new kinds of spaces, necessary to respond to the climate emergency and injustices related to current development processes.

Building on Lefebvre, Milián Bernal et al. set out how power manifests in the design of abstract space, i.e., the space designed by architects and planners typically for abstract users that does not acknowledge them and actively ignores and erases diversity and the impacts of exploitation and extractive sourcing origins. The authors argue that for tackling the climate emergency, a clear decoupling of architectural thinking from capitalist practices is required. This can be expressed by the appropriation of space and how there are other ways of contributing to design and planning by those excluded from formal processes, as has been done in the past (though typically unrecognised in history). Importantly, as the authors describe and illustrate with case studies, these spaces encourage co-created and shared knowledge and learning together for what the community needs, acting as spaces fostering social and political action and mobilisation to achieve much-needed alternative social and circular economies.

Myllymäki's design case study in Chapter 4 can be interpreted as an insurgent spatial practice given its speculative proposal for appropriation of space by local communities, users and nature. Chapter 5 further illustrates how this has been done around the world in different places by different communities in real-world case studies. They illustrate how communities self-organised to create spaces that meet their needs, and they shared knowledge and learned together how to create these spaces, typically with little to no resources other than those embedded in the community. Big challenges are the negotiations of use of the space, and perseverance is needed from the community against market forces and capitalist-driven development processes, often leading to temporary use spaces only and facing the risk of eviction over time. Instead of abstract spaces, insurgent spatial practices help create the differentiated spaces that meet communities' needs and desires as they themselves instigated and enacted them. It elevates their local knowledge, instead of that imposed by architects paid for by corporate interests. Ultimately, changing architectural thinking and practices that foster social and environmental justice challenges the current status quo of how spaces are produced and can inspire others (communities, architects, planners, policy-makers etc.) elsewhere for alternative approaches that focus on equity and situated approaches that support sustainability and democratic processes.

In common throughout the four chapters in Part 1 is the call for changing design and decision-making processes and a shift in mindsets and values. This includes a shift in architectural thinking as a social activity, elevating situated knowledge and ways of knowing by local communities and accepting uncertainty but also slowness: it takes more time to study the site and needs, deeply understand, discuss, co-design, refine, resource, learn together, refine again, construct, check it works, use, adapt, etc. In itself, these are principles that go against capitalist-focused development where the mantra is 'time is money'. Architecture as a social activity, where architects are one of several stakeholders among the communities and users impacted by the design, can however create challenges and barriers to true inclusivity where architects attempt inclusive democratic processes but they themselves are often not representing a diverse body of people compared to the communities they design for and with in society. When we design with partial knowledge or knowledge that may systematically disadvantage vulnerable communities exposed to higher risks, it risks erasing others' voices and perpetuating past injustices; this is a recurring theme discussed in Part 3. Clearly the architecture profession needs to diversify itself to be able to represent those whom it designs for and with and to renew itself with different perspectives rather than continuing the exploitative status quo.

This section of the book also shows how that these new processes and thinking create a new kind of architecture and aesthetics: accepting working with uncertainty, imperfection, constructed or self-organised by people themselves. Accepting changing aesthetics is a crucial part of the shift needed, which may be unconventional for its aesthetics today, but is not unappealing, yet it profoundly challenges architecture's culture that continues to prioritise a certain aesthetics, style and form, and 'the new', regardless of its disastrous implications.

While experimental practices can inspire others to protect and act, ultimately they may have limited power for change if our societies remain coupled to practices and systems of power that are unsustainable and that lead to injustices. This is where architects' leadership and ethical positioning and advocating for what is right may be necessary – despite them having collectively failed to do so, so far. It will require individual and collective introspection and reflection on how we have been so resistant to act in the interests of nature and the many. Design for the collective good is a common theme in all the chapters, and while this requires a shift in thinking, all chapters illustrated the potential for this shift to lead to a more just, inclusive society ready to face the climate and biodiversity crises.

Given the longevity of the places and spaces we design, the importance of getting their design and construction and long-term use right is crucial, as it otherwise locks in negative impacts for decades. With the same notion, creating positive impacts with our design choices also creates long-lasting positive impacts for decades to come. But to do so means we need to design to exceed regulatory standards and narrow definitions of what sustainability or design for the climate emergency is, as they rarely acknowledge broader socio-economic implications and injustices related to development.

The chapters in Part 1 all embody a need to challenge the status quo. That means working outside the norm. The importance of embedding different values, mindsets, processes and practices in architecture education from the early years is crucial and subject of discussion in Part 2.

2

ARCHITECTS AND CLIMATE JUSTICE

Jonathon Taylor, Salla Jokela, Markus Laine, and Lauren Ferguson

Introduction

Objects built by humans now weigh more than all life on earth (Elhacham et al., 2020). In Europe, buildings are the single-largest consumers of energy, responsible for 40% of all energy consumption and 36% of greenhouse gas (GHG) emissions, while only 1% of buildings are retrofitted each year (European Commission, 2019); construction and demolition is the largest waste stream in the EU, accounting for 32% of all waste produced (European Commission, 2018); and poorly designed buildings may 'lock-in' poor energy consumption or environmental performance for lifespans which may extend beyond the dangerous levels of warming projected by 2100 (Intergovernmental Panel on Climate Change, 2021). Few industries have contributed to global environmental unsustainability more than building design and construction, and within this sector, architects – estimated to generate 162 average American lifestyles worth of carbon emissions if they build just three typical buildings in their career (Hurst, 2021) – have a disproportionate negative impact. In the face of global crises, including a changing climate, biodiversity loss, land use change, and unsustainable use of resources, there is an urgent need to improve the sustainability of buildings by reducing energy and resource consumption, GHG emissions, and waste production. However, despite the overwhelming evidence of the role of buildings on emissions and climate change, the majority of buildings are designed and constructed to meet – rather than exceed – energy efficiency building regulations. Architectural form and economic considerations continue to take precedence over minimising environmental impact.

Climate change has been described as the greatest global health threat of the twenty-first century (Romanello et al., 2021). It poses a serious threat to human health and wellbeing due to increases in extreme weather such as heatwaves and heavy rainfall, weather-related hazards such as flooding, and because of the disruptions to the ecosystems and human systems on which people rely. Housing, at its very core, should provide protection against these elements. Dwellings and their indoor environmental quality play a significant role in population health and wellbeing, modifying the degree

DOI: 10.4324/9781003293903-3
This chapter has been made available under a CC-BY-NC-ND 4.0 license.

to which occupants are exposed to environmental health hazards. Housing, therefore, has an important role not just in mitigating climate change, but also in providing protection against more frequently occurring extreme weather. However, the quality of living environments and their resilience to environmental hazards can vary substantially across sociodemographic groups (Ferguson et al., 2020; Tonne et al., 2018) with disadvantaged populations typically living in poorer housing. The driving factors which can lead to poor indoor environments are often systemic and out of the control of occupants, and can include building design and quality, the local outdoor environment, maintenance, and the actions of neighbours (Ferguson et al., 2021). Housing is, therefore, also important for addressing parallel challenges to social sustainability that may lead to increased health inequalities as the climate changes, including inequity, social cohesion, inclusion, and justice (Dempsey et al., 2011).

Architects, architectural engineers, developers, and the building construction sector have a critical role to play in responding to these interlinked environmental, health, and social sustainability challenges. This chapter begins by describing why climate change is a health and equity issue, why low-income populations are more vulnerable to negative health impacts from their surrounding environment, and how this leads to health inequalities between socioeconomic groups. It then describes current socioeconomic inequalities in GHG and air pollution emissions from housing, as well as inequalities in local and indoor exposure to climate and air pollution hazards, illustrating how – even within the same city – housing is an area of climate and environmental injustice. We then conclude by describing how decisions by architects can entrench and perpetuate inequalities, and why greater leadership and agency from architects is required to mitigate climate change and reduce climate inequity.

Health, inequalities, and buildings

Health is not simply the absence of disease or infirmity, but rather a state of physical, mental, social, and spiritual wellbeing (World Health Organization, 1948). While the causes of disease are related to an individual's characteristics and behaviours, the social, economic, and physical environments in which people are born, live, learn, work, and age are considered even more important (Rose, 1992). These are often referred to as Social Determinants of Health, and can include, for example, housing conditions, the characteristics of the neighbourhoods they inhabit, their working conditions and income, and their access to medical care – all circumstances which are often determined by factors outside the control of an individual. Social determinants of health are themselves shaped by Structural Determinants of Health, or the governing processes, the social and economic policies, and subsequent decisions that impact housing, neighbourhoods, education, working conditions, and income by affecting the distribution of health-related resources throughout society.

Structural and social determinants of health are a root cause of health inequalities, or when certain groups within the population face a disproportionate burden of poor health (Diderichsen et al., 2001). Social contexts lead to social stratification, where individuals are assigned to different social positions. Social position can then determine the degree of exposure to health-damaging conditions, the risk of underlying health problems, and a lack of resources that can make individuals more vulnerable. Finally, the stratification also leads to differences in consequences from this exposure, with greater health, social, and economic impacts amongst these vulnerable groups. This is often referred to as 'triple jeopardy' in public health, where low-income communities are 1) more exposed to environmental hazards, 2) are more susceptible to poor health outcomes from this exposure due to high rates of pre-existing health issues, and therefore 3) face greater health issues once exposed, reinforcing a vicious circle of low income and poor health.

Living environments are a key social determinant of health. One reason is because they are the environment in which people spend the majority of their time: up to 90% in the UK, for example (Lader et al., 2006). Another reason is the quality of homes. Health-affecting exposures in homes, include for example low or high indoor temperatures, damp and mould, indoor air pollution, noise, and overcrowding, which may have various physical and mental health consequences for occupants (World Health Organization, 2018). Living environments are themselves shaped by structural determinants such as building regulations, urban planning policy, and housing costs. Example of the roles of social and structural determinants of health may be found in the COVID-19 situation (see Box 2.1).

BOX 2.1 COVID-19, HOUSING, AND THE SOCIAL AND STRUCTURAL DETERMINANTS OF HEALTH

Covid-19 has highlighted inequality worldwide (Patel et al., 2020), with many countries presenting disproportionately high Covid-19 mortality rates in lower income groups, due to increased exposure risk, greater underlying health vulnerabilities, and limited financial resilience. Buildings also modify this risk: high rates of within-household transmission have been observed during the Covid-19 outbreak (SAGE, 2020a), with overcrowding, larger household sizes, and multi-generational living associated with an elevated risk of mortality (SAGE, 2020b). Evidence that the coronavirus can be transmitted through airborne pathways has led to calls to improve how we design building ventilation systems (Morawska et al., 2020), and it has been suggested multi-unit accommodation and dense housing developments may contribute to transmission (Bray, Gibson, & White, 2020; Kenway et al., 2020). As with Covid-19, the impacts of the changing climate will not be borne equally.

Climate justice and housing

The concept of climate justice derives from the fact that climate change will have disproportionate negative social, economic, and public health impacts on disadvantaged communities due to differences in both exposure to climate hazards and vulnerability to the effects (Klinsky & Mavrogianni, 2020). Even within the same city, region, or country, low-income populations are at greater risk of exposure from climate and environmental hazards (Environmental Protection Agency, 2021) and more vulnerable to negative outcomes due to a greater sensitivity and lower adaptive capacity to these hazards; this can act to exacerbate existing inequalities. In addition, while low-income populations bear a disproportionate burden from the negative effects of the changing climate, higher income populations produce a disproportionate amount of GHG emissions (Grunewald et al., 2017). Given the substantial carbon emissions from the residential sector, its varying resilience to extreme conditions, and role as a health determinant, housing is an important component of climate justice. This section summarises inequalities in local environments, housing design and performance, focusing on emissions and exposure to various climate and air pollution-related hazards.

Thermal inequality

Inequalities in indoor cold temperatures across income groups has been long recognised via the concept of fuel poverty, where households are unable to afford to heat their homes to an adequate temperature. In cold and temperate regions, building regulations mean that new homes are easier to heat, reducing the rates of fuel poverty for occupants. However, the changing climate means the emergence of inequalities in heat exposure in such areas, for which regulatory guidance continues to be lacking for renovations and new builds.

Heat exposure is an acute health risk, particularly in individuals older than 65 years and those with pre-existing health conditions (Kovats & Hajat, 2008; M. Li et al., 2015). Climate change is increasing annual average temperatures globally, and the frequency and severity of heatwave events is growing. Global heat-related mortality was estimated to reach a record high of 345,000 deaths in 2019, an increase of 80.6% from the 2000–2005 average (Romanello et al., 2021) and between 1991 and 2018, an estimated 37% of global heat-related deaths were attributable to anthropogenic climate change (Vicedo-Cabrera et al., 2021). Increased exposure to heat is also thought to pose a range of risks to mental health-related hospital admissions and suicidality (Mullins & White, 2019; Obradovich et al., 2018).

Living in urban areas is also a risk factor for heat-related mortality (Basu, 2002; Kovats & Hajat, 2008). Human-made surfaces, such as pavements and buildings, absorb and re-emit heat from the sun more than natural surfaces, while human activities such as air conditioning, heating buildings, vehicles, and industry can introduce additional heat into urban environments. Reduced coverage of natural surfaces also means less evapotranspiration from vegetation, which would otherwise help to cool the urban air. As a result, cities tend to be hotter than the surrounding rural areas – an effect known as the Urban Heat Island (UHI), and can lead to relatively higher heat exposure and heat-related mortality in urban centres (Milojevic et al., 2017; Ruuhela et al., 2020). The temperature difference can be significant – for example, the UHI means urban centres in cities such as London, New York, Manchester, and Birmingham are around 2–4°C hotter on average relative to the surrounding rural area, although in extreme cases this may be 5–10°C higher (Heaviside et al., 2017). In London, it has been estimated that up to half of heat-related mortality is attributable to the UHI (Heaviside & Macintyre, 2019).

The distribution of urban heat exposure risk is not shared equally between socioeconomic populations in many urban areas. Green and leafy areas can be relatively cooler than denser, more built-up urban areas (Murage et al., 2020), but the provision of green space is not equally distributed amongst neighbourhoods of different socioeconomic status (Chen et al., 2020; Hoffimann et al., 2017; Pham et al., 2012). In the USA, the historical policy of disinvestment and denial of services to specific communities – usually home to large numbers of ethnic minorities – may be a contributor to the land cover differences that cause urban heat exposure disparities (Hoffman et al., 2020; Wilson, 2020). A study of the 175 largest urban areas in the USA found that 95% of these cities had higher land surface temperatures in areas inhabited by people of color, with similar results for households living below the poverty line (Hsu et al., 2021). Other studies in the USA have shown racial and income disparities in urban heat exposure in Phoenix, Arizona (Harlan et al., 2006; Jenerette et al., 2006), New Jersey (Solecki et al., 2011), and Richmond, Virginia (Saverino et al., 2021). This phenomenon is not just in the USA, but worldwide. For example, low-income communities have been found to live in areas with higher UHI temperatures in the West Midlands region of the UK (Macintyre et al., 2018). A study that linked data obtained from satellites with census data in 25 cities around the world found 72% of cities had increased land surface temperatures in low-income neighbourhoods (T. Chakraborty et al., 2019). This has led to the concept of thermal inequity, where marginalised communities are more exposed to high outdoor temperatures.

Buildings can modify exposure to heat, depending on their built form, fabric, ventilation performance, and any passive or active cooling measures employed to reduce heat exposure (Taylor et al., 2015, 2018). Here, too, there can be inequalities in heat exposure. In England, where domestic air conditioning is rare, evidence indicates that low-income households overheat more than non-low-income households due to the characteristics of buildings and the behaviour of the occupants (Lomas et al., 2021). There is evidence that, in hotter countries, the outdoor and indoor temperature exposure disparities are a likely contributing factor in the increased morbidity or mortality in deprived populations during periods of hot weather (Medina-Ramón et al., 2006; O'Neill et al., 2005; Semenza et al., 1996; Xu et al., 2020). This is in part due to access to air conditioning, where low-income households cannot afford to install, operate, or repair faulty units, resulting in greater indoor temperature exposures. As a consequence, low-income populations can be exposed to higher temperatures, while also the least equipped to adapt to and mitigate the effects of heat. As Klinenberg (1999) concluded in his study of the 1995 Chicago heatwave, the structural and social drivers of poverty and resource inequality had an even greater role in heat-associated mortality than absolute temperature exposures.

Flooding and rainfall

In many parts of the world, climate change is projected to lead to increased periods of heavy rainfall and greater flood risks, while those close to the sea may be at risk of rising sea levels and storm surges. Flooding presents immediate health risks from events such as drowning, injuries, and hypothermia, medium-term risks such as poisoning, and complications from injury, and long-term risks such as disability and poor mental health (Du et al., 2010). Indoor damp from flooding or rainwater leaks is associated with a number of respiratory problems in occupants (World Health Organization, 2009). Despite this, in many countries such as the UK, demand for housing means dwellings continue to be built on functional flood plains.

The risk of flooding is also not evenly shared. Studies have shown that areas with greater exposure to flooding have disproportionate numbers of disadvantaged residents, for example in the USA (J. Chakraborty et al., 2014, 2019; Smiley, 2020) and Europe (Hall & Bailey, 2021; Walker & Burningham, 2011). This may in part be driven by cost, as homes in flood-risk areas may be more affordable; similarly, the lack of green space in low-income communities may restrict drainage and increase surface flooding risk. There are also differences in the social vulnerability to floods (Tapsell et al., 2002). Recovery can be slower in lower income groups, as they are less likely to be insured or have the support of financial resources during an emergency, are less likely to be able to continue in work if displaced, and face a disproportionate risk of poor physical and mental health outcomes (Blaikie et al., 2005).

Buildings and neighbourhoods can be designed to be more resilient to flood events (Bennett, 2010), however, some houses are more prone to extended periods of post-flood dampness than others (Taylor et al., 2012). Indoor damp from a range of different moisture sources has been shown to be more prevalent in low-income houses in England (MHCLG, 2020), and mould growth is higher in homes with lower socioeconomic status residents in Europe (Braubach & Fairburn, 2010). As a consequence, low-income populations may have a greater exposure to flooding and moisture damage, while also having a reduced ability to adapt or recover following exposure.

Air pollution

Air pollution exposure is one of the greatest environmental health threats globally. It is associated with a number of physical health outcomes such as respiratory (Atkinson et al., 2016) and cardiovascular problems (Arden Pope et al., 2011), birth defects (Padula et al., 2013), childhood asthma cases and sudden infant deaths (National Institute for Health and Care Excellence, 2013), as well as mental health outcomes such as adult depression (Braithwaite et al., 2019), schizophrenia (Horsdal et al., 2019), and cognitive impairment and increased risk of dementia (Delgado-Saborit et al., 2020).

Within many cities worldwide, levels of outdoor air pollution can vary spatially, for example due to intra-urban density of road traffic, energy production, and industry. Evidence indicates that low socioeconomic communities often live in areas with higher levels of outdoor air pollution, including studies from Europe (Brainard et al., 2002; Brunt et al., 2017; Fairburn et al., 2019; Fecht et al., 2015; Milojevic et al., 2017; Samoli et al., 2019), North America (Clark et al., 2017; Colmer et al., 2020; Pinault et al., 2016), and Asia (Choi et al., 2016; V. O. Li et al., 2018). These disparities may arise due to various factors, for example, underinvestment and redlining in the USA, or because land or housing prices close to industry or busy roads may be more affordable, attracting purchase by local governments for social housing and low-income individuals.

Housing modifies exposure to air pollution, either by impacting the degree to which outdoor air pollution can infiltrate indoors, or by the rate of removal of pollution produced indoors from activities such as cooking or smoking (Ferguson et al., 2020). There is increasing evidence of higher levels of indoor air pollution exposures in low-income homes (Ferguson et al., 2020), and the causes of this disparity may be due to a range of different driving factors such as exposure to higher outdoor air pollution levels, the quality of housing and ventilation, and differences in occupant ventilation and indoor pollution generating behaviours (Ferguson et al., 2021). While occupants can influence their own behaviours, many of these drivers of unequal exposures are due to social determinants related to housing, neighbourhoods, and community.

Carbon and air pollution emissions

The above illustrates how exposure to climate and environment-related hazards can be unequally distributed across the built environment and housing, and how this can impact the health of the most vulnerable. As the climate changes, so too will exposure to climate hazards, and by modifying exposures to these hazards, the role of housing as a determinant of health will increase.

However, the inequalities in exposure within living environments are in stark contrast with the production of GHG and pollution from buildings across socioeconomic groups. In Europe, around 11% of housing GHG come from the highest 1% of emitters (Ivanova & Wood, 2020). This is due to these high emitters owning and occupying larger homes, owning multiple residences, and a higher proportion of energy-intensive household appliances such as air conditioning (Nielsen et al., 2021). Socioeconomic differences in energy-efficient retrofit can also exist within countries. In the UK, energy-efficient retrofits are more common in low-income neighbourhoods than in middle- and high-income neighbourhoods (Hamilton et al., 2014) due to higher rates of retrofit in government-provided social housing.

Similarly, the emission of air pollution through the burning of household fuels appears to vary by socioeconomic status. While households in countries with high scores on the United Nations' Human Development Index (HDI) have generally better means to access clean fuels, domestic burning of solid fuels can still lead to appreciative levels of household air pollution exposure and health risks and residential wood combustion is a significant source of air pollution in countries with lower HDI scores (Olsen et al., 2020). Domestic wood burning is now the primary source of particulate air pollution in the UK (DEFRA, 2021) despite only 8% of the population burning solid fuels, of which 46% are from the wealthiest quarter of the population (Kantar, 2020).

The contrast between exposure to climate and environmental hazards, and energy efficiency and air pollution emissions, is illustrated for England in Figure 2.1. It shows how climate justice is not just a global issue, but also exists within countries, and is impacted by the design and performance of the building stock. Housing – intended to protect occupants from the climate – may actually exacerbate exposures in vulnerable groups due to their design and quality.

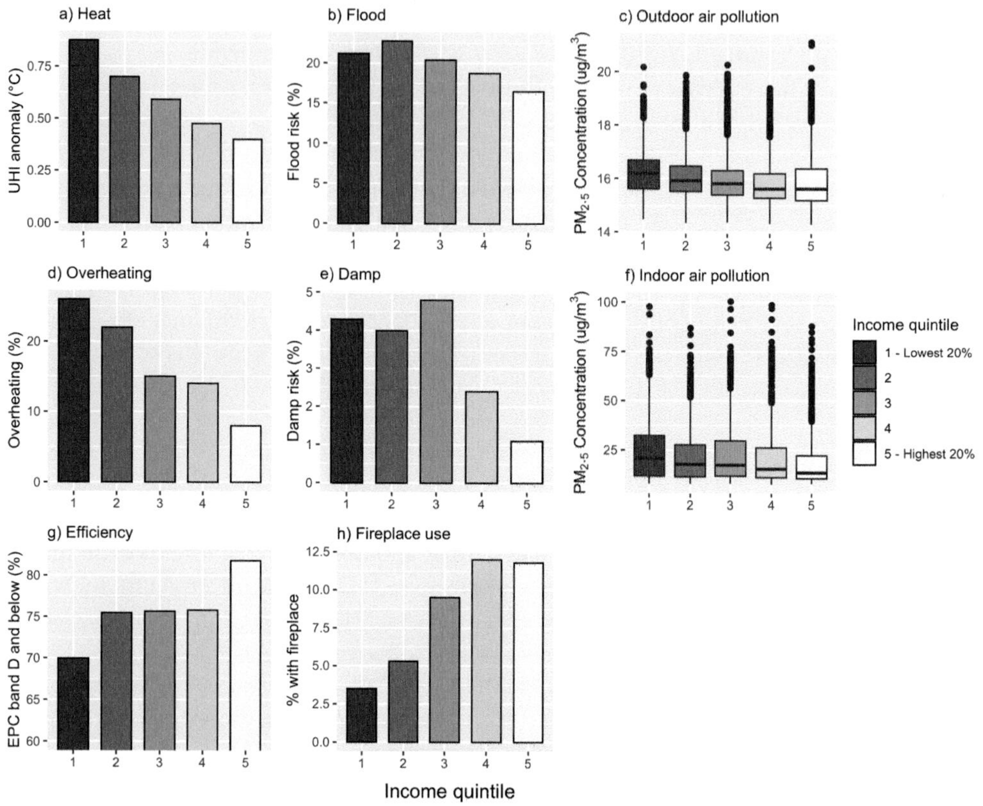

FIGURE 2.1 Exposure to climate hazards in the English built environment, by deprivation quintile where 1 is most deprived and 5 is least deprived. (a) Average outdoor temperature anomaly in the West Midlands, UK, during a heatwave period by LSOA Indices of Multiple Deprivation (IMD), adapted from Macintyre et al., 2018; (b) Percentage of areas at risk of flooding in England, by LSOA IMD, adapted from Hall and Bailey, 2021; (c) Distribution of outdoor air pollution levels by LSOA IMD in London, adapted from Ferguson et al., 2021; (d) Percentage of houses that were found to exceed overheating criteria, by household income quintile, adapted from Lomas et al., 2021; (e) Percentage of homes with evidence of damp in England, by household income, obtained from the 2017–2018 English Housing Survey (MHCLG, 2020); (f) Distribution of indoor air pollution levels by LSOA IMD in London, adapted from Ferguson et al., 2021; (g) Dwellings with an Environmental Impact Rating of C or below in England, by household income quintile (MHCLG, 2020); and (h) Households that burn solid fuels in a fireplace for secondary heating in England (MHCLG, 2020) by household income quintile.

Architectural agency and climate justice

It is clear that if we are to mitigate climate change to achieve the international goals set out in the 2015 Paris Climate Agreement, we need significant and transformative changes to how buildings are designed, constructed, operated, adapted, and disassembled. Due to the importance of housing as a determinant of health, climate adaptations are also required to mitigate the profound public health risks posed by the changing climate. Without action, climate change will worsen health inequalities, where the poor are more exposed to climate hazards, while the wealthy are less exposed and better able to adapt their living environments.

Buildings are therefore important examples of the intersection between environmental and social sustainability, with a profound impact on planetary health – or the health of the human population and the natural systems on which they depend (Whitmee et al., 2015). But, how can architects and architectural engineers help address issues of climate injustice and protect the health of people in response to climate hazards?

Design for sustainability

There is an urgent need to reduce carbon emissions from building construction and operation, and many countries are enacting regulations to increase the efficiency of new or renovated buildings. However, built environment professionals face various complex challenges, including balancing efficiency with other competing priorities. Energy efficiency is too often an afterthought, and much greater emphasis on energy efficiency is needed by all in the building sector. Indeed, the benefits of increased energy efficiency of homes also has significant co-benefits for population health and wellbeing (Thomson et al., 2013; Willand et al., 2015).

Beyond just energy efficiency, though, there needs to be a shift towards sufficiency in the built environment – or having enough to live well without consuming beyond the limits of sustainability. The production of new building materials is a significant source of greenhouse gas (GHG) emissions. Embodied carbon in building construction accounts for an average of 20–25% of their lifecycle GHG emissions (Röck et al., 2020), meaning that a huge number of GHG emissions occur prior to the building even being occupied. The production of concrete is estimated to be responsible for 5% of global GHG emissions and is a disproportionate component of construction and demolition waste in the EU. The most energy-efficient building is one already constructed, and the priority should be to make use of existing buildings – either through retrofit, adaptive reuse, transformation, or materials re-use. The latter would be easier if the buildings were routinely designed for disassembly (Rios et al., 2015). Architects should advocate for this becoming the norm, designing only buildings that can be dismantled for re-use as part of a circular approach.

Building performance often depends on the adopted system boundary within which performance is defined and assessed. Energy performance is not only determined by design choices, with occupant behaviours playing a critical role; indeed, new low energy buildings may consume more energy than older buildings due to occupants worrying less about energy conservation (Sunikka-Blank & Galvin, 2012). Design is often considered to end when the building nears completion, but there is a need to extend this temporal boundary into Post Occupancy Evaluation of buildings, to ensure that buildings are performing as designed, and to feed lessons forward into future designs (Whyte & Gann, 2001). Architects who are already familiar calculating the carbon footprints of their buildings should go further to consider the carbon handprint, which is used to express the carbon benefits, such as storing carbon in materials such as wood, or by generating surplus renewable energy production. This can allow developers to claim 'net zero carbon' buildings by offsetting emissions using activities such as tree-planting. While offsetting can help contribute to net zero strategies, too often it can be used as greenwashing and an opportunity to justify vanity projects when the real focus should instead be on reducing emissions as much as possible. Offsetting can also have unintended negative consequences, for example impacting the ecological and social resilience of landscapes from nature-based offsetting practices.

Architects can be involved in every phase of planning, design, building and occupancy, meaning they have significant scope to shape building performance. At the very beginning, through the client brief and planning stage, architects are in a unique position to liaise with the client to emphasise the importance of energy efficiency, sufficiency and environmental sustainability particularly amongst wealthier private clients, who are more likely to invest in energy-saving technologies (Umit et al., 2019). Architects also have the obvious opportunity to embed energy efficiency into every element of the design from the very beginning – from geometry to the selection of materials – either directly, or subversively by not offering unsustainable options to clients. There is a need for architects in positions of power and influence to show greater leadership and ambition in terms of sustainable construction and to support sustainable design using evidence and building performance feedback (Kimpian et al., 2021), and for all architects to feel empowered to resign if clients choose to continue with an unsustainable design.

Design for current and future climates and hazards

Environmental design, or designing buildings by addressing the surrounding environmental parameters, is often conducted according to assumptions or requirements that do not reflect the actual use(rs) or future climate. Indeed, anthropogenic climate change has already led to observable changes in the environmental parameters used to design buildings, and within the lifespan of a building constructed today there are a wide range of possible future parameters dependent on carbon emissions pathways. These surrounding environmental parameters may also change within urban areas. Take, for example, the issue of increasing summertime temperatures and heatwave events worldwide. Many building regulations worldwide now require calculations to show that the overheating risk of dwellings is below threshold levels, often requiring dynamic thermal simulation conducted using standard assumptions and example weather files. There are now a multitude of different extreme weather files for modelling buildings that can be used in building design to understand a building's robustness to future thermal performance: e.g., Design Summer Years (DSYs) representing hot but not extreme weather conditions, Typical Hot Years (THYs), Hot Summer Years (HSYs), Extreme Meteorological Years (XMYs) (Herrera et al., 2017). These are often available for baseline periods representing the current climate, as well as for future climates under different emission scenarios epochs, and with probabilities of occurrence.

These weather files are often developed using weather data from local weather stations, for example those located at airports, meaning that they do not typically capture the spatial variations in temperature across an urban area. Therefore, these weather files may underestimate the actual urban climate conditions, with potential consequences for buildings located within hotter parts of the city. Given that low-income communities often live in the hottest part of the city, this may result in a systematic underestimation of outdoor temperatures, and insufficient design adaptations to reduce indoor temperatures for housing in these communities.

Failure to design climate-resilient housing will have significant implications for climate justice. Those that can afford it will adapt their housing, for example installing air conditioning systems, or move to better adapted housing. Those who cannot afford to adapt, pay for energy costs or maintenance for active systems, or leave, or may have to remain in poorly performing housing if they cannot afford to pay for energy costs, exasperating exposures and worsening inequalities. Architects have a unique opportunity to 'stress test' building designs for current and future climate scenarios (Puckett & Gethering, 2019) to ensure climate resilient housing for all, and equitable climate adaptation solutions should be embedded within the building design. Failure to do so can help to entrench and perpetuate exposure inequalities.

Design for occupants

An essential element of architectural design is accounting for the needs of occupants, but these can be diverse and also change over the lifetime of a building. Given that occupants and functions may change, buildings need to be designed to be adaptable, enabling sustainable retrofit rather than replacement (Saarimaa & Pelsmakers, 2020). There is also a need to go beyond standard requirements for climatic design, but to evaluate the design and performance of buildings for occupancy scenarios that reflect the true range of actual uses of the buildings.

Considering again the issue of increasing hot weather and heatwaves, where occupant behaviour can play a significant role in the overheating risk of a building (Mavrogianni et al., 2014), building regulations for overheating contain a standard set of assumptions based on occupant behaviour, for example the number of occupants, their time spent at home, their use of equipment, and the actions that they take to cool their homes during hot weather. However, low-income households may have significantly different behaviours: for example, evidence from London indicates that they spend more time at home, are more likely to live in overcrowded dwellings, spend more time using cooking equipment, and may be less willing to open their windows due to security, pollution, or noise concerns (Ferguson et al., 2021). Such behaviours can significantly increase overheating risk, meaning that buildings that may be designed to meet overheating criteria for a 'standard' occupant may systematically underestimate internal gains and overestimate ventilation in homes occupied by low-income communities. Technological solutions to heat exposure, such as air conditioning systems have requisite installation and running costs that may not be accessible to a household living in energy poverty (Sanchez-Guevara et al., 2019).

By recognising that occupants may behave in diverse ways that may not reflect standard assumptions, and that a dwelling will likely be home to a number of different occupants over its lifespan, architects can enable a building to meet the needs of a range of different future occupants. This can include designing for diverse user scenarios (Kimpian et al., 2021), ensuring that indoor environments reach a high indoor environmental standard, minimise ongoing costs for active cooling systems, and prioritising passive solutions and equitable community solutions. The desired outcomes in terms of climate and health are difficult to achieve if the end users' behaviour contradicts the assumptions made in housing regulations or by the designers. For example, if users have a poor understanding of the purpose and functioning of new technologies and spatial solutions, they may be unable to use these appropriately. Employing a Soft Landing's approach in the design of new and transformed buildings can help ensure the buildings perform as designed (Way & Bordass, 2009); Soft Landing is a framework that aims to improve the performance of buildings and satisfaction of users by undertaking specific actions from the early stages of the design process to building completion and building occupation (Kimpian et al., 2021).

Design for the collective good

Building design can be conceived as a social activity, where architects are just one of multiple stakeholders (Green & Sergeeva, 2020; Laine et al., 2018). Support from the local community is important when improvements to existing housing are introduced or new residential buildings are designed. Common ways of ensuring the support of the local community include raising awareness of the possibilities of new technologies and solutions and involving users in planning processes (Gianfrate et al., 2017). The problem with these approaches is that they easily turn into box-ticking exercises with little real impact. Residents' participation – especially in disadvantaged areas – may be hindered by cognitive constraints caused by financial challenges, health conditions, or other chronic problems. Furthermore, many people in these areas may lack confidence in, and motivation for, taking part in planning processes or trying to change the status quo in response to top-down initiatives, because they have been marginalised and disempowered by previous planning practices and structural injustices. In some cases, the norms, values, and language of the experts driving the planning processes and publicity campaigns are also incompatible with those of residents (Sjöblom et al., 2020). Low levels of diversity in architecture schools and the profession may restrict the ability of architects to communicate with different groups within society.

While architects are generally trained to consider the interconnectedness of spatial design and social life, a more holistic view of the relationship between architecture and society enables addressing the inequities in housing impacts on climate and health more effectively. Instead of merely enabling public participation or raising awareness of new technologies and solutions, it is important to gain an understanding of social cohesion and the ways in which it enables residents to take an active role in fostering more sustainable living. According to Fonseca et al. (2018), social cohesion is '[t]he ongoing process of developing well-being, sense of belonging, and voluntary social participation of the members of society, while developing communities that tolerate and promote a multiplicity of values and cultures, and granting at the same time equal rights and opportunities in society.' In other words, when people feel that they want to be a part of a community and their participation in it is supported by laws, regulations, values and norms, they are motivated and better able to contribute to a change towards an equitable and sustainable society.

One way of fostering social cohesion in today's diverse urban communities is to design spaces that enable simultaneous functions and activities, supporting a peaceful co-existence of different people. Richard Sennett (Sennett, 2019) refers to these kinds of spaces as an 'open city', which he distinguishes from more closed and homogeneous urban forms, such as gated communities. According to him, buildings should be 'incomplete' in the sense that they can be flexibly changed, inviting people to think of living in a city as an open-ended path which involves coping with complexity. This could be done in an old or new urban structure, where urban space could be used differently. An example of the former is Barcelona Superblocks (Mueller et al., 2020), where a walking friendly, pleasant, and safe community space is created by using the street scape of nine city blocks, while car traffic is guided to the periphery. This is combined with increasing urban greenery by creating a network of pocket parks within the Superblock area.

In the so-called Nordic Superblock (Sjöblom et al., 2020) (2.2), a new urban structure, sustainable energy infrastructure, community spaces and shared services can be planned from scratch and sustainability goals can be combined in a meaningful way (Meadowcroft et al., 2019). The core idea of the Nordic Superblock is to share resources and to have maximum spaces, functions and services with minimal cost and environmental impact. This is achieved in various ways that challenge the conventional planning ideas. Firstly, residents share spaces, functions, and services – for example, gyms, saunas, laundry facilities, workshops, and guest rooms. Secondly, residents also share common energy, green and blue infrastructures. Thirdly, the location of each shared space, function and service and user routes and meeting points (i.e., social infrastructure) are carefully planned to help develop a sense of community. Finally, the Nordic Superblock entity can purchase services together, allowing residents to benefit from economies of scale. This approach allows for reduced emissions, and the equal access to services is intended to allow for improved community resilience – for example, when facing a changing climate or another pandemic.

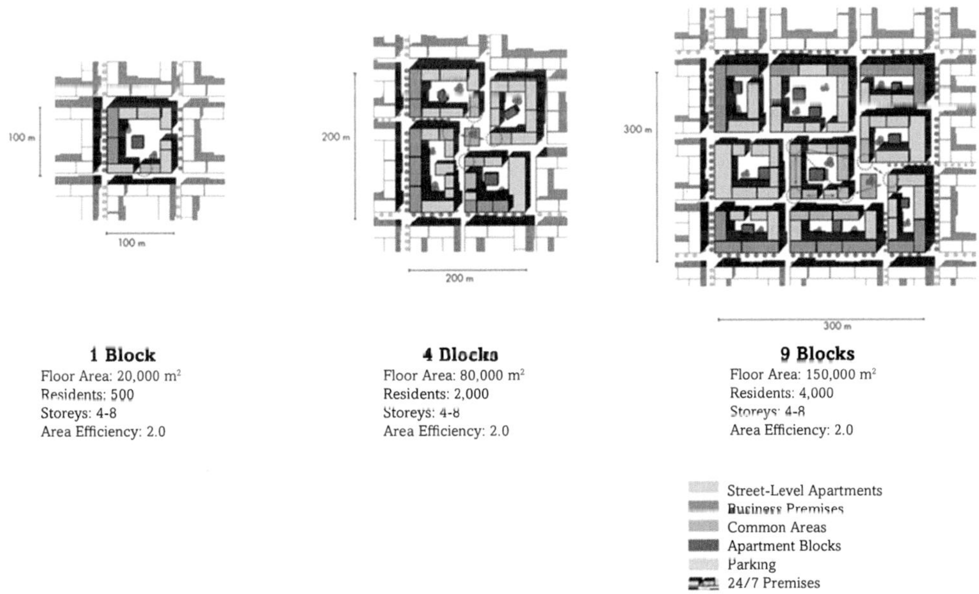

FIGURE 2.2 An illustration of the Nordic Superblock in three scales, depending on the site and purpose. Credit: Jolma Architects and TUPA Architecture.

Education and protection of function

The quality of the design can lock-in poor energy performance or indoor environmental quality for the lifetime of the building. However, despite the substantial consequences for poor design on carbon emissions and public and planetary health, there is little regulatory oversight for architects. Not every EU country protects the 'function' of an architect, which prevents those without registration from practicing architecture (Visscher & Meijer, 2008). Protecting the function of an architect and setting and assessing competency requirements for sustainable design is an important step towards improving building performance. This needs to be done in parallel with standards in other professions that are responsible for the design and development of the built environment.

Education also has a critical role to play in providing architects with the core skills to design and retrofit sustainable buildings, however architecture departments in many universities have been slow to respond to the climate crisis with many continuing to prioritise aesthetics over sustainable design. Architecture schools need to do more to teach students about the climate crisis, and to embrace zero carbon as part of their professional identities (Green & Sergeeva, 2020). Beyond just new construction, schools need to provide skills in energy-efficient retrofit, which can pose a number of challenges and potential unintended consequences if done improperly (Shrubsole et al., 2014). Providing future architects with the knowledge and skills to address the climate crisis, and agency through professional certification, empowers them to assume a leadership role in the drive to improve building sustainability. Many architecture students enter the profession seeking to positively impact society and improve cities (Till, 2020), and acknowledging the role of buildings as a social determinant of health and an area of climate inequality is an important way of doing so. However, architects cannot do this alone, and architectural education should provide students with the capability to conduct interdisciplinary collaboration with communities, engineers, and social scientists that can support the design of socially and environmentally sustainable buildings.

Conclusions

This chapter has examined the disproportionate role of housing on GHG emissions and exposure to climate and environmental hazards, and how inequalities in building quality and performance are an area of climate injustice. Evidence from various locations indicates that low-income populations are more exposed to climate hazards, such as heat, flooding, or air pollution, and more vulnerable to negative health outcomes from these exposures. Rather than providing protection, much of the housing inhabited by low-income communities exacerbates these exposures due to poor quality or climate resilience, with differential exposures and vulnerabilities manifesting themselves as environmental health inequalities. At the same time, low-income housing tends to be more energy efficient or sufficient than housing from higher-income groups, meaning that they are contributing less GHG emissions.

With their key role in the design and specification of buildings, architects and architectural engineers have a critical role in mitigating the climate crisis. We argue that their design decisions also have an important role in addressing or perpetuating social inequalities and climate injustice in the building stock. Energy-efficient and climate-resilient buildings are no longer an option, but an environmental and moral responsibility and an issue of justice. To be truly sustainable and equitable, those working in the buildings sector need to account for climate mitigation through not just energy efficiency, but sufficiency, as well as adaptation and resilience to local current and future climates and hazards, and the social cohesion and community of the people they are building for.

KEY TAKEAWAYS FROM CHAPTER 2

- There is an urgent need to reduce GHG emissions from buildings and adapt them to the changes in climate change-related hazards that will occur in the future.
- Often, low-income populations are more exposed to climate-related environmental hazards despite being the least responsible for GHG emissions – a concept known as climate injustice.
- There is increasing evidence of climate injustice in residential indoor environments, with differences in the emissions from housing as well as indoor exposure to climate-related environmental hazards between population groups. This is both due to housing and environmental quality, as well as the financial resources to adapt.
- Designing to regulatory standards will not address this injustice. Rather, homes should be designed to be as energy efficient as possible and stress-tested for future climates in order to provide resilient housing for a range of occupants with different means and abilities.

Bibliography

Arden Pope, C., Burnett, R. T., Turner, M. C., Cohen, A., Krewski, D., Jerrett, M., Gapstur, S. M., & Thun, M. J. (2011). Lung cancer and cardiovascular disease mortality associated with ambient air pollution and cigarette smoke: shape of the exposure-response relationships. *Environmental Health Perspectives*, *119*(11), pp. 1616–1621. https://doi.org/10.1289/ehp.1103639 (Accessed: 9 November 2023)

Atkinson, R. W., Analitis, A., Samoli, E., Fuller, G. W., Green, D. C., Mudway, I. S., Anderson, H. R., & Kelly, F. J. (2016). Short-term exposure to traffic-related air pollution and daily mortality in London, UK. *Journal of Exposure Science and Environmental Epidemiology*, *26*(2), pp. 125–132. https://doi.org/10.1038/jes.2015.65 (Accessed: 9 November 2023)

Basu, R. (2002). Relation between elevated ambient temperature and mortality: a review of the epidemiologic evidence. *Epidemiologic Reviews*, *24*(2), pp. 190–202. https://doi.org/10.1093/epirev/mxf007 (Accessed: 9 November 2023)

Bennett, O. (2010). *Reducing Flood Risk*. UK Parliament.

Blaikie, P., Cannon, T., Davis, I., & Wisner, B. (2005). At Risk: Natural Hazards, People's Vulnerability and Disasters. In *At Risk*. Routledge. https://doi.org/10.4324/9780203974575 (Accessed: 9 November 2023)

Brainard, J. S., Jones, A. P., Bateman, I. J., Lovett, A. A., & Fallon, P. J. (2002). Modelling environmental equity: access to air quality in Birmingham, England. *Environment and Planning A: Economy and Space*, *34*(4), pp. 695–716. https://doi.org/10.1068/a34184 (Accessed: 9 November 2023)

Braithwaite, I., Zhang, S., Kirkbride, J. B., Osborn, D. P. J., & Hayes, J. F. (2019). Air pollution (particulate matter) exposure and associations with depression, anxiety, bipolar, psychosis and suicide risk: a systematic review and meta-analysis. *Environmental Health Perspectives*, *127*(12). https://doi.org/10.1289/EHP4595 (Accessed: 9 November 2023)

Braubach, M., & Fairburn, J. (2010). Social inequities in environmental risks associated with housing and residential location – a review of evidence. *European Journal of Public Health*, *20*(1), pp. 36–42. https://doi.org/10.1093/EURPUB/CKP221 (Accessed: 9 November 2023)

Bray, I., Gibson, A., & White, J. (2020). Coronavirus disease 2019 mortality: a multivariate ecological analysis in relation to ethnicity, population density, obesity, deprivation and pollution. *Public Health*, 185, pp. 261–263. https://doi.org/10.1016/j.puhe.2020.06.056 (Accessed: 25 January 2025)

Brunt, H., Barnes, J., Jones, S. J., Longhurst, J. W. S., Scally, G., & Hayes, E. (2017). Air pollution, deprivation and health: Understanding relationships to add value to local air quality management policy and practice in Wales, UK. *Journal of Public Health*, *39*(3), pp. 485–497. https://doi.org/10.1093/pubmed/fdw084 (Accessed: 9 November 2023)

Chakraborty, J., Collins, T. W., & Grineski, S. E. (2019). Exploring the environmental justice implications of Hurricane Harvey flooding in Greater Houston, Texas. *American Journal of Public Health*, *109*(2), pp. 244–250. https://doi.org/10.2105/AJPH.2018.304846 (Accessed: 9 November 2023)

Chakraborty, J., Collins, T. W., Montgomery, M. C., & Grineski, S. E. (2014). Social and spatial inequities in exposure to flood risk in Miami, Florida. *Natural Hazards Review*, *15*(3), 04014006. https://doi.org/10.1061/(ASCE)NH.1527-6996.0000140 (Accessed: 9 November 2023)

Chakraborty, T., Hsu, A., Manya, D., & Sheriff, G. (2019). Disproportionately higher exposure to urban heat in lower-income neighborhoods: a multi-city perspective. *Environmental Research Letters*, *14*(10), 105003. https://doi.org/10.1088/1748-9326/AB3B99 (Accessed: 9 November 2023)

Chen, Y., Yue, W., & La Rosa, D. (2020). Which communities have better accessibility to green space? An investigation into environmental inequality using big data. *Landscape and Urban Planning*, *204*, 103919. https://doi.org/10.1016/J.LANDURBPLAN.2020.103919 (Accessed: 9 November 2023)

Choi, G., Heo, S., & Lee, J. T. (2016). Assessment of environmental injustice in Korea using synthetic air quality index and multiple indicators of socioeconomic status: a cross-sectional study. *Journal of the Air and Waste Management Association*, *66*(1), pp. 28–37. https://doi.org/10.1080/10962247.2015.1107657 (Accessed: 9 November 2023)

Clark, L. P., Millet, D. B., & Marshall, J. D. (2017). Changes in transportation-related air pollution exposures by race-ethnicity and socioeconomic status: outdoor nitrogen dioxide in the United States in 2000 and 2010. *Environmental Health Perspectives*, *125*(9), 097012. https://doi.org/10.1289/EHP959 (Accessed: 9 November 2023)

Colmer, J., Hardman, I., Shimshack, J., & Voorheis, J. (2020). Disparities in PM2.5 air pollution in the United States. *Science*, *369*(6503), pp. 575–578. https://doi.org/10.1126/SCIENCE.AAZ9353 (Accessed: 9 November 2023)

DEFRA. (2021). *Emissions of air pollutants in the UK*. Department for Environment, Food & Rural Affairs. www.gov.uk/government/statistics/emissions-of-air-pollutants/emissions-of-air-pollutants-in-the-uk-summary (Accessed: 9 November 2023)

Delgado-Saborit, J. M., Guercio, V., Gowers, A. M., Shaddick, G., Fox, N. C., & Love, S. (2020). A critical review of the epidemiological evidence of effects of air pollution on dementia, cognitive function and cognitive decline in adult population. *Science of The Total Environment*, 143734. https://doi.org/10.1016/j.scitotenv.2020.143734 (Accessed: 9 November 2023)

Dempsey, N., Bramley, G., Power, S., & Brown, C. (2011). The social dimension of sustainable development: defining urban social sustainability. *Sustainable Development*, *19*(5), pp. 289–300. https://doi.org/10.1002/sd.417 (Accessed: 9 November 2023)

Diderichsen, F., Evans, T., & Whitehead, M. (2001). The Social Basis of Disparities in Health. In T. Evans (Ed.), *Challenging Inequities in Health: From Ethics to Action*. Oxford University Press. https://doi.org/10.1093/acprof:oso/9780195137408.003.0002 (Accessed: 9 November 2023)

Du, W., FitzGerald, G. J., Clark, M., & Hou, X. Y. (2010). Health impacts of floods. *Prehospital and Disaster Medicine*, *25*(3), 265–272. https://doi.org/10.1017/S1049023X00008141 (Accessed: 9 November 2023)

Elhacham, E., Ben-Uri, L., Grozovski, J., Bar-On, Y. M., & Milo, R. (2020). Global human-made mass exceeds all living biomass. *Nature*, *588*(7838), 442–444. https://doi.org/10.1038/s41586-020-3010-5 (Accessed: 9 November 2023)

Environmental Protection Agency. (2021). *Climate Change and Social Vulnerability in the United States: A Focus on Six Impacts*. U.S. Environmental Protection Agency. www.epa.gov/cira/social-vulnerability-report (Accessed: 9 November 2023)

European Commission. (2018). Report from the commission to the European Parliament, the Council, the European Economic and Social Committee and the Committee of the Regions on the implementation of EU waste legislation, including the early warning report for Member States at risk of missing the 2020 preparation for re-use/recycling target on municipal waste.

European Commission. (2019). *Energy performance of buildings directive*. https://ec.europa.eu/energy/topics/energy-efficiency/energy-efficient-buildings/energy-performance-buildings-directive_en (Accessed: 9 November 2023)

Fairburn, J., Schüle, S. A., Dreger, S., Hilz, L. K., & Bolte, G. (2019). Social inequalities in exposure to ambient air pollution: A systematic review in the WHO European region. *International Journal of Environmental Research and Public Health*, *16*(17). https://doi.org/10.3390/ijerph16173127 (Accessed: 9 November 2023)

Fecht, D., Fischer, P., Fortunato, L., Hoek, G., De Hoogh, K., Marra, M., Kruize, H., Vienneau, D., Beelen, R., & Hansell, A. (2015). Associations between air pollution and socioeconomic characteristics, ethnicity and age profile of neighbourhoods in England and the Netherlands. *Environmental Pollution*, *198*, pp. 201–210. https://doi.org/10.1016/j.envpol.2014.12.014 (Accessed: 9 November 2023)

Ferguson, L., Taylor, J., Davies, M., Shrubsole, C., Symonds, P., & Dimitroulopoulou, S. (2020). Exposure to indoor air pollution across socio-economic groups in high-income countries: a scoping review of the literature and a modelling methodology. *Environment International*, *140*, 105748.

Ferguson, L., Taylor, J., Zhou, K., Shrubsole, C., Symonds, P., Davies, M., Dimitroulopoulou, S., Shrubsole, C., Symonds, P., Davies, M., & Dimitroulopoulou, S. (2021). Systemic inequalities in indoor air pollution exposure in London, UK. *Buildings and Cities*, *2*(1), pp. 425–448. https://doi.org/10.5334/BC.100 (Accessed: 9 November 2023)

Fonseca, X., Lukosch, S., & Brazier, F. (2018). Social cohesion revisited: a new definition and how to characterize it. *Innovation: The European Journal of Social Science Research*, *32*(2), pp. 231–253. https://doi.org/10.1080/13511610.2018.1497480 (Accessed: 9 November 2023)

Gianfrate, V., Piccardo, C., Longo, D., & Giachetta, A. (2017). Rethinking social housing: behavioural patterns and technological innovations. *Sustainable Cities and Society*, *33*, pp. 102–112. https://doi.org/10.1016/J.SCS.2017.05.015 (Accessed: 9 November 2023)

Green, S. D., & Sergeeva, N. (2020). The contested privileging of zero carbon: plausibility, persuasiveness and professionalism. *Buildings and Cities*, *1*(1), pp. 491–503. https://doi.org/10.5334/BC.49 (Accessed: 9 November 2023)

Grunewald, N., Klasen, S., Martínez-Zarzoso, I., & Muris, C. (2017). The trade-off between income inequality and carbon dioxide emissions. *Ecological Economics*, *142*, pp. 249–256. https://doi.org/10.1016/j.ecolecon.2017.06.034 (Accessed: 9 November 2023)

Hall, M., & Bailey, P. (2021). *Social deprivation and the likelihood of flooding*. Environment Agency.

Hamilton, I. G. I., Shipworth, D., Summerfield, A. J., Steadman, P., Oreszczyn, T., & Lowe, R. (2014). Uptake of energy efficiency interventions in English dwellings. *Building Research & Information*, *42*(3), pp. 255–275. https://doi.org/10.1080/09613218.2014.867643 (Accessed: 9 November 2023)

Harlan, S. L., Brazel, A. J., Prashad, L., Stefanov, W. L., & Larsen, L. (2006). Neighborhood microclimates and vulnerability to heat stress. *Social Science & Medicine*, *63*(11), pp. 2847–2863. https://doi.org/10.1016/J.SOCSCIMED.2006.07.030 (Accessed: 9 November 2023)

Heaviside, C., & Macintyre, H. (2019). The impact of the Urban Heat Island on heat related mortality in a European city, and potential benefits of adaptation measures. *Geophysical Research Abstracts*, *21*, 2019–15169.

Heaviside, C., Macintyre, H., & Vardoulakis, S. (2017). The urban heat island: implications for health in a changing environment. *Current Environmental Health Reports*, *4*(3), pp. 296–305. https://doi.org/10.1007/S40572-017-0150-3/FIGURES/1 (Accessed: 9 November 2023)

Herrera, M., Natarajan, S., Coley, D. A., Kershaw, T., Ramallo-González, A. P., Eames, M., Fosas, D., & Wood, M. (2017). A review of current and future weather data for building simulation, *38*(5), pp. 602–627. https://doi.org/10.1177/0143624417705937 (Accessed: 9 November 2023)

Hoffimann, E., Barros, H., & Ribeiro, A. I. (2017). Socioeconomic inequalities in green space quality and accessibility – evidence from a southern European city. *International Journal of Environmental Research and Public Health*, *14*(8), 916 https://doi.org/10.3390/IJERPH14080916 (Accessed: 9 November 2023)

Hoffman, J. S., Shandas, V., & Pendleton, N. (2020). The effects of historical housing policies on resident exposure to intra-urban heat: a study of 108 US urban areas. *Climate*, *8*(1), p. 12. https://doi.org/10.3390/CLI8010012 (Accessed: 9 November 2023)

Horsdal, H. T., Agerbo, E., McGrath, J. J., Vilhjálmsson, B. J., Antonsen, S., Closter, A. M., Timmermann, A., Grove, J., Mok, P. L. H., Webb, R. T., Sabel, C. E., Hertel, O., Sigsgaard, T., Erikstrup, C., Hougaard, D. M., Werge, T., Nordentoft, M., Børglum, A. D., Mors, O., … Pedersen, C. B. (2019). Association of childhood exposure to nitrogen dioxide and polygenic risk score for schizophrenia with the risk of developing schizophrenia. *JAMA Network Open*, *2*(11), e1914401. https://doi.org/10.1001/jamanetworkopen.2019.14401 (Accessed: 9 November 2023)

Hsu, A., Sheriff, G., Chakraborty, T., & Manya, D. (2021). Disproportionate exposure to urban heat island intensity across major US cities. *Nature Communications*, *12*(1), pp. 1–11. https://doi.org/10.1038/s41467-021-22799-5 (Accessed: 9 November 2023)

Hurst, W. (2021, November 10). Typical architect generates '162 American lifestyles' worth of carbon emissions. *Architects' Journal*. www.architectsjournal.co.uk/news/typical-architect-generates-162-american-lifestyles-worth-of-carbon-emissions?tkn=1 (Accessed: 9 November 2023)

Intergovernmental Panel on Climate Change. (2021). Climate Change 2021: The Physical Science Basis. Contribution of Working Group I to the Sixth Assessment Report of the Intergovernmental Panel on Climate Change (V. Masson-Delmotte, P. Zhai, A. Pirani, S. L. Connors, C. Péan, S. Berger, N. Caud, Y. Chen, L. Goldfarb, M. I. Gomis, M. Huang, K. Leitzell, E. Lonnoy, J. B. R. Matthews, T. K. Maycock, T. Waterfield, O. Yelekçi, R. Yu, & B. Zhou, Eds.). Cambridge University Press.

Ivanova, D., & Wood, R. (2020). The unequal distribution of household carbon footprints in Europe and its link to sustainability. *Global Sustainability*, *3*. https://doi.org/10.1017/SUS.2020.12 (Accessed: 9 November 2023)

Jenerette, G. D., Harlan, S. L., Brazel, A., Jones, N., Larsen, L., & Stefanov, W. L. (2006). Regional relationships between surface temperature, vegetation, and human settlement in a rapidly urbanizing ecosystem. *Landscape Ecology*, *22*(3), pp. 353–365. https://doi.org/10.1007/S10980-006-9032-Z (Accessed: 9 November 2023)

Kantar. (2020). *Burning in UK Homes and Gardens*. Department for Environment, Food, and Rural Affairs.

Kenway, P. et al. (2020) *People and places in London most vulnerable to COVID-19 and its social and economic consequences*. New Policy Institute.

Kimpian, J., Hartman, H., & Pelsmakers, S. (2021). *Energy / People / Buildings: Making Sustainable Architecture Work*. RIBA Publishing.

Klinenberg, E. (1999). Denaturalizing disaster: a social autopsy of the 1995 Chicago heat wave. *Theory and Society*, *28*(2), pp. 239–295.
Klinsky, S., & Mavrogianni, A. (2020). Climate justice and the built environment. *Buildings and Cities*, *1*(1), pp. 412–428. https://doi.org/10.5334/bc.65 (Accessed: 9 November 2023)
Kovats, R., & Hajat, S. (2008). Heat stress and public health: a critical review. *Annu. Rev. Public Health*. www.annualreviews.org/doi/abs/10.1146/annurev.publhealth.29.020907.090843 (Accessed: 9 November 2023)
Lader, D., Short, S., & Gershuny, J. (2006). *The Time Use Survey, 2005, How we spend our time* (Issue August). Office for National Statistics (ONS).
Laine, M., Helamaa, A., Kuoppa, J., & Alatalo, E. (2018). Bricolage in collaborative housing in Finland: combining resources for alternative housing solutions. *Housing, Theory and Society*, *37*(1), pp. 101–117. https://doi.org/10.1080/14036096.2018.1492438 (Accessed: 9 November 2023)
Li, M., Gu, S., Bi, P., Yang, J., & Liu, Q. (2015). Heat waves and morbidity: current knowledge and further direction – a comprehensive literature review. *International Journal of Environmental Research and Public Health*, *12*(5), pp. 5256–5283. https://doi.org/10.3390/IJERPH120505256 (Accessed: 9 November 2023)
Li, V. O., Han, Y., Lam, J. C., Zhu, Y., & Bacon-Shone, J. (2018). Air pollution and environmental injustice: are the socially deprived exposed to more PM2.5 pollution in Hong Kong? *Environmental Science and Policy*, *80*, pp. 53–61. https://doi.org/10.1016/j.envsci.2017.10.014 (Accessed: 9 November 2023)
Lomas, K. J., Watson, S., Allinson, D., Fateh, A., Beaumont, A., Allen, J., Foster, H., & Garrett, H. (2021). Dwelling and household characteristics' influence on reported and measured summertime overheating: a glimpse of a mild climate in the 2050's. *Building and Environment*, *201*, 107986. https://doi.org/10.1016/j.buildenv.2021.107986 (Accessed: 9 November 2023)
Macintyre, H. L., Heaviside, C., Taylor, J., Picetti, R., Symonds, P., Cai, X.-M. M., & Vardoulakis, S. (2018). Assessing urban population vulnerability and environmental risks across an urban area during heatwaves – implications for health protection. *Science of the Total Environment*, 610–611, pp. 678–690. https://doi.org/10.1016/j.scitotenv.2017.08.062 (Accessed: 9 November 2023)
Mavrogianni, A., Davies, M., Taylor, J., Chalabi, Z., Biddulph, P., Oikonomou, E., Das, P., & Jones, B. (2014). The impact of occupancy patterns, occupant-controlled ventilation and shading on indoor overheating risk in domestic environments. *Building and Environment*, *78*, pp. 183–198. https://doi.org/10.1016/j.buildenv.2014.04.008 (Accessed: 9 November 2023)
Meadowcroft, J., Banister, D., Holden, E., Langhelle, O., Linnerud, K., & Gilpin, G. (2019). *What next for sustainable development? Our common future at thirty*. Edward Elgar Publishing.
Medina-Ramón, M., Zanobetti, A., Cavanagh, D. P., & Schwartz, J. (2006). Extreme temperatures and mortality: assessing effect modification by personal characteristics and specific cause of death in a multi-city case-only analysis. *Environmental Health Perspectives*, *114*(9), pp. 1331–1336.
MHCLG. (2020). *English Housing Survey, 2018: Housing Stock Data*. UK Data Service. https://doi.org/10.5255/UKDA-SN-8670-1 (Accessed: 9 November 2023)
Milojevic, A., Niedzwiedz, C. L., Pearce, J., Milner, J., MacKenzie, I. A., Doherty, R. M., & Wilkinson, P. (2017). Socioeconomic and urban-rural differentials in exposure to air pollution and mortality burden in England. *Environmental Health*, *16*(1), pp. 1–10. https://doi.org/10.1186/s12940-017-0314-5 (Accessed: 9 November 2023)
Morawska, L. et al. (2020). 'How can airborne transmission of COVID-19 indoors be minimised?', *Environment International*, 142, p. 105832. https://doi.org/10.1016/J.ENVINT.2020.105832 (Accessed: 25 January 2025)
Mueller, N., Rojas-Rueda, D., Khreis, H., Cirach, M., Andrés, D., Ballester, J., Bartoll, X., Daher, C., Deluca, A., Echave, C., Milà, C., Márquez, S., Palou, J., Pérez, K., Tonne, C., Stevenson, M., Rueda, S., & Nieuwenhuijsen, M. (2020). Changing the urban design of cities for health:

the superblock model. *Environment International*, *134*, 105132. https://doi.org/10.1016/J.ENVINT.2019.105132 (Accessed: 9 November 2023)

Mullins, J. T., & White, C. (2019). Temperature and mental health: evidence from the spectrum of mental health outcomes. *Journal of Health Economics*, *68*, 102240. https://doi.org/10.1016/J.JHEALECO.2019.102240 (Accessed: 9 November 2023)

Murage, P., Kovats, S., Sarran, C., Taylor, J., McInnes, R., & Hajat, S. (2020). What individual and neighbourhood-level factors increase the risk of heat-related mortality? A case-crossover study of over 185,000 deaths in London using high-resolution climate datasets. *Environment International*, *134*, 105292. https://doi.org/10.1016/J.ENVINT.2019.105292 (Accessed: 9 November 2023)

National Institute for Health and Care Excellence. (2013). *Smoking: Harm reduction.*

Nielsen, K. S., Nicholas, K. A., Creutzig, F., Dietz, T., & Stern, P. C. (2021). The role of high-socioeconomic-status people in locking in or rapidly reducing energy-driven greenhouse gas emissions. *Nature Energy*, pp. 1–6. https://doi.org/10.1038/s41560-021-00900-y (Accessed: 9 November 2023)

Obradovich, N., Migliorini, R., Paulus, M. P., & Rahwan, I. (2018). Empirical evidence of mental health risks posed by climate change. *Proceedings of the National Academy of Sciences*, *115*(43), pp. 10953–10958. https://doi.org/10.1073/PNAS.1801528115 (Accessed: 9 November 2023)

Olsen, Y., Nøjgaard, J. K., Olesen, H. R., Brandt, J., Sigsgaard, T., Pryor, S. C., Ancelet, T., Viana, M. del M., Querol, X., & Hertel, O. (2020). Emissions and source allocation of carbonaceous air pollutants from wood stoves in developed countries: a review. *Atmospheric Pollution Research*, *11*(2), pp. 234–251. https://doi.org/10.1016/j.apr.2019.10.007 (Accessed: 9 November 2023)

O'Neill, M. S., Zanobetti, A., & Schwartz, J. (2005). Disparities by race in heat-related mortality in four US cities: the role of air conditioning prevalence. *Journal of Urban Health: Bulletin of the New York Academy of Medicine*, *82*(2), pp. 191–197. https://doi.org/10.1093/jurban/jti043 (Accessed: 9 November 2023)

Padula, A. M., Tager, I. B., Carmichael, S. L., Hammond, S. K., Yang, W., Lurmann, F., & Shaw, G. M. (2013). Ambient air pollution and traffic exposures and congenital heart defects in the San Joaquin Valley of California. *Paediatric and Perinatal Epidemiology*, *27*(4), pp. 329–339. https://doi.org/10.1111/ppe.12055 (Accessed: 9 November 2023)

Patel, J. A. et al. (2020). Poverty, inequality and COVID-19: the forgotten vulnerable. *Public Health*, 183, p. 110. https://doi.org/10.1016/J.PUHE.2020.05.006 (Accessed: 25 January 2025)

Pham, T. T. H., Apparicio, P., Séguin, A. M., Landry, S., & Gagnon, M. (2012). Spatial distribution of vegetation in Montreal: an uneven distribution or environmental inequity? *Landscape and Urban Planning*, *107*(3), pp. 214–224. https://doi.org/10.1016/J.LANDURBPLAN.2012.06.002 (Accessed: 9 November 2023)

Pinault, L., Crouse, D., Jerrett, M., Brauer, M., & Tjepkema, M. (2016). Socioeconomic differences in nitrogen dioxide ambient air pollution exposure among children in the three largest Canadian cities. *Health Reports*, *27*(7), pp. 3–10.

Puckett, K., & Gethering, W. (2019). *Design for Climate Change*. RIBA Publishing. https://doi.org/10.4324/9780429347641 (Accessed: 9 November 2023)

Rios, F. C., Chong, W. K., & Grau, D. (2015). Design for disassembly and deconstruction – challenges and opportunities. *Procedia Engineering*, *118*, pp. 1296–1304. https://doi.org/10.1016/J.PROENG.2015.08.485 (Accessed: 9 November 2023)

Röck, M., Saade, M. R. M., Balouktsi, M., Rasmussen, F. N., Birgisdottir, H., Frischknecht, R., Habert, G., Lützkendorf, T., & Passer, A. (2020). Embodied GHG emissions of buildings – the hidden challenge for effective climate change mitigation. *Applied Energy*, *258*, 114107. https://doi.org/10.1016/J.APENERGY.2019.114107 (Accessed: 9 November 2023)

Romanello, M., McGushin, A., Di Napoli, C., Drummond, P., Hughes, N., Jamart, L., Kennard, H., Lampard, P., Solano Rodriguez, B., Arnell, N., Ayeb-Karlsson, S., Belesova, K., Cai, W.,

Campbell-Lendrum, D., Capstick, S., Chambers, J., Chu, L., Ciampi, L., Dalin, C., ... Hamil ton, I. (2021). The 2021 report of the Lancet Countdown on health and climate change: code red for a healthy future. *The Lancet*. https://doi.org/10.1016/S0140-6736(21)01787-6 (Accessed: 9 November 2023)

Rose, G. (1992). *Rose's Strategy of Preventive Medicine*. Oxford University Press.

Ruuhela, R., Votsis, A., Kukkonen, J., Jylhä, K., Kankaanpää, S., & Perrels, A. (2020). Temperature-related mortality in Helsinki compared to its surrounding region over two decades, with special emphasis on intensive heatwaves. *Atmosphere*, *12*(1), p. 46. https://doi.org/10.3390/ATMOS12010046 (Accessed: 9 November 2023)

Saarimaa, S., & Pelsmakers, S. (2020). Better living environment today, more adaptable tomorrow? Comparative analysis of Finnish apartment buildings and their adaptable scenarios. *Yhdyskuntasuunnittelu*, *58*(2 SE-Artikkelit). https://doi.org/10.33357/ys.89676 (Accessed: 9 November 2023)

SAGE (2020a). *COVID-19 housing impacts. For SAGE meeting 10 September 2020*. Scientific Advisory Group for Emergencies.

SAGE (2020b). *Housing, household transmission and ethnicity: For SAGE meeting 26 November 2020*. Scientific Advisory Group for Emergencies.

Samoli, E., Stergiopoulou, A., Santana, P., Rodopoulou, S., Mitsakou, C., Dimitroulopoulou, C., Bauwelinck, M., de Hoogh, K., Costa, C., Marí-Dell'Olmo, M., Corman, D., Vardoulakis, S., & Katsouyanni, K. (2019). Spatial variability in air pollution exposure in relation to socioeconomic indicators in nine European metropolitan areas: a study on environmental inequality. *Environmental Pollution*, *249*, pp. 345–353. https://doi.org/10.1016/j.envpol.2019.03.050 (Accessed: 9 November 2023)

Sanchez-Guevara, C., Núñez Peiró, M., Taylor, J., Mavrogianni, A., & Neila González, J. (2019). Assessing population vulnerability towards summer energy poverty: case studies of Madrid and London. *Energy and Buildings*, *190*. https://doi.org/10.1016/j.enbuild.2019.02.024 (Accessed: 9 November 2023)

Saverino, K. C., Routman, E., Lookingbill, T. R., Eanes, A. M., Hoffman, J. S., & Bao, R. (2021). Thermal inequity in Richmond, VA: the effect of an unjust evolution of the urban landscape on urban heat islands. *Sustainability*, *13*(3), p. 1511. https://doi.org/10.3390/SU13031511 (Accessed. 9 November 2023)

Semenza, J. C., Rubin, C. H., Falter, K. H., Selanikio, J. D., Flanders, W. D., Howe, H. L., & Wilhelm, J. L. (1996). Heat-related deaths during the July 1995 heat wave in Chicago. *New England Journal of Medicine*, *335*(2), pp. 84–90. https://doi.org/10.1056/nejm199607113350203 (Accessed: 9 November 2023)

Sennett, R. (2019). *Building and dwelling: ethics for the city*. Penguin Books.

Shrubsole, C., Macmillan, A., Davies, M., & May, N. (2014). 100 Unintended consequences of policies to improve the energy efficiency of the UK housing stock. *Indoor and Built Environment*, *23*(3), pp. 340–352. https://doi.org/10.1177/1420326X14524586 (Accessed: 9 November 2023)

Sjöblom, J., Kuoppa, J., Laine, M., & Alatalo, E. (2020). Crafting a planning issue with citizens in the context of planning competition: a case of 'Nordic Superblock.' *Journal of Urban Design*, *26*(1), pp. 117–131. https://doi.org/10.1080/13574809.2020.1832886 (Accessed: 9 November 2023)

Smiley, K. T. (2020). Social inequalities in flooding inside and outside of floodplains during Hurricane Harvey. *Environmental Research Letters*, *15*(9), 0940b3. https://doi.org/10.1088/1748-9326/ABA0FE (Accessed: 9 November 2023)

Solecki, W. D., Rosenzweig, C., Parshall, L., Pope, G., Clark, M., Cox, J., & Wiencke, M. (2011). Mitigation of the heat island effect in urban New Jersey. *Global Environmental Change Part B: Environmental Hazards*, *6*(1), pp. 39–49. https://doi.org/10.1016/J.HAZARDS.2004.12.002 (Accessed: 9 November 2023)

Sunikka-Blank, M., & Galvin, R. (2012). Introducing the prebound effect: The gap between performance and actual energy consumption. *Building Research & Information*, *40*(3), 260–273. https://doi.org/10.1080/09613218.2012.690952 (Accessed: 9 November 2023)

Tapsell, S. M., Penning-Rowsell, E., Tunstall, S., & Wilson, T. (2002). Vulnerability to flooding: health and social dimensions. *Philosophical Transactions of the Royal Society*, *360*, pp. 1511–1525.

Taylor, J., Biddulph, P., Davies, M., Ridley, I., Mavrogianni, A., Oikonomou, E., & Lai, K. M. M. (2012). Using building simulation to model the drying of flooded building archetypes. *Journal of Building Performance Simulation*, *51*(2), pp. 182–195. https://doi.org/10.1080/19401493.2012.703243 (Accessed: 9 November 2023)

Taylor, J., Mavrogianni, A., Davies, M., Das, P., Shrubsole, C., Biddulph, P., & Oikonomou, E. (2015). Understanding and mitigating overheating and indoor PM2.5 risks using coupled temperature and indoor air quality models. *Building Services Engineering Research and Technology*, *36*(2), pp. 275–289. https://doi.org/10.1177/0143624414566474 (Accessed: 9 November 2023)

Taylor, J., Symonds, P., Wilkinson, P., Heaviside, C., Macintyre, H., Davies, M., Mavrogianni, A., & Hutchinson, E. (2018). Estimating the influence of housing energy efficiency and overheating adaptations on heat-related mortality in the West Midlands, UK. *Atmosphere*, *9*(5), Article 5. https://doi.org/10.3390/atmos9050190 (Accessed: 9 November 2023)

Thomson, H., Thomas, S., Sellstrom, E., & Petticrew, M. (2013). Housing improvements for health and associated socio-economic outcomes. *Cochrane Database of Systematic Reviews*, *2013*(2). https://doi.org/10.1002/14651858.CD008657.pub2 (Accessed: 9 November 2023)

Till, J. (2020). Architecture After Architecture. In *Architects After Architecture*. Routledge.

Tonne, C., Milà, C., Fecht, D., Alvarez, M., Gulliver, J., Smith, J., Beevers, S., Ross Anderson, H., & Kelly, F. (2018). Socioeconomic and ethnic inequalities in exposure to air and noise pollution in London. *Environment International*, *115*, pp. 170–179. https://doi.org/10.1016/j.envint.2018.03.023 (Accessed: 9 November 2023)

Umit, R., Poortinga, W., Jokinen, P., & Pohjolainen, P. (2019). The role of income in energy efficiency and curtailment behaviours: findings from 22 European countries. *Energy Research & Social Science*, *53*, pp. 206–214. https://doi.org/10.1016/J.ERSS.2019.02.025 (Accessed: 9 November 2023)

Vicedo-Cabrera, A. M., Scovronick, N., Sera, F., Royé, D., Schneider, R., Tobias, A., Astrom, C., Guo, Y., Honda, Y., Hondula, D. M., Abrutzky, R., Tong, S., Coelho, M. de S. Z. S., Saldiva, P. H. N., Lavigne, E., Correa, P. M., Ortega, N. V., Kan, H., Osorio, S., … Gasparrini, A. (2021). The burden of heat-related mortality attributable to recent human-induced climate change. *Nature Climate Change*, *11*(6), pp. 492–500. https://doi.org/10.1038/s41558-021-01058-x (Accessed: 9 November 2023)

Visscher, H. J., & Meijer, F. M. (2008). The meaning of the protection of the architects title in European countries. In R. Haigh, & D. Amaratunga (Eds.), *Building Resilience* (pp. 1287–1298). School of the built environment, University of Salford, UK. www.irbnet.de/daten/iconda/CIB11461.pdf (Accessed: 25 January 2025)

Walker, G., & Burningham, K. (2011). Flood risk, vulnerability and environmental justice: evidence and evaluation of inequality in a UK context: *Critical Social Policy*, *31*(2), pp. 216–240. https://doi.org/10.1177/0261018310396149 (Accessed: 9 November 2023)

Way, M., & Bordass, W. (2009). The Soft Landings Framework: For better briefing, design, handover and building performance in-use. Building Services Research and Information Association (BSRIA).

Whitmee, S., Haines, A., Beyrer, C., Boltz, F., Capon, A. G., De Souza Dias, B. F., Ezeh, A., Frumkin, H., Gong, P., Head, P., Horton, R., Mace, G. M., Marten, R., Myers, S. S., Nishtar, S., Osofsky, S. A., Pattanayak, S. K., Pongsiri, M. J., Romanelli, C., … Yach, D. (2015). Safeguarding human health in the Anthropocene epoch: Report of the Rockefeller Foundation-Lancet

Commission on planetary health. *The Lancet*, *386*(10007), pp. 1973–2028. https://doi.org/10.1016/S0140-6736(15)60901-1 (Accessed: 9 November 2023)

World Health Organization. (1948). *Constitution of the World Health Organization*.

World Health Organization. (2009). *WHO Guidelines for Indoor Air Quality: Dampness and Mould*.

World Health Organization. (2018). *Housing and health guidelines*. World Health Organization. www.who.int/sustainable-development/publications/housing-health-guidelines/en/

Whyte, J., & Gann, D. M. (2001). Closing the loop between design and use: post-occupancy evaluation. *Building Research and Information*, *29*(6), pp. 460–462. https://doi.org/10.1080/09613210110072683 (Accessed: 9 November 2023)

Willand, N., Ridley, I., & Maller, C. (2015). Towards explaining the health impacts of residential energy efficiency interventions – a realist review. Part 1: Pathways. *Social Science & Medicine (1982)*, *133*, pp. 191–201. https://doi.org/10.1016/J.SOCSCIMED.2015.02.005 (Accessed: 9 November 2023)

Wilson, B. (2020). Urban heat management and the legacy of redlining. *Journal of the American Planning Association*, *86*(4), pp. 443–457. https://doi.org/10.1080/01944363.2020.1759127 (Accessed: 9 November 2023)

Xu, R., Zhao, Q., Coelho, M. S. Z. S., Saldiva, P. H. N., Abramson, M. J., Li, S., & Guo, Y. (2020). Socioeconomic inequality in vulnerability to all-cause and cause-specific hospitalisation associated with temperature variability: a time-series study in 1814 Brazilian cities. *The Lancet Planetary Health*, *4*(12), pp. e566–e576. https://doi.org/10.1016/S2542-5196(20)30251-5 (Accessed: 9 November 2023)

3

THE NEW AESTHETICS OF CIRCULAR ARCHITECTURE

Antti Lehto and Sanna Lehtinen

Introduction

Currently, the field of architecture is struggling to meet the challenges of the ongoing climate emergency. One potential new direction for architecture is circularity. The greater underlying shift is the circular economy, which has gained a growing foothold in policymaking, academic research and practical projects. This framework focuses on narrowing, slowing and closing resource loops in the construction sector by avoiding new construction, extending lifecycles and reusing building materials (Geissdoerfer et al., 2017). The concept of the circular economy has been criticised for being vaguely defined and for retaining the consumption model of global capitalist economics (Corvellec et al., 2022). Different schools of thought regarding the profoundness of change have been identified (Reike et al., 2018). From these schools, we take for our study the transformational model of circular economy (Reike et al., 2018), which directly addresses the change needed at the foundations of production. If viewed in conjunction with this larger societal shift, a revolution in the construction sector and architecture can occur. It will not only reduce the environmental impact of buildings but it will, and should, be a visible and genuinely design-changing transformation.

A circular approach to *construction* can be described as "the dynamic total of associated processes, materials, and stakeholders that accommodate circular flows of building materials and components" (Geissdoerfer et al., 2017). However, architecture is more than mere buildings, as the component of art has unarguably been embedded in the concept. Here, aesthetics has a central role, as it refers to the human experience that interlinks the physical environment with abstract thinking. The aesthetics of sustainable architecture have been discussed by few other scholars (e.g., Lee, 2011; Hosey, 2012; Koźmińska, 2019a). Regarding circularity, handbooks and architectural publications are even more scarce, and the ones available typically highlight the aesthetic outcomes, which are valuable as such, but leave the philosophical dimension rather undeveloped (Gorgolewski, 2017; Heisel et al., 2022; ter Steege, 2021).

DOI: 10.4324/9781003293903-4
This chapter has been made available under a CC-BY-NC-ND 4.0 license.

This chapter is structured similarly to Winters (2007): in other words, theory tells us *how best to build*, and practice means *to design and build*, whereas the philosophy of architecture attempts to answer the question *what architecture is*. The latter is a classic question in the philosophy of architecture, yet it has not been adequately discussed in relation to the emerging circular architectural practice and theory. We ask two questions in this chapter: (1) what is the conceptual framework of circular architecture, as informed by philosophical aesthetics; and (2) what are the implications of the resulting framework for the current theory and practice of circular construction?

Seven design strategies for circular construction

In its current form, the theory of circular construction is a multifaceted array of design strategies. For instance, Eberhardt et al. (2020) have identified 16 building and engineering strategies in their literature review based on articles on both theory and practice in the industry. Remaining on a more general level in the context of product design, Bocken et al. (2016) recognise nine strategies. Nordby et al. (2008) describe six strategies, focusing on the salvageability of materials. In the literature, most authors are not suggesting that all strategies should be implemented within a single building.[1] These three studies have been used as the basis of this chapter because they approach the circular economy of construction from different perspectives.[2] Based on the articles, we further synthesised the theory into seven strategies (see Table 3.1), as described below. The resulting synthesis is not all-inclusive, but summarises the strategies and removes overlaps, creating an overall picture of the current theory of circular construction.

1 *Refusing and reducing*

Refusing and reducing (1) *unmaterial* practices as a new mindset for design, critically examining the role of the architecture profession in resource consumption (Till & Schneider, 2012), and (2) efficiency through sharing and optimisation (Ellen McArthur Foundation, 2015).[3] In the first case, the need for construction is minimised through intervention on a systems level, such as reprogramming of use. This requires a major shift in architectural practice and education as well as legislative changes to prevent unnecessary demolition. We also include in this strategy *adaptive reuse*, that is, utilising and transforming an existing building or most of it. In the second case, the need for resources is decreased through using less material for a purpose, such as lean structures, or sharing spaces for various users.

2 *Design for Disassembly*

Design for Disassembly (DfD) is a central strategy in circular construction that allows the recovery of building materials after the end of life of the building, manifested in reversible construction techniques and joints, as well as careful considerations of different layers in buildings (Riegler-Floors & Hillebrandt, 2019). Other strategies, such as generality and durability of materials, as well as adaptability, have been identified as success factors for DfD (Akinade et al., 2017).

TABLE 3.1 Overview of seven design strategies for circular construction.

Eberhardt et al. (2020)	Bocken et al. (2016) *(Note: Bocken et al. divide strategies into two categories of 1–6 "slowing loops" and 6–9 "closing loops.)*	Nordby et al. (2008)	Synthesis for this chapter
1. Assembly/disassembly. 2. Material selection/substitution. 3. Adaptability/flexibility. 4. Modularity. 5. Secondary materials. 6. Prefabrication. 7. Standardisation. 8. Optimised shapes/dimensions. 9. Durability. 10. Component and material optimisation. 11. Accessibility. 12. Reusing existing building components/materials. 13. Layer independence. 14. Material storage. 15. Symbiosis/sharing. 16. Short use.	1. Design for attachment and trust. 2. Design for reliability and durability. 3. Design for ease of maintenance and repair. 4. Design for upgradability and adaptability. 5. Design for standardisation and compatibility. 6. Design for dis- and re-assembly (slowing loops). 7. Design for a technological cycle. 8. Design for a biological cycle. 9. Design for dis- and re-assembly (closing loops).	1. Limited material selection. 2. Durable design. 3. High generality. 4. Flexible connections. 5. Suitable layering. 6. Accessible information.	1. Refusing and reducing. 2. Design for disassembly. 3. Reuse of materials. 4. Adaptability and flexibility. 5. Durable design and repairability. 6. Building for attachment and trust. 7. Material selection and generality of materials.

3 *Reuse of materials*

The strategy of reusing materials enables a major potential in saving embodied carbon emissions (Rasmussen et al., 2019; Bertin et al., 2022). The approach is closely connected to the strategies of design for disassembly (DfD), material selection and generality.

4 *Adaptability and flexibility*

The discourse on adaptability/flexibility in architecture aims at answering changing user needs in different time scales, caused by a change in society. The strategy belongs to a longer discourse in architecture, derived from the idea of counteracting obsolescence in architecture (Abramson, 2016, 79–106). In circular design, adaptability and the potential to extend the lifespan and slow resource flows have been discussed by several authors (e.g., Geldermans et al., 2019; Gosling et al., 2013; Askar et al., 2021). We also include the capacity of a building to adapt to climate conditions in this approach (Pelsmakers et al., 2020).

5 *Durable design and repairability*

Durability and repairability are means to extend the lifespan of a building and its materials, thus avoiding technical, functional, or aesthetic obsolescence (Bocken et al., 2016).

The strategy manifests in materials that withstand repeated use for generations, and encourage the application of the value retention processes of a circular economy and the business models supporting it. In addition, robust solutions, such as avoiding the increase of technical systems integrated into buildings, may be placed under this category.[4]

6 *Building for attachment and trust*

Building for attachment and trust aims at prolonging the end of life of buildings or materials. The strategy is derived from smaller-scale consumer product design (Bocken et al., 2016; Chapman 2005), and encourages the adaptation of other circular strategies, such as repairability. It includes a social dimension that is seldom directly expressed by addressing the potential in creating an emotional bond between product and user.

7 *Generality of materials*

With high generality, the potential for the future reuse of materials used in a building is increased. We combine here the strategies of choosing few materials, utilising standard dimensions, and preferring standardised elements for maximising the reusability of materials (Nordby et al., 2008).[5] Leaving out surface finishes when possible or using non-toxic and natural finishes increases their future usability (Cheshire, 2019). This strategy has parallels with the approaches of material storage and modularity/standardisation (Eberhardt et al., 2020).

Four changes necessary to practice of circular architecture

Present architectural practice is based on the linear utilisation of virgin materials: "Architects are agents for the adding of more stuff to the world, acting at the end of a chain of extraction that draws raw materials from the earth and turns them into the largest of objects" (Till & Schneider, 2012, 38). Several barriers to adapting a circular economy in the construction sector, in general, have been identified (Kanters, 2020; Hart et al., 2019). Interestingly, some recent pioneering works, such as the office buildings Kristian August Gate 13 in Oslo (Mad arkitekter, 2021), K118 in Winterthur (baubüro in situ, 2021), and the laboratory building Biopartner 5 in Leiden (Popma ter Steege Architecten, 2021) provide valuable insights into the implementation of circularity, particularly regarding reuse of materials, the design process, and overcoming barriers.[6] To be able to comprehensively understand implementation of circular architecture, we need to identify changes in the practice. We identified four fundamental changes in architectural practice in the literature.

1 *Accepting scarcity*

Scarcity, besides directly affecting the material conditions of new architecture, creates significant changes in architectural practice. As a concept, scarcity has been presented 1) as physical exhaustion due to unprecedented quantities of materials extracted and 2) as a scenario based on the ease of available natural resources and energy intensity of processing, costs and transport (Organisation for Economic Co-operation and Development

[OECD], 2019, 163–178). The concept has also been framed as a social, political and economic construct (Till & Schneider, 2012). Following Till and Schneider's line of thought, but limiting it to the sphere of the working framework of designers, that circular design also deals with both factual and self-constructed scarcity. In the first, the designer works with the actual limited availability or applicability of materials.[7] In the second, scarcity is a more artificially constructed limitation caused by several considerations, such as availability, salvageability, CO2 emissions, or aesthetic potential. This is the method utilised by designers in pioneering cases.[8] As for aesthetic considerations, there is great potential for how scarcity can be turned into a positive driver for new types of expression.

2 *Accepting new roles and aesthetics*

The role of the architect consists of more than the design of the project. To some extent, architects also adopt roles such as being 1) connectors between actors; 2) mediators of business and economic knowledge; 3) challengers of the linear mindset, making circularity tangible (Dokter et al., 2021), and taking on 4) a new kind of leadership (Kanters, 2020). The agency of the designer is thus broad, multifaceted and flexible. The *aesthetic knowhow* consists of tacit knowledge of "what works". This practice-based aesthetic knowledge will change with the paradigm shift if circularity becomes the norm of architectural production and design. The intersubjective thinking about what architects consider beautiful will change.

3 *New design processes*

Circularity, and particularly reuse, will have a significant effect on the design process. The overall process is expected to become more complex and extended in scope, and the importance of cooperation and underwriting similar values among actors will be highlighted (Dokter et al., 2021). Reuse causes a shift towards a means-oriented approach, informed by the availability of materials (Massmünster, 2022, 36–37). Uncertainty increases because the materials at hand at the moment are not often known for certain, thus calling for added flexibility in design (Dokter et al. 2021, 697; Koźmińska 2019b). Uncertainty and scarcity open paths for "adhocism" as a way to achieve a creative and unified combination for a specific purpose (Jencks & Silver, 2013; see also Rowe & Koetter 1983, 102–103)[9]. Both of these pairs of authors refer to the anthropologist Claude Lévi-Strauss:

> [...] the "bricoleur" is still someone who works with his hands and uses devious means compared to those of a craftsman... The bricoleur is adept at performing a large number of diverse tasks; but, unlike the engineer, he does not subordinate each of them to the availability of raw materials and tools conceived and procured for the purpose of the project. His universe of instruments is closed and the rules of his game are always to make do with "whatever is at hand".
>
> *Claude Lévi-Strauss (1966), as quoted in* Adhocism: The Case for Improvisation *(Jencks and Silver, 2013, 16)*

In *adhocism*, the bricoleur uses pre-existing subsystems, staying within the boundaries of the existing resources "with whatever resources are available" (Jencks & Silver 2013, 17).

However, adhocism in reuse does not imply an attitude where anything is acceptable. Instead, utilising reused materials calls for a deepened understanding of the physical properties of the building parts, and for conceiving material relations in a novel way (Massmünster, 2022). The necessity to create unique architectonic qualities for each design task, such as fragility, tension, or haptic materiality, does not dissolve when turning towards reused materials. This question of architectural meaning-making and perceptibility in our aesthetic framework will be returned to later in this chapter.

4 *New circular business models and innovations*

Architectural design is expected to be affected by new circular economic activity and innovations. An increasing share of materials will be provided by manufacturers utilising value retention cycles, also called "r-imperatives" (e.g. reuse, remanufacture, or refurbish). Designers are expected to support the economics by making building parts detachable and accessible. In addition, material and methodological innovations, such as digital design or new biobased materials, are to be expected (Cheshire, 2019; Heisel et al., 2022).

An aesthetic framework for circular architecture

The previous sections described theoretical and practical viewpoints, and here the focus is on how these changes affect the essence of architecture. We will first discuss accounts of functionalism, phenomenology, semiology, and applied aesthetics in architecture, in the light of circularity.

Functionalist aesthetics

The current theory of circular construction tends to focus on the technical aspects. Could it be that aesthetics should have nothing to do with architecture, in order to offer solutions to the climate emergency? Here, the concept of circular architecture would then be comparable to the "austere" functionalism in architecture in the early twentieth century: "scientific solutions to the problems brought into focus" (Winters, 2007, 47).[10] Buildings could be simply conceived of as manipulating devices of material flows, and the form would follow, in addition to serving other functional needs, the *circular function* of material flows. However, excluding aesthetics is not as straightforward as it first appears. As Winters points out, infrastructure, such as bridges, can be aesthetically appreciated, even though the original intention would have been mostly concerned with fulfilling a specific logical or measurable purpose.

Functionalist aesthetics has an appeal as a "solving the problem" attitude, but philosophically at least two problems arise. First, we can agree with the ambiguity of *function*, as discussed by Forty (2000, 174–195). A circular function may consist of a wide range of rather unspecific means, and while narrowing it down to a specific strategy is helpful (reuse, for instance) the function is practically always compromised. A building is never 100% circular, regardless of whether it is measured as the amount of secondary material, or as the capacity to act as a material bank (Akanbi et al., 2018). Secondly, it is hard to include design strategies that focus on human behaviour and emotions, such as "building for attachment and trust", within the functionalist approach.

To conclude, functionalist aesthetics has potential in its demand for making circularity an essential and form-giving property of a building, but its limitations makes it far from satisfactory in defining circular architecture.

Cognitive aesthetics

The *constructive relativism* of the American philosopher Nelson Goodman, as described by Capdevila-Werning (2014), offers a particularly suitable path for the field of architecture to follow. Goodman's essay *How Buildings Mean* (1985), explains how the meaning of symbols in architecture is relative, allowing several simultaneous interpretations. This is why, according to him, it is more appropriate to ask *when* architecture is, rather than what (Capdevila-Werning 2014, 104). Goodman positions his constructive relativism between "absolutism" that suggests a single interpretation and the radical relativism of deconstructivism (Goodman, 1985, 650; Winters, 2007, 90). For Goodman, "a building is a work of art only insofar as it signifies, means, refers, symbolizes in some way" (Goodman, 1985, 643).

According to Goodman, meanings are created through modes of *denotation* (e.g. quoting another building by using a similar element); *exemplifying* (e.g. a structure exemplifies when it is distinguishable from the other non-structural elements), *articulation* (e.g. emphasising a building's features); *expression* (e.g. an environmentally-friendly building that truly possesses the properties, creating metaphors for being a green building). Goodman's thinking allows for an advanced conceptual framework in defining the condition for circular architecture: it enables meaning-making of its possessed circular qualities. Furthermore, doing this deepens our understanding of the sustainable properties embedded in the building. According to Capdevila-Werning (2014), the unique potential of architecture for Goodman is in "worldmaking":

> Architecture can offer new insights in a way that no other discipline can provide and, in so doing, create unique versions of the world that, in their turn, can influence or contribute to other versions. By symbolising in several different ways, architecture can mould or reshape our perception and reorganize our understanding of the worlds or world-versions that constitute reality.
>
> *(Capdevila-Werning, 2014, 104)*

Herein lies an essential aspect. Goodman's cognitive aesthetics allows us to describe how circular architecture is capable of symbolising sustainability. In the "versions of the world", that is the aesthetic experience, a novel understanding of material and energy flows is possible. Architecture can both point towards the condition of the world (climate emergency) and also create a vision of an alternative future.

The downside of Goodman's semantic approach is its lack of recognising aesthetic judgment and psychological aspects (Robinson, 2013).[11] To overcome this dilemma, we next look towards phenomenological accounts of architecture.

Representational model

Another way to conceive art in buildings is through representation. The philosopher Karsten Harries, in his work, *The Ethical Function of Architecture* (1997), criticises the strands of modernism in which aesthetic values are considered, if at all, to be only secondary. Using the metaphor of the "decorated shed" (coined by Venturi, Scott Brown & Izenour), Harries further develops Ruskin and Pevsner's idea that a building must be designed both in relation to and in opposition to functional needs (Harries, 1997, 70–96). For Harries, for a building to be counted as architecture, it should represent (1) the idea of the building and (2) the materials. Here, architecture necessitates a mutual relationship between appearance and purpose, articulated by ornamentation that is understood as non-purposeful elements in general.

For circular architecture, the first prerequisite of representing ideas is unfitting, particularly if an idea is understood as a building type (e.g. school or apartment building) that supports a certain function. The urgency of climate change necessitates fluid concepts, in which building types allow change of both function and adaptation to climate. Harries's second prerequisite, material re-presentation, is more suitable for circular architecture. According to Harries (1997, 121), materials, and the way they are connected, should represent the essential characteristics, and even sources, of materials. This is applicable considering several design strategies, such as design for disassembly, and reuse. The mundaneness of the secondary material flows in a contemporary city is somewhat contrary to the divinity of Heideggerian wood, stone, and clay.[12] To discuss this further, viewpoints from the field of everyday aesthetics are included.

Everyday aesthetics of imperfection and care

The reuse of building materials raises the subject of the ageing of materials and imperfection. These aspects have been discussed by Yuriko Saito (2007; 2017; 2017b). Saito's concept of *Green Aesthetics* is also relevant as a proposed design strategy that is informed by aesthetic theory (Saito, 2007). Saito identifies a) minimalism; b) durability and longevity; c) fittingness, appropriateness or site-specificity; d) contrast between past and present; e) perceivability of nature's function; f) health; and g) a caring and sensitive attitude (humbleness). So far, c), for example, has not been acknowledged amongst the circular design strategies. This has links to critical regionalism (Frampton, 1983, 21) as part of the still-existing postmodern discourse. It needs to be reconsidered whether fittingness to the site is something to be taken into consideration, especially if material reuse is considered a priority. This would be a major paradigm change which could have potentially even catastrophic consequences. Site-specificity is one example of how we see that Saito's *Green Aesthetics* opens the way to discussing more general questions about 1) which values to put first: lowering the CO_2-level at the expense of aesthetic sustain ability? And 2) what is the role of site-specificity in the context of different temporalities of buildings?

A proposal for the aesthetics of circular architecture

We argue that circular architecture must fulfil the following aesthetic necessities.

1) Meaning-making and cognition: circular architecture enables meaning-making on the embedded circular properties of a building. Thus, circularity is perceivable, enabling a possibility to understand flows of material and energy. Interpretations are relational, evolving over time.

2) Sensuosity: in addition to meaning-making, circular architecture has sensuous qualities, which enable aesthetic judgment and value. Nevertheless, it is agreed that aesthetic value may change over time; circular architecture should be open to transformations.
3) Material value: circular architecture utilises the circularity of material on all scales, from buildings to single parts, both in performative and aesthetic meanings. Recognising the aesthetic values in materials, even challenging ones, turns them into a positive factor, which also takes into account the effect of ageing.

An overarching theme of the aesthetic necessities is time. That is to say that a standpoint on the temporality of a building forms the basis of the philosophy: circular architecture always entails a considered relation with time, translated to aesthetics in various ways. A meaningful building with sensuous qualities may avoid obsolescence and be taken care of. The full adoption of the aesthetics necessities for circular architecture can *nudge* general opinion towards appreciating circular solutions and sustainability in architecture in general (Lehtinen, 2021).

To highlight the proposed framework, we take Peter Zumthor's Swiss Pavilion at the Hannover Expo (2000) as an example (Zumthor & Durisch, 2014). As illustrated in Figure 3.1, a temporary wooden structure could possess new symbolic interpretations of "carbon and material storage" and denote temporality by exemplifying a detachable structural system while a visitor's understanding of the infinity of natural resources may reinforce the experience (1: meaning and cognition). The building also forms a strong material presence and acoustics (2: sensuosity) and materials are possible to detach and reuse, and the ageing of the wooden material due to sunlight and rainwater are visible and aesthetically pleasing (3: material value and ageing).

To further describe some aesthetic trajectories, five tentative dimensions were created that oscillate between the philosophy of aesthetics, architectural theory and practice.

Five dimensions of circular aesthetics

Through the aesthetical framework, we discovered five potential dimensions for circular architecture. This tentative taxonomy of the dimensions is not intended to be conclusive but aims to further demonstrate the way in which the proposed conceptual framework connects to practice and theory and opens trajectories for architectural thinking.

1 *Unmaterialised aesthetics*

The avoidance of waste demands a re-thinking of aesthetic understanding. The ecological value of the natural resources embedded and already in use in the built fabric is not currently completely understood (Rosen, 2019, 22). This will by necessity affect our aesthetic perceptions of it and support reuse rather than demolition. A wetland, for example, can be far from a conventionally appreciated landscape, but it may become aesthetically appreciable through the perceiver's knowledge of its meaning for an ecosystem (Carlson, 2000; Saito, 2007, 70). This translates to human environments, such as in the case of many urban environments characterised by challenging beauty.

FIGURE 3.1 Swiss Pavilion at the Hannover Expo (Zumthor 2000). Photo: Roland Halbe

Non-materialised aesthetics in the context of architecture sounds paradoxical, but critical re-thinking, as an "unmaterial architectural practice" (Till & Schneider, 2012) has its aesthetic trajectories. This may lead to investing in a network of places within the existing built fabric or adaptive reuse, as illustrated by Figure 3.2.[13]

FIGURE 3.2 Walzwerk project, Münchenstein, Switzerland by baubüro in Situ (2014). Photo: Martin Zeller

2 Aesthetics of temporality

This aesthetic strategy emphasises the selection and reuse of materials and components but also includes choices that lead to increased adaptability and flexibility of the designs. The relation to time is manifested in circular architecture in terms of lifespans on different scales. This concerns whole buildings in general but also the various materials used for their construction, and the way they are joined together.

As well as the general notion of understanding buildings as constantly evolving things, even the most long-lasting buildings, conceived of as "permanent", are the result of continuous care-taking and maintenance. An extreme example may be seen in Japanese shrines, in which temporality and permanence coexist. The structures are temporal and changeable when needed, yet make the existence of the overall structure to be experienced (and conceived of) as permanent. Aesthetics thus take an active stance on either the permanence or impermanence of the built structure.

Besides maintainability and the potential for disassembly, the aesthetics of temporality also concerns materials and the ageing of their appearance, which is a perceptually unfolding process that is easier to link directly to aesthetics. Different materials stand in different relations to time (Harries, 1997, 229; Kalakoski, 2016). Aged surfaces include associated thoughts and images (Saito 2017, 182). According to Saito (2007), natural materials age in a dignified way, whereas those with a more industrial character are more vulnerable, such as easily scratched glossy surfaces. Saito notes also the contrast between the past and present. The concept has been most notably regulated in architectural practice by the Venice Charter (1964), which aligns conservation principles to make explicit the old and the subsequent newer parts in buildings. This principle needs to be reconsidered, because in circular architecture the components and materials are derived from various sources, with a gamut of values (Huuhka & Vestergaard, 2020).

Emphasising materials or putting them first in the design process does not mean that designers are forced to accept imperfections as such. A part of the new design paradigm is to deal with imperfection (Saito 2007, 186–188). Quilting, such as in traditional Japanese *boro* materials, uses repetitiveness and acceptance of imperfections as central methods. In the way that old scraps of cloth are mended by patching and turned into new garments in traditional crafts, similar methods of celebrating the aesthetic idea of accepting imperfection can be used in circular architecture (Gorgolewski, 2017, 42)[14]. While it is tempting to consider only surface qualities, circularity may be another kind of imperfection, such as non-optimal shapes or proportions deriving from the reuse of larger components. Even if these kinds of aesthetics are unconventional, they are not automatically aesthetically unappealing: weird conditions may become a means of expression.[15]

3 *Critical and serendipitous aesthetics*

By critical and serendipitous aesthetics, we refer to the strategy of reusing materials and components in a way that plays with the meanings of their unexpected collisions and contrasts, creating collage-like interpretations[16] and/or utilising waste as a source for the material (Figure 3.3). The aesthetics result in the new design paradigm, utilising the reuse, scarcity and unpredictability of the design process as a method.

FIGURE 3.3 Office building K.118 in Winterthur (baubüro in situ, 2021). Photo: Martin Zeller.

Already in Western antiquity, it was common to reuse building parts, called *spolia*, and this took place in the pre-industrial as well as vernacular building contexts (Kalakoski & Huuhka, 2018). Installing old, looted columns in a new building, for example, could symbolise war victory. In the industrial period, the reuse of building materials may have had similar symbolic meanings of power over preceding times and ideologies. *Contemporary spolia* emphasises authenticity, structurality, intention and perceptibility of material, with a strong aesthetic relation to temporality and associations (Kalakoski & Huuhka, 2018, 201–203).[17] In Kalakoski and Huuhka's study, not all means of reuse fit within the concept of contemporary spolia, such as when the reused materials are unperceivable. In circular architecture, using elements that will not be directly perceivable, or with smaller contrast between old and new (Goodman's *articulation* in previous section), may lead to different, yet equally potential aesthetic qualities, such as a certain spatiality or atmosphere. In these, the perceptibility of the reuse or its intention is more challenging. Reuse for new functions ("*superuse*"), or turning waste into a material ("*upcycling*"), are similarly less stringent concepts in terms of authenticity, yet essential to increase circular material flows.

Critical and serendipitous aesthetics, and their practical implications, lead to interesting parallels with postmodern architecture, in which the transfer and multi-interpretability of meanings were mostly utilised in an eclectic manner.[18] For circular architecture, sustainability is a serious effort, in which the play of meanings is not an end in itself. However, the re-invention of critical connotations in architecture, particularly critical ones manifesting material overconsumption and justified by the ethics of sustainability, is a relevant path for circular architecture.

4 *Aesthetics of care*

The aesthetics of care coincides with the design strategy of building for attachment and trust, in which "users and products flourish within long-lasting empathic partnership" (Chapman, 2005, 21). Care has been proposed as an essential factor for the longevity and survival of products (Hosey, 2012, 106). The theme of care has recently been conceptualised in aesthetic theory following theories mainly formulated in feminist care ethics (Saito, 2022; Saito, 2020; Lehtinen, 2020). Activities to take care of and maintain the spaces of living have connections with urban-scale design and planning. As a classic example, Jane Jacobs explained how the physical and social aspects of neighbourhoods are intertwined, either encouraging or discouraging caretaking and positive social control (Jacobs, 1962). In architectural design, care and maintenance are still currently considered a burden, especially if it leads to aesthetic compromises.[19] This is problematic from the perspective of circular architecture. Change in attitudes is however possible, and even likely, as cases from similar change already testify (Saito, 2007).

There might, however, be reason to worry about a possible conflict between attachment, practices of care, and transience. Are we able to create an attachment to transient architecture?[20] The other concern comes from the economic investment in transient architecture. According to Abramson (2016), the pay-back time often influences the amount of investment, and is reflected in design objectives and parameters. With short lifespans, the investments in materials are smaller, leading to the predominant use of cheaper, poor quality and noncircular materials. This conflicts with *durability* and *material selection*, for which several authors (e.g. Eberhardt, 2020; Nordby et al., 2008) have pointed out

FIGURE 3.4 A visualisation of Bürogebäude Hortus, Herzog & de Meuron (2022).

the need for ecological, durable and well-ageing materials that have potential in the cascade use of a circular economy.

5 *Aesthetics of generality*

This aesthetic strategy is directly linked to the high generality of available materials. This depends on production on the systemic level, because the quantity and availability of general materials and components drive this aesthetic. Aesthetics of generality is based on recognising major material flows in an urban area or in certain buildings that are going to be demolished, and/or by designing the building as a material bank. By conceiving the functional, aesthetic and ecological potential, the general tectonic principles appear and lead to aesthetic ramifications. This aesthetic strategy in design strives to maximise the building's capacity for reuse, as illustrated in Figure 3.4. It also links the adaptability/flexibility design strategy to the high generality of materials. Taken to an extreme, this strategy leads to a narrowing of aesthetic choices and even a certain flattening of the overall aesthetic appearance.[21] The strategy emphasises instead the function of buildings as climate change mitigating devices.

Conclusion

In this article, we described the philosophy and aesthetics of circular architecture by reflecting the most notable accounts of the aesthetics of architecture against the emerging theory and practice. As a result, we proposed a conceptual framework that creates the following conditions for circular architecture: (1) reconstructive meaning-making, (2) sensuous qualities in aesthetic judgment, and (3) recognition of the value and ageing of materials. Here, the first point of meaning-making refers to cognitive aesthetics and the possibility to understand the world through architecture. The second and third points complement the semiotics approach by adding aesthetic engagement and judgment to the framework.

As to the question of how aesthetics should inform the theory of circular construction, one outcome of this chapter is that the environmental performance of buildings is not a satisfactory end in itself. The theory of circular construction could be complemented by the three conditions for the aesthetics of circular architecture, as presented in this chapter. The first point, reconstructive meaning-making, explains how the aesthetics of circular architecture can accelerate wider social change towards sustainable development. The second point, sensory qualities and aesthetic judgment are needed to guide technology-oriented theoretical discourse in a holistic direction. Beauty, which makes buildings architecture, is an essential prerequisite for long-term sustainability. The third point, the value and ageing of materials, are in themselves familiar themes in practical architectural design. In the discourse, however, there has been less talk about the effect of the aesthetic appreciation of materials on decision-making in design. In addition, the third point emphasises the aesthetic value of the materials alongside the use value and monetary value, which are typically highlighted in the circular economy. To illustrate the way in which our conceptual framework relates to practice (and theory), we demonstrated the reciprocal relationship using five tentative aesthetic dimensions of circular architecture, namely, *non-materialised aesthetics*, *aesthetics of temporality*, *critical and serendipitous aesthetics*, *aesthetics of care* and *the aesthetics of generality*. The proposed dimensions demonstrate how both the "language" and essence of circular architecture is varied and they encourage further investigation by scholars. Utilised in design practice and education, the dimensions may help in clarifying creative thinking by relating the central concepts to their philosophical roots, or act as a tool for interpretation: to understand how certain forms have materialised out of circular economy conditions.

The assumption taken for this chapter, that the circular economy has true change-making potential, has been rightly criticised. The rapidity of urban growth and the resulting need for construction materials in many areas of the world exceed available secondary materials, while the alternative, bio-based cascading, would require a major shift in land use in order to avoid damaging ecosystems. Nevertheless, circular architecture is one of the most powerful and optimistic approaches because its theory includes a wide-ranging set of strategies for reacting to the climate emergency. An underlying factor, and necessity, is how designers, and society in general, value natural resources, and how this translates into architecture.

KEY TAKEAWAYS FROM CHAPTER 3

- The circular economy, particularly if it leads to a larger societal shift, will have a significant effect on both practice and theory of architecture.
- The chapter creates a conceptual framework for defining circular architecture through philosophical aesthetics.
- To highlight the reciprocal relationship between practice and theory, we propose five tentative aesthetic dimensions for circular architecture, namely, non-materialised aesthetics, aesthetics of temporality, critical and serendipitous aesthetics, aesthetics of care and aesthetics of generality.

Notes

1 For example, Eberhardt et al. (2020) suggest focusing on the strategies that have the biggest potential in reducing building-related climate impacts.
2 Here, some research is left aside (e.g., Akhimien et al., 2021; Hamida et al., 2023; Geldermans, 2016), because it is similar to those selected, as well as earlier accounts on architectural theory related to consumption (e.g. Zanotto, 2020) and focus on the current theoretical discourse.
3 We include the utilisation of industrial by-products in the optimisation of material use. For an example, see Upcycle Studios apartment building in Copenhagen (Lendager Group 2018), for utilising left-over concrete from an adjacent infrastructure project. It has to be mentioned that efficiency is a markedly disputed concept for achieving long-term sustainability goals, as the approach may lead to doing "less harm", yet bypassing root causes and hindering spatial flexibility.
4 The strategy is parallel with the strategy of DfD for its ease of access to materials.
5 Nordby et al. (2008) also mention high spatial generality, which allows the whole building's functions to change. We situate it in the adaptability/flexibility categories.
6 Here, a common nomination for a large-scale implementation of circularity (particularly reuse), appears to be a combination of (1) indomitable work by the architect and other central actors, and (2) rather exceptionally and financially supported objectives set by the client.
7 For example, architects involved in providing aid for natural disasters, or in developing countries, are obliged to use limited material sources.
8 Currently, as the markets of reused materials are practically non-existent, hands-on methods to acquire materials are required. In the future, digital platforms may enhance supply-demand processes, leading to an increasing range of available secondary materials.
9 See also Jencks (1977) on *adhocism*, participatory design and their aesthetics. Whereas Jencks and Silver somewhat detach "bricolage" from the original idea of Lévi-Strauss by referring to a direct material meaning, Rowe and Koetter are more devoted to the main point of the two distinct modes of scientific thought, drawing the idea conceptually to the sphere of the architect's work and politics: thinking both scientifically and primitively with a "real-life" attitude to problem-solving.
10 Functionalism can be divided into aesthetical and austere functionalism, in which the latter denies aesthetics altogether (Winters, 2013). However, even if aesthetics would not have been the focus of architectural designers, "they're still worth looking at", as put by Winters (2013), making the claim of aesthetic indifference rather unconvincing.
11 As an example, any building using detachable joints could be counted as circular architecture as long as it creates meanings. Thus, a temporary kiosk at a festival could fulfil this requirement, no matter how hastily done, or of whatever material quality.
12 In addition, where we can agree with Harries is in his notion of the two types of beauty, of which the non-Platonic one is of "ephemeral beauty" (Harries 1997, 241). It is something that we relate to the nature of impermanence of a buildings, essential in circular architecture.
13 While these practices are not, in the end, completely unmaterial, because a certain "look", or "brand" might be needed, through refurbishment or adaptive reuse, to create tangibility for the users, yet it is something very unmaterial compared to a whole new building, the conventional symbol of economic power.
14 Example of this approach can be seen in experiments on reused materials, such as using shingles from repurposed thin/plate profiles (Vandkunsten Architects, 2017, 33).
15 Parts from large-scale infrastructure projects, for example, have been used as the loadbearing structure of a private house (Gorgolewski, 2017, 171–174).
16 For definitions of collage (several material sources are detectable) and bricolage (the origin of material is less relevant), see Kalakoski and Huuhka, 2018, 206.
17 See Meier (2016) on early examples of contemporary architecture and reuse. While Meier stresses locality and place in his notions on these examples in contemporary spolia, the serendipitous aesthetics of circular architecture we propose open wider possibilities for architectural expression and meaning-making.
18 See e.g. Alan Colquhoun on the "intellectual" eclecticism of Michael Graves to the "parodistic" of Robert Venturi (Colquhoun 1981).
19 The preference for uniform surfaces, or hidden connections, may lead to more difficult or even impossible conditions (such as using glue in joining to achieve "slim" details) for repairing and refurbishing buildings.

20 With living creatures, there does not seem to be this ethical dilemma. Any dog owner knows that acknowledging that the pet's passing away sooner or later does not affect attachment, but does this kind of involvement transfer to objects?
21 For example, industrial housing construction in Europe in the 1970s led to a similar outcome.

Bibliography

Abramson, D. M., 2016. *Obsolescence.* The University of Chicago Press.

Akanbi, Oyedele, L. O., Akinade, O. O., Ajayi, A. O., Davila Delgado, M., Bilal, M. & Bello, S. A., 2018. Salvaging building materials in a circular economy: A BIM-based whole-life performance estimator. *Resources, Conservation and Recycling*, 129, 175–186.

Akhimien, N., Latif, E. & Hou, S. S., 2021. Application of circular economy principles in buildings: A systematic review. *Journal of Building Engineering*, 38, 102041.

Akinade, O. O., O. Oyedele, L.O. Ajayi, S. O., Bilal, M., Alaka, H. A., Owolabi, H. A., Bello, S. A., Jaiyeoba, B. E. & Kadiri, K. O. 2017. Design for Deconstruction (DfD): Critical success factors for diverting end-of-life waste from landfills. *Waste Management* (Elmsford), 60, 3–13.

Askar, R., Bragança, L. & Gervásio, H., 2021. Adaptability of buildings: A critical review on the concept evolution. *Applied Sciences*, 11(10), 4483. https://doi.org/10.3390/app11104483

Baubüro in situ, 2021. K.118 – *Kopfbau Halle 118.* www.insitu.ch/projekte/196-k118-kopfbau-halle-118. (Accessed 5 April 2023)

Bertin, I., Saadé, M., Le Roy, R., Jaeger, J.-M. & Feraille, A., 2022. Environmental impacts of Design for Reuse practices in the building sector. *Journal of Cleaner Production, 349*, 131228.

Bocken, N. M. P., Ingrid, d. P., Conny, B. & v. d. Grinten, B., 2016. Product design and business model strategies for a circular economy. *Journal of Industrial and Production Engineering*, 33(5), pp. 308–320.

Capdevila-Werning, R., 2014. *Goodman for Architects.* Vol. 10. Routledge.

Carlson, A., 2000. *Aesthetics and the Environment: The Appreciation of Nature, Art and Architecture.* Routledge.

Chapman, J., 2005. *Emotionally Durable Design: Objects, Experiences and Empathy.* Earthscan.

Cheshire, D., 2019. *Building Revolutions: Applying the Circular Economy to the Built Environment.* RIBA Publishing.

Colquhoun, A., 1981. *From Bricolage to Myth, or How to Put Humpty-Dumpty Together Again.* MIT Press.

Corvellec, H., Stowell, A. & Johansson, N., 2022. Critiques of the circular economy. *J Ind Ecol.*, 26, 421–432. https://doi-org.libproxy.tuni.fi/10.1111/jiec.13187

Dokter, G., Thuvander, L. & Rahe, U., 2021. How circular is current design practice? Investigating perspectives across industrial design and architecture in the transition towards a circular economy. *Sustainable Production and Consumption*, 26, 692–708.

Eberhardt, L. C. M., Birkved, M. & Birgisdottir, H., 2020. *Building Design and Construction Strategies for a Circular Economy.* Architectural Engineering and Design Management.

Ellen MacArthur Foundation, 2015. *Towards a Circular Economy: Business Rationale for Accelerated Transition.* Ellen MacArthur Foundation.

Forty, A. (2000) *Words and Buildings: A Vocabulary of Modern Architecture.* Thames & Hudson.

Frampton, K., 1983. "Towards a Critical Regionalism: Six Points for an Architecture of Resistance", in Foster, H. (ed.), *The Anti-Aesthetic: Essays on Post-Modern Culture.* Bay Press.

Geissdoerfer, M., Savaget, P., Bocken, N. M. P. & Hultink, E. J., 2017. The circular economy – a new sustainability paradigm?. *Journal of Cleaner Production*, 143, 757–768.

Geldermans, R.J., 2016. Design for change and circularity – accommodating circular material & product flows in construction. *SBE16 Tallinn and Helsinki Conference; Build Green and Renovate Deep*, pp. 301–311.

Geldermans, B., Tenpierik, M. & Luscuere, P., 2019. Circular and flexible infill concepts: Integration of the residential user perspective. *Sustainability*, 11(1), 216–235.
Goodman, N., 1985. How buildings mean. *Critical Inquiry*, *11*(4), 642–653.
Gorgolewski, M., 2017. *The Architecture of Reuse.* Wiley-Blackwell.
Gosling, J., Sassi, P., Naim, M. & Lark, R., 2013. Adaptable buildings: A systems approach. *Sustainable Cities and Society*, 7, 44–51.
Hamida, M. B., Remøy, H., Gruis, V., & Jylhä, T., 2023. Circular building adaptability in adaptive reuse: Multiple case studies in the Netherlands. *Journal of Engineering, Design and Technology*, 23(1), 161–183. https://doi.org/10.1108/JEDT-08-2022-0428
Harries, K., 1997. *The Ethical Function of Architecture.* The MIT Press.
Hart, J. et al., 2019. Barriers and drivers in a circular economy: the case of the built environment. *s.l., 26th CIRP Life Cycle Engineering (LCE) Conference.*
Heisel, F., Hebel, D. E. & Webster, K., 2022. *Building Better – Less – Different: Circular Construction and Circular Economy: Fundamentals, Case Studies, Strategies.* Walter de Gruyter GmbH.
Hosey, L., 2012. *The Shape of Green: Aesthetics, Ecology, and Design.* Island Press.
Huuhka, S. & Vestergaard, I., 2020. Building conservation and the circular economy: A theoretical consideration. *Journal of Cultural Heritage Management and Sustainable Development*, 10.1, 29–40.
Jacobs, J., 1962. *The Death and Life of Great American Cities.* Jonathan Cape.
Jencks, C. A., 1977. Post Modern Architecture. In: *The Language of Post-Modern Architecture.* London: Academy Editions, pp. 87–102.
Jencks, C. & Silver, N., 2013 [1972]. *Adhocism: The Case for Improvisation.* Expanded and updated edition edition. MIT Press.
Kalakoski, I., 2016. *Defining Requirements for Appreciation of Patina.* Trondheim, Nordisk Arkitekturforskning, pp. 27–40.
Kalakoski, I. & Huuhka, S., 2018. Spolia revisited and extended: The potential for contemporary architecture. *Journal of Material Culture*, 187–213.
Kanters, J., 2020. Circular building design: An analysis of barriers. *Buildings*, 10(4).
Koźmińska, U., 2019a. The Aesthetics of Sustainable Architecture: Reused Materials. *The 6th International Conference on Architecture and Built Environment.* S.Arch.
Koźmińska, U., 2019b. Circular design: reused materials and the future reuse of building elements in architecture. Process, challenges and case studies. *IOP Conference Series: Earth and Environmental Science*, Issue 225.
Lee, S., 2011. *Aesthetics of Sustainable Architecture.* 010 Publishers.
Lehtinen, S., 2020. "Buildings as Objects of Care in the Urban Environment", in Zoltan S. & Max R. (ed.) *Aesthetics in Dialogue: Applying Philosophy of Art in a Global World.* Peter Lang, pp. 223–236.
Lehtinen, S., 2021. "Aesthetic Sustainability", in Krieg C. P. & Toivanen, R. (eds.), *Situating Sustainability: A Handbook of Contexts and Concepts: A Handbook of Contexts and Concepts.* Helsinki University Press, pp. 255–267.
Lévi-Strauss, C., 1966. *The Savage Mind.* University of Chicago Press.
Mad arkitekter (2021). *Kristian August gate 13.* www.mad.no/prosjekter/kristian-august-gate-13 (Accessed 5 April 2023).
Massmünster, 2022. "K118 Reportage", in Stricker, E., Brandi, G., & Sonderegger, A., (ed.) *Re-Use in Construction: A Compendium of Circular Architecture.* Park Books, pp. 10–11.
Meier, H.-R., 2016 [2011]. "Spolia in Contemporary Architecture: Searching for Ornament and Place", in. Brilliant, R. & D. Kinney, (ed.) *Reuse Value: Spolia and Appropriation in Art and Architecture from Constantine to Sherrie Levine.* Routledge, pp. 223–236.
Nordby, A. S., Hakonsen, F., Bjørn, B. & Hestnes, A. G., 2008. Salvageability; implications. *Nordic Journal of Architectural Research*, 20(3), 29–42.

OECD, 2019. *Global Material Resources Outlook to 2060: Economic Drivers and Environmental Consequences*. OECD Publishing.
Pelsmakers, S., Poutanen, J. & Saarimaa, S. 2020. "(Hybrid) architecture in and over time", in M. Pedersen Zari, P. Connolly & M. Southcombe (eds), *Ecologies Design: Transforming Architecture, Landscape, and Urbanism*. Routledge Research in Sustainable Urbanism, Routledge, pp. 268–275.
Popma ter Steege Architecten, 2021. Biopartner 5. https://ptsa.nl/biopartner-5/ (Accessed 5 April 2023)
Rasmussen, Birkved, M. & Birgisdóttir, H., 2019. Upcycling and Design for Disassembly – LCA of buildings employing circular design strategies. *IOP Conference Series. Earth and Environmental Science*, 225(1), 12040.
Reike, D., Vermeulen, W. J. V. & Witjes, S., 2018. The circular economy: New or Refurbished as CE 3.0? – Exploring Controversies in the conceptualization of the circular economy through a focus on history and resource value retention options. *Resources, Conservation & Recycling*, 135, 246–264.
Riegler-Floors, P. & Hillebrandt, A., 2019. "Detachable Connections and Constructions", in Hillebrandt A., Riegler-Floors P., Rosen A. & Seggewies J., (eds). *Manual of Recycling: Buildings as Source of Materials*. Detail Business Information GmbH, pp. 42–57.
Rosen, A., 2019. "Dismantling, Recovery and Disposal in Construction", in Hillebrandt A., Riegler-Floors P., Rosen A. & Seggewies J., (ed.). *Manual of Recycling: Buildings as Source of Materials*. Detail Business Information GmbH, pp. 16–31.
Robinson, J., 2013. "Goodman", in Gaut, B. Lopes, D. (eds). *The Routledge Companion to Aesthetics*. Routledge, pp. 179–189.
Rowe, C. & Koetter, F., 1983. *Collage City*. The MIT Press.
Saito, Y., 2007. *Everyday Aesthetics*. Oxford University Press.
Saito, Y., 2017. *Aesthetics of the Familiar: Everyday Life and World-Making*. Oxford University Press.
Saito, Y., 2017b. The Role of Imperfection in Everyday Aesthetics. Contemporary Aesthetics. (Journal Archive): 15(15).
Available at: https://digitalcommons.risd.edu/liberalarts_contempaesthetics/vol15/iss1/15
Saito, Y., 2020. "Aesthetics of Care", in Somhegyi, Z. & Ryynänen, M. (eds.). *Aesthetics in Dialogue. Applying Philosophy of Art in a Global World*. Peter Lang.
Saito, Y., 2022. *Aesthetics of Care: Practice in Everyday Life* (1st edn). Bloomsbury Publishing.
Till, J. & Schneider, T., 2012. Invisible agency. *Architectural Design*, 82(4), 38–43.
Vandkunsten Architects, 2017. *Rebeauty: Nordic Built Component Reuse*. Tegnestuen Vandkunsten A/S.
Winters, E., 2007. *Aesthetics and Architecture*. Continuum.
Winters, E., 2013. 'ARCHITECTURE' in Gaut, B. Lopes, D. (eds). *The Routledge Companion to Aesthetics*. Routledge, pp. 627–638.
Zanotto, F., 2020. *Circular Architecture: A Design Ideology*. AUID Series.
Zumthor, P. & Durisch, T. (2014) *Peter Zumthor : Buildings and Projects / Vol. 3, 1998–2001*, Scheidegger & Spiess.

4

WHO DO WE CREATE ARCHITECTURE FOR, AND WHY?

Mirjami Myllymäki, Sofie Pelsmakers and Laura Uimonen

The Capitalocene

We live in a time in which humans are changing the earth's biosystems through the burning of fossil fuels on a gigantic scale. The geological stratigraphic layers are changed by the abundance of plastic and nuclear waste, the (de)growth of forests and so on. This phenomenon is referred to as the Anthropocene, i.e., the period where humans are the dominant force in shaping the environment and climate ('the age of man as geological force', Moore, 2019). But this is a deceptive term (Crist, 2013; Haraway, 2016; Moore, 2019), because not *all* humans are to blame for this exploitation of nature and our planet for the creation of endless accumulation of capital (i.e., capitalism). Instead, *some* humans (usually white, male, and wealthy) do the exploitation while *other* humans are exploited and face the consequences of this imbalance of power (Ghosh, 2021; Moore, 2019). As such, the planetary crisis is better understood through the concept of the Capitalocene, which is the inextricable connection between the maximisation of profit and the suppression and domination of humans, their cultures, non-humans and nature. This exploitation goes back to the invasion, colonisation and devastation of other countries (Ghosh, 2021; Moore, 2019) and this fate was 'meted out to indigenous peoples, to the Irish, to virtually all women, to African slaves, to colonial peoples around the world' (Moore, 2019). This cheapening of life and work, and abuse of raw materials and people (through slavery or low-cost labour) makes possible the endless accumulation of capital for a minority of people, sequentially creating grave injustices within the world that we live in.

This planet-transforming process is near exhaustion: people are starting to resist this reality; raw materials are becoming depleted; soil, water, and air on Earth are toxified (affecting the very people and nature the system exploits). Climate change makes unfettered growth more difficult and more expensive. All of this is not only destabilising the planet but also questioning the very system and values that are at the cause of it all. It also highlights the challenges we face: that our governments' promised climate neutrality targets far into the future cannot be just a technological fix if our societies remain coupled

DOI: 10.4324/9781003293903-5
This chapter has been made available under a CC-BY-NC-ND 4.0 license.

to endless consumption, production and systems of power that create these 'embodied injustices' (Tucker, 2021), such as those described earlier.

This chapter explores the implications for (and of) architecture in the Capitalocene, followed by concepts of radical inclusivity, biophilia and topophilia, leading to a restorative approach that is illustrated with a case study.

Architecture in the Capitalocene

Architecture, which includes not only the built product but also the process and production of architectural activities (Mangold, 2010), is at the heart of the exploitative profit-making global structure, giving (quite literally) a facade to this growth-oriented system. Typically, surplus capital and profits have been required to create architecture, which in turn enabled the safe storing and growth of capital in real estate. This continuous production in the name of profit has led to an extractive and exploitative built environment and industry with embedded injustices and suffering, and is thereby complicit to the planetary crisis (e.g., vast extraction and waste of resources both nearby and far away; pollution of land, water and air, manipulation and exploitation of nature and different communities). Architecture has customarily celebrated and therefore legitimised 'the activity of those in power, often at the expense of large numbers of disadvantaged others' (Ward, 1996, 27). Architecture has been complicit in the exploitation of people at different stages of its process, production and end product (e.g., unsafe construction sites; negative impacts on health and well-being; displacement, exclusion and inequalities of people; unfit end solutions for their purpose). Not only is architecture the spatialisation of capitalism (Till, 2020), but by extension it is typically also the spatialisation of exclusivism and privilege, originating by the privileged, and working for *some* humans but not for the majority of *other* humans, let alone nature and non-humans.

All of this has been partly made possible by the 'architecture-as-art' view – a symbol of class and social superiority, including the social superiority of the architect – that is still the prevailing discourse, affecting architects' internal values. This view and the associated activities are disconnected from social and moral imperatives (Ward, 1996, 28), further perpetuating capitalist abuse of nature and *the other*. Similarly, the 'architecture-as-science' view was also used to reinforce power and authority by the dominant culture. This happened because attempts to put social issues at the heart of architecture, and the scientific background to those social issues, were quickly silenced or completely misrepresented until relatively recently (Ward, 1996). This silencing is exemplified by William Morris, an advocate for socialism of the nineteenth century Arts and Crafts Movement, who realised that reawakening of the lost crafts could potentially put all people in charge of building production again, thereby empowering them (Ward, 1996, 31–32). Another unfairly silenced example is Hannes Meyer, the director of the Bauhaus in the late 1920s, who believed that 'the making of architecture was essentially a social process, involving not just one individualistic designer but the whole community' (Ward, 1996, 33). Morris and Meyer's shared bravery was derived from an idea capable of redirecting some of the monetary flows and distribution of power away from the cultural elite holding most capital and authority. They moved quickly to deflect any revolution of this kind. Instead of advancing the social and moral imperatives in architecture, the agenda of 'architecture-as-science' declared itself as the liberation of people, diagnosing people's needs without consulting them. Thus, architects became 'cultural gatekeepers', only advancing the elite's interests (Ward, 1996, 35). It paradoxically argued that 'the social process of architecture could only be achieved through its suppression' (Ward, 1996, 36).

The role of architecture in its complicity in the planetary crisis and oppression through the capital-driven structures of power are now being revealed and questioned. Instead, 'architecture-as-social-process' is needed (Ward, 1996, 35–36), recognising that to empower people and community, means to challenge the capitalist profit-seeking structures, i.e., the same structures which are also at the heart of the societal and planetary crises. Tackling one cannot be done without the other, and clearly points to architects' moral obligation. Crist (2013) argues that to face the global challenges, our priority should be to pull back and scale down, and 'welcoming limitations of our numbers, economies, and habitats for the sake of a higher, more inclusive freedom and quality of life'. However, our current architectural values and culture, obsessed with 'starchitects' is unfit to address the big societal and planetary challenges we face (Fraser, 2015, 26). As such, it is our duty to create conditions and practices to address these challenges while decoupling from the unequal terms of power, wealth and opportunity that come with extreme profit-seeking structures. Buildings designed for profit typically have a built-in obsolescence, i.e., they are not designed to be permanent (Ward, 1996, 49). Profit-seeking commonly leads to cheap, standardised, mass-produced solutions instead of adaptable, diverse, user-driven approaches that meet people's needs and conditions (Mangold, 2010; Ward, 1996). People directly impacted by development are also generally not involved in its creation and production. Solidarity is often omitted from design, leading to universalising and interpreting others' needs (Ward, 1996, 65), without contact with the people it serves (Ward, 1996, 57; Yusof and Kozlowski, 2015, 78–103).

Moreover, nature and non-human species rarely feature in a profit-seeking world, unless it is to exploit them. Urbanisation, where the majority of people live in urban areas, has led to an increasing lack of access to connection with nature and a lack of day-to-day experience with nature from one generation to the next (Beery et al., 2015, 2023). The concern is that this disconnection may affect how we think about the natural world and our connection with it as humans, and lead to its destruction (Beery et al., 2015). In the past, humans lived close to nature, and they saw themselves as part of and not separate from nature. We have now migrated from rural to urban areas and from outside to internal environments (Tabb, 2021). This indoor life affects health and well-being and detaches humans from the natural world to which they belong (Tabb, 2021), leaving its exploitation and consumption mostly unchallenged, leading us further into the climate and biodiversity crisis. This is why Haraway (2016) argues it is time for the Chthulucene, in which human beings are not the only important actors as centred in the Anthropocene and Capitalocene but are completely intertwined with all other species on our planet.

Radical inclusivity, biophilia and topophilia

It is clear that architecture's priority should be empathy, solidarity and communality with humans and non-humans and nature, rejecting the current false boundaries between nature and humans and endless profit-seeking exploitation. This is closely aligned with principles of radical inclusivity, biophilic design and topophilia – each briefly unfolded below and illustrated by a design case study.

Radical inclusivity is defined by Nawratek and Nawratek (2015, 21) as 'a possibility to use the building in various ways, by various users, human and not human actors'. Radical inclusivity means creating user-adaptable architecture that 'moves away from a polished unapproachable architecture towards one that is inviting, inclusive and un-intimidating for people to interact with' (Cole, 2015, 122). The idea is that even an egoistic programme and building can also work for the common good and should include social infrastructure; e.g., a building with private functions yet with open accessible spaces for (or shared with) the public (Nawratek and Nawratek, 2015). However, the word inclusivity is questioned by Tonkiss (2015), given that 'different people want different things' and that this can create conflict when trying to design for everybody so that it risks working for nobody. Instead Tonkiss argues to aim for 'non-exclusionary' or 'anti-exclusionary spaces' – but the terms are based on the same intention, i.e. 'the same space being used by different sets for different purposes at different times' (Nawratek and Nawratek, 2015, 21). This description is also used to characterise hybrid building approaches (Pelsmakers et al., 2020), however radical inclusivity explicitly centres nature, non-humans and social inclusivity.

Gaining empathy for different users starts in architecture education, where internal values are first established. Encouragingly, empathetic teaching approaches and strategies in architecture education are being formulated (see Chapter 7 in Nisonen et al.; Nawratek, 2021), suggesting trading the conventional individualistic master-apprentice design studio model for active peer-learning discussions and workshops with teachers as facilitators in a blended learning model (see also Donovan and Pelsmakers, 2019; Pelsmakers et al., 2019).

Such educational environments create the space where new architectural thinking, approaches and ideas can be freely explored. Myllymäki (2021) did exactly that with her masters in sustainable architecture design thesis *On How To Reimagine Lost Spaces Through The Concept Of A Table* and challenged exploitative, profit-seeking and exclusionary architecture. She imagined a more equal future for local people and other species; to not fall prey to the capital developer-led programmes (i.e., capital-led) but instead be community (end-user) led. The project is set in the outer suburbs of Paris, on the remains of a forgotten railway line called the Petite Ceinture. Focusing on this existing place and neighbourhood in Paris, the case helped to understand how simple everyday needs of every person has to be considered as a starting point for social, cultural and environmental change. From this deep understanding, a communal kitchen shared with neighbours and urban citizens becomes a prototypical temporary, movable infrastructure. A food loop of farming food in the city from seed to community growing, and enjoying, and composting, creates fertile soil for socially binding actions, supporting a radically inclusive and restorative living culture.

Myllymäki's thesis (2021) is as much an exploration of an existing place, as it is a design thesis ending with imaginative visualisations of a possible future on-site. The existing social and physical structures are carefully studied in order to preserve and enhance through making components with minimal redirection and construction, instead of rewriting and reclaiming a place for the sake of capital gain and self-fulfilment. Myllymäki's design is subject to at-the time available local, recycled materials and building parts (discussed in Chapter 3), as well as the community needed to create a place inside and out, leaving the architect as a sort of 'mediator of the community' (as discussed in

Chapter 5), as she herself puts it: 'when we let go of the idea that every part of a city should be resolved, planned, and built to its maximum capacity, we start to uncover a resource already fulfilling its greatest potential in solitude, governed by the people who then need it the most' (Myllymäki, 2021, 153).

Radical inclusivity principles are connected to the principles of biophilia and topophilia (Tuan, 1974), of which biophilia is the most established one. In architecture, biophilia[1] is generally interpreted to mean people's innate connection to nature (Ryan, 2014), whether visually or physically. However, the typical biophilia interpretation in architecture is reductive. Instead, the connection to nature was distinctly intended to be an *emotional connection* of humans to other living organisms, related to our evolutionary adaptability and survival (Beery et al., 2015). For decades Wilson (2016) has delivered the vital importance of understanding species, and not only biologically but socially formed ecosystems, in which taking care of the pack is essential to the whole society's survival. Biophilia centres around the notion that our appreciation of the natural world evolved to support the survival and health of our ancestors.

Biophilic design includes, e.g., access to good natural daylight and ventilation and views of nature, and this supports physical and mental health and well-being (Griffin, 2004). For example, visual connection to nature can lead to reduced stress, improved mood and relaxation and increased restorative benefits with increased biodiversity (Ryan et al., 2014). Non-visual connection to nature in the built environment relates to e.g., air flow and scents and nature sounds (water, rustling of trees), and has been shown to lower stress and blood pressure and improve mental health. Simultaneous exposure to visual and non-visual connections to nature leads to greater well-being benefits. Access to nature also encourages physical activity and helps the sleep cycle (Tabb, 2021). Biophilia has developed into the biophilic design of housing and urban design and further to Biophilic Urbanism, Biophilic Cities and a social movement; the different characteristics of biophilic design at different scales are summarised in Table 4.1.

Similarly, in Myllymäki's design, the location of the Petite Ceinture itself attests how amazing nature is at spreading even when forced to retreat from the cityscape into fallows left untouched and undisturbed, in turn creating a retreat for animals and humans alike. In line with biophilic and radical inclusivity principles, Myllymäki proposes preservation of all wildlife, making sure not to disrupt any important layers of the ecosystem, introducing only lightweight infrastructures like a protective veil, freely welcoming a diversity of life-forms into the space. These ecology of repair principles extend to the restorative safeguarding of a major green corridor linking to local waterways and vast forests outside Paris. This was done, again not by building, but the opposite: restoring it by leaving nature alone and by elevating the corridor as an important biodiverse, inclusive cycling and walking route through central Paris, and enabling coexistence of people and different endangered species. This also brings different people closer into nature, affecting health and well-being, reconnecting humans to everyday nature to which they belong (Tabb, 2021). As described before, a reconnection to nature can encourage environmental concern and more sustainable lifestyles that reject exploitative patterns (Beery et al., 2015).

The difference between Biophilic Design and Biophilic Urbanism (Tabb, 2021, 29) is where the former creates human–nature interactions and the latter includes the social aspects (e.g., the mix of uses, social infrastructure, public space etc.), but also includes a broader environmental dimension (e.g., open spaces, parks, corridors) and a transportation dimension (e.g. alternative modes of transport, reduced impact of cars and reduced greenhouse gases, increased clean fuel use etc.). Beatley (2020) simply states that 'biophilic cities are just and equitable cities; cities that exemplify and commit to a Just Biophilia, a recognition that contact with nature is a right deserved by all irrespective of income, race or status', highlighting that radical inclusivity as defined earlier is intrinsic to biophilic design principles. Furthermore, biophilia challenges the design processes in a positive, more inclusive way by changing the role of the planner and power relations of urban planning. This is achieved through meetings and discussions between citizens for shared understanding of the human connection to nature, and the need to bring nature back into citizens' daily lives as a collective goal.

However, at present, the institutional level of planning sustainability is usually broad and separated in sectorised governance areas, instead of being considered in the social and cultural context at the macro and micro levels. Moreover, the current biophilic efforts in architecture and urban planning fail to escape a human-centric perspective, where nature only exists to serve, soothe, and satiate people's stomachs, lungs, and minds. Generally, the focus is on what nature can do for humans and not for the benefit of non-humans or nature alone. Holistic approaches are lacking, because sustainability strategies such as the greening and rewilding of areas, and participation projects with local communities, are typically planned and organised separately (Hajer, 1995; Vazquez, 2022). For example, stormwater projects, green walls and pollinator meadows appear to the city as technical products while participation in the allotment gardens are managed by social projects, green management, or the garden society.

Further, collected user data are generally simplified and generalised in a few categories, so do designers really know for whom they are planning? Despite recent attempts to tackle social inequalities and impacts of planning on the lived environment (through soft-GIS and user-friendly planning) connections to everyday emotions of the community and genuine inclusivity in planning have often been neglected. Yet community visioning and true democratic processes are crucial and are acknowledged in the 1994 Aalborg Charter, which has over 2,500 regional signatories in the EU (Aalborg Charter, 1994).

At the heart of Myllymäki's case study (2021) is empathy towards a place and the people realising this. Through 'the art of noticing' (Tsing, 2015), Myllymäki discovers that empathy and radical inclusivity (acceptance) of people are essential to restorative architecture. Tsing (2015, 20–29) defines 'the art of noticing' as where the designer notices unintentional patterns, the interplay of temporal rhythms and scales in the divergent life-forms, and where the designer reorients their attention to understanding that making worlds is not limited to humans. Myllymäki's design concept is accordingly realised in collaboration with, and dedication to, existing life forms on-site, with an objective to let a discouraged community rebuild itself, and enabling different life forms to co-exist (see Figures 4.1 and 4.2). Urban fallows, like disused rail and tunnels, are often seen as wounds in the urban structure, yet they can be a refuge and solace for often marginalised communities and left-alone nature. In the Petite Ceinture people had painted murals on the retaining walls and planted seeds next to the disused wooden tracks to ground themselves into the place (see Figures 4.1 and 4.2).

Myllymäki's (2021) project 'becomes an investigation of the people. The people, whose presence and minds are needed in order to make sense of a place [...] above all

FIGURE 4.1 The existing stone retaining walls of the Petite Ceinture and the new wooden frame create an alleyway in between for gardens and leisure. Illustration of two 'flaneurs' on a morning walk along the Petite Ceinture with its murals (painted unobserved) and left-alone nature (Myllymäki, 2021).

recognizing the value in the Petite Ceinture, the people, and their efforts' (Myllymäki, 2021, 153). To investigate people and their needs, Myllymäki created personas based on deep contextual (desktop) research[2] – see Figures 4.3 and 4.4. This process is closely aligned with Havik's (2014, 2019) threefold-step framework of description, transcription,

FIGURE 4.2 Illustration of portable seedbeds on the rails (Myllymäki, 2021).

and prescription. Her proposal of a literary view on the experience, use and imagination were applied during the pandemic. Specifically, Myllymäki follows the framework taking the selective position of a designer addressing reality critically and understanding social spatial practices and moreover, using narratives offering a passage to new ways of living (Havik, 2014, 27–32, 107, 149–150).

The notion of personas (or characters) is used for dual purposes; firstly, the empathetic views of the area are discovered seeing the place through the eyes of another, e.g., from the viewpoint of different local inhabitants. The second purpose is that even nature and existing structures can be studied as 'characters', connected to the urban social memory of the place (Havik, 2019, 104–106, 115–119). After the *descriptive step* (i.e., collection and analysis of detailed material, atmospheric and other contextual observations through 'the art of noticing'), the social aspects of the site analysis are foregrounded through a *transcriptive step*. For example, empathic and inclusive understanding is created from

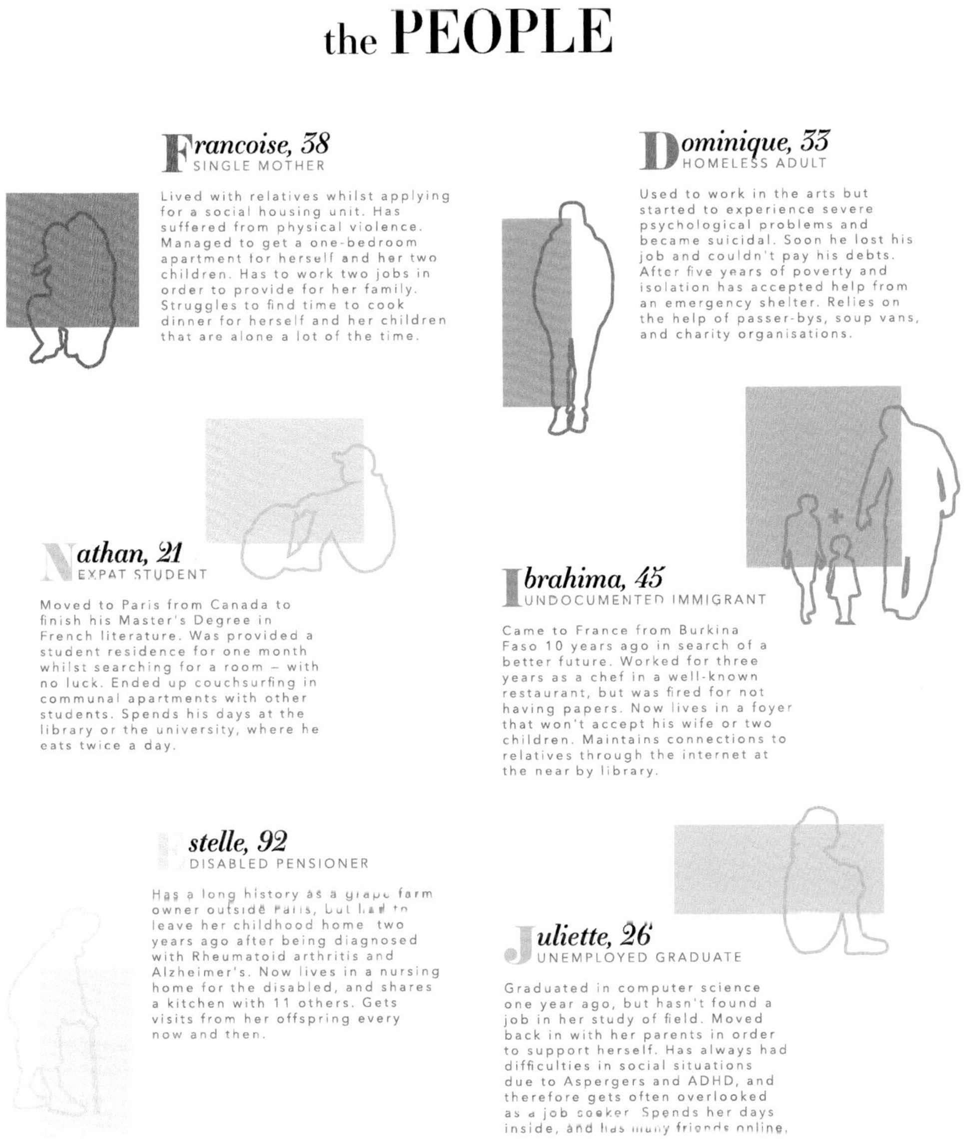

FIGURE 4.3 Introduction of the six human personas used in Myllymäki's (2021) thesis.

interviews and observations of social activity of different user groups but also from literary narratives and personas (characters) – see Figure 4.3. Both in student projects and in real life this helps to see how places are lived, used and remembered by inhabitants. In the *prescription* of the framework, fictional scenarios are used to imagine possible futures and spatio-temporal structures for the current and prospective users (see Figures 4.4 and 4.5).

Through the art of noticing, Myllymäki (2021) was then able to recognise common unmet needs shared by the neighbourhood personas, embedded in real stories of vulnerable local people, who find themselves lacking a complete home. Myllymäki uses personas to navigate through her proposed programme and later design iterations. The same personas guide the designer every step of the way and help her to communicate her design proposal for the place. She goes in depth to explore the meanings of food and the activities around food for different people, and allows those ideas to give birth to the architecture instead of the other way around. As a result, Myllymäki envisions

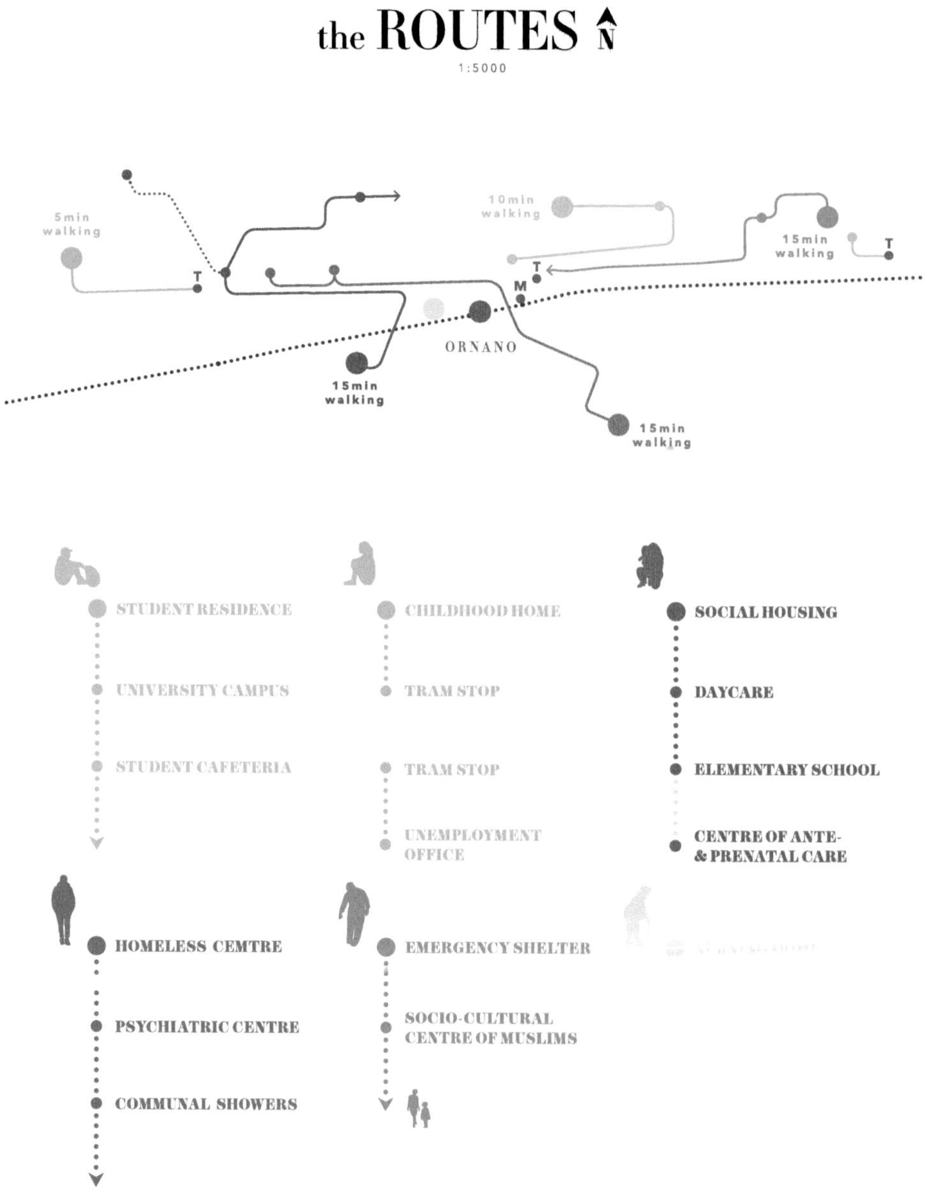

FIGURE 4.4 Imaginary scenarios of the six human personas and their weekly routes mapped around the main site (Ornano) (Myllymäki, 2021).

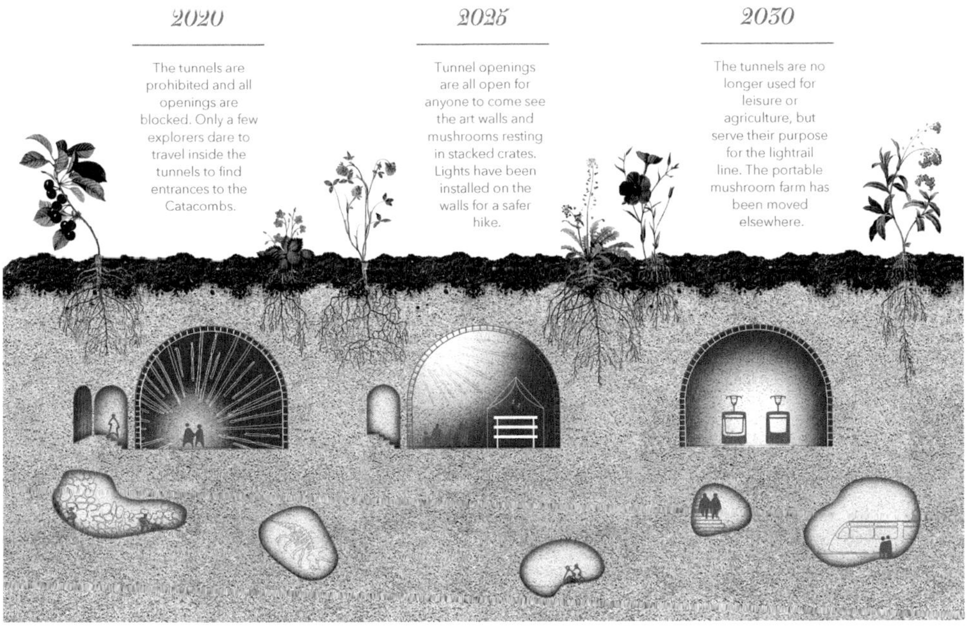

FIGURE 4.5 Potential temporal scenarios for proposed future action inside the tunnels of the Petite Ceinture, with space given over to mushrooms (a seventh non-human persona) and bringing humans and nature closer together (Myllymäki, 2021).

a meeting place revolving around food, welcoming everyone with an appetite to feel loved, refuelled and fulfilled. Moreover, the major binding ties of the ecosystems are food webs (Wilson, 2016, 106), and links back to the appreciation of the natural world that evolved to support the survival and health of our ancestors, which is at the heart of biophilic principles.

Nature, often similarly homeless inside the city, is a seventh persona, and it is never neglected in Myllymäki's work. For example, white button mushrooms revive a piece of lost culture in twenty-first-century urban life and are given a home to grow in abandoned railway tunnels (see Figures 4.5 and 4.6), opening an almost invisible yet straight-in-our-stomach kind of connection between natural processes and human life. The biophilic design aspects that are included in the case study are listed in Table 4.1.

While buildings can create public biophilic spaces such as rooftop gardens, green walls at street level and courtyards, we need to think more radically: not of a building with some greenery, but of a fundamental rewilding of our cities (i.e. greenery with some buildings located within), so as to connect natural areas together and expand biodiversity, while ensuring that humans are part (again) of nature (and not simply have a daily interaction with it). The importance of biophilic cities and buildings goes beyond human health and supporting humans that thrive; it is also of crucial importance for non-humans to thrive by providing habitats, movement corridors and sources for food for different species (Beatley, 2020).

TABLE 4.1 Examples of biophilic design characteristics at different scales; summarised from Tabb (2021). Text in italics signifies all the biophilic principles at different scales as present in Myllymäki's case study.

Biophilic Buildings	*Biophilic Streets*	*Biophilic Blocks*
– *Passive design* – Renewables – *Food growing* – *Views to nature* – *Sun/daylight access* – *Natural ventilation* – Natural materials – *Durability, adaptability* – *Landscaping*	– *Accessibility* – *Pedestrian movement* – *Edible landscapes* – *Protective trees and vegetation* – *Social infrastructure (sidewalks, benches etc)* – *Reduced night lights* – *Car free connections* – *Good solar access / protection*	– Clustered blocks – *Human scale* – *Social infrastructure* – *Free courtyards* – *Native planting* – District heat or cooling (e.g., geothermal) – *Recycling* – *Play areas* – *Food growing*
Biophilic Neighbourhood	*Biophilic Communities*	*Biophilic Eco regions*
– *Play areas* – Schools – *Gardens and community gathering places* – *Mixed use* – *Views and walking access to nature* – *Native planting* – *Sense of place* – *Car free zones, walking networks*	– *Mixed use* – *Urban agriculture* – *Recycling* – *Sustainable urban drainage systems* – *Car free environment* – *Social infrastructure (e.g. schools, library)* – *Community Renewable energy provision*	– *Integrate development with natural areas and topography* – *Connections to nature and of nature and water* – *Habitat protection* – Flood resilience – Renewable energy – Connects to other regions

However, protecting local biodiversity is only one part: we must also protect more abstract and larger-scale notions of nature. It is also about how cities and their ecologies are linked at (bio)regional scale but also globally and to reduce their climatic and ecological impacts: 'they work to protect more distant nature at the same time that they work to protect, regenerate and celebrate local ecologies and biodiversity' (Beatley, 2020).

The notion of biophilia has been further expanded into topophilia ('fondness of place') and includes affiliation of all living and non-living elements of nature (e.g., stone, water), the latter is usually excluded in the definition of biophilia, though often implicitly present. Topophilia is also acquired through culture-based learning by experiencing the environment.

Whether biophilia and topophilia have an underlying evolutionary basis or not remains open for debate, but the benefits of a day-to-day connection to nature (whether living and non-living elements) are well documented, and create 'a rationale at a deep human level for how place attachment may be a powerful tool in creating environmental concern and interest in local sustainability efforts' (Beery et al., 2015). Connecting urban citizens with nature and their local place may bring about a greater self-awareness of our destructive actions and hence may lead to actions towards more sustainable living, rejecting exploitation and the systems that support it.

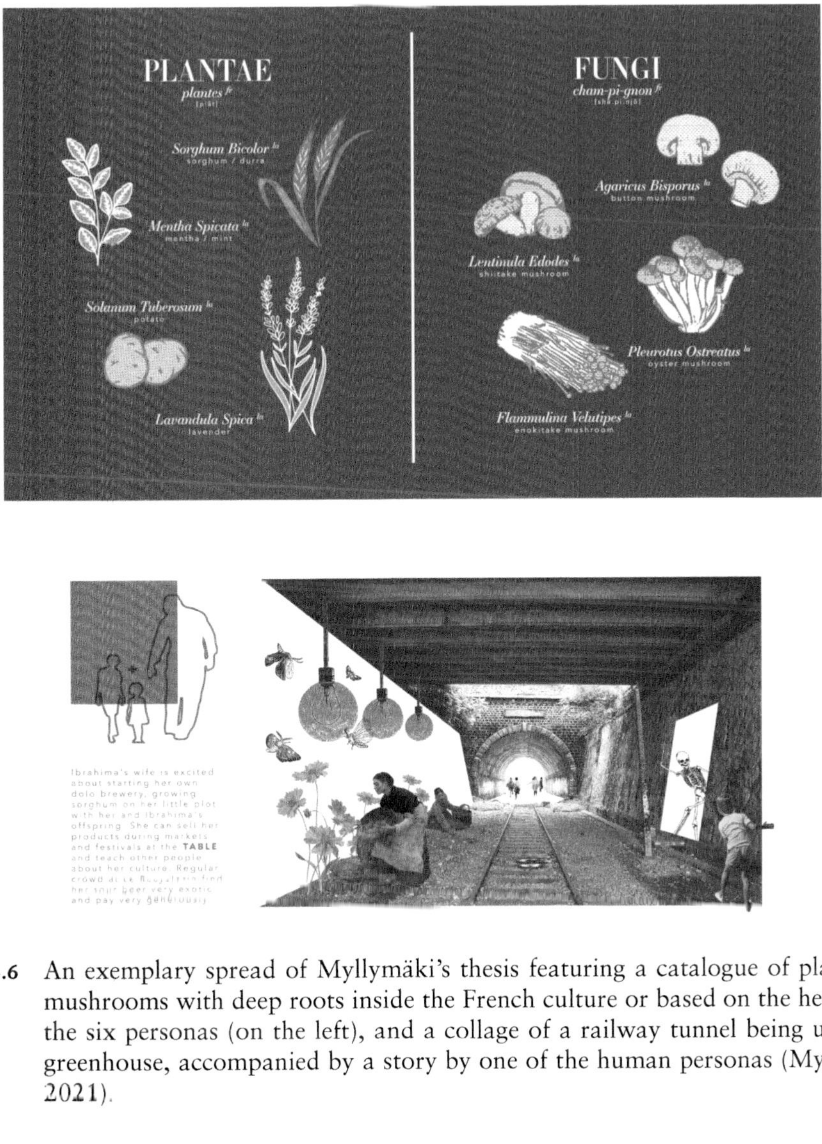

FIGURE 4.6 An exemplary spread of Myllymäki's thesis featuring a catalogue of plants and mushrooms with deep roots inside the French culture or based on the heritage of the six personas (on the left), and a collage of a railway tunnel being used as a greenhouse, accompanied by a story by one of the human personas (Myllymäki, 2021).

Changing our internal architecture values towards ecologies of repair

It is clear that the principle of empathy relates to the philosophy of biophilia and its original idea of humans' inherent emotional connection to nature. Empathy asks the designer to stand in the shoes of others, especially the most vulnerable in society and those with no voice (including nature and non-humans). The absence (or presence) of empathy in architectural thinking goes beyond professional and official roles and is innately connected to architects' inter-subjective 'internal values', i.e., shaped by the profession's traditions and worldviews that have been complicit in exploitation.

Architects' internal values need to expand the concept of nature from objective biological and ecological notions into socially and culturally shared practices involving people in restorative practices (Blanco-Wells, 2021; Söderlund, 2019; Söderlund and Newman; 2015; Tabb, 2021). Such restorative practices, also referred to as ecologies of repair

(Blanco-Wells, 2021), brings biophilia and radical inclusivity together by connecting social and ecological issues to (a)mend what is seen or felt as broken. Repair and restoring are not conceived from a moralising or romantic point of view but offer an inclusive social perspective and encompass material and non-material circumstances for the design of infrastructures as part of a network of complex ecosystems (Blanco-Wells, 2021, 2–10). Especially in a crisis, such as climate change, the crisis itself represents a possibility for collective rethinking and change.

Myllymäki's (2021) proposed restorative environments occur at different scales: from mushroom and food growing to composting and communal kitchen, to a biodiverse protected cycling loop that provides a haven for a diversity of life-forms. Myllymäki offers an alternate reality to the Capitalocene and sees it as the architect's responsibility to 'design out' social and ecological injustices by eliminating profit-seeking in its construction and operation. Instead, as an architect Myllymäki envisions a natural cycle profiting nature and humans in symbiosis. For all cooking and fertilising purposes, the kitchen and garden rely on leftovers, bio-waste from daily bypassers, and any surplus food past its best-before date from nearby restaurants or markets, completed solely by organic processes and local hands. In return people, some lacking capital needed for engaging in a formal type of leasing, or a decent meal and a family to share it with, receive fresh produce grown on their very own seed beds, creating a feeling of belonging and accomplishment enjoyed in tune with the whole community. Since there are no rooms for rent or paywalls to encounter, only seedbeds to retain by *anyone* willing, people give value to the system by lending their time, talent and empathy, with radical inclusivity at its heart. Myllymäki's restorative approach further manifests itself in approachable, accessible and user-adaptable architecture, which is in line with Cole's (2015, 122) earlier described notion of radical inclusivity.

Moreover, radical inclusivity is extended to nature and the climate by avoiding any extractive practices and carefully designing components that serve several functions and living forms, as per Nawratek and Nawratek's (2015) definition of radical inclusivity. For example, the piloti structure adapts through seasons, guiding rainwater towards thirsty crops, able to shelter different lifeforms (e.g., people, animals, and vegetables) through the coldest months, and providing shade during hot summers (see Figure 4.7). Existing unused, gloomy tunnels house mushrooms, provide natural refrigeration and allow bats to hibernate undisturbed. Similarly, the open kitchen hosts events ranging from communal dinners to pop-up juiceries, to social workshops, and accidental encounters (see Figures 4.5 and 4.9). It is a place where different lifeforms can simply be among one another, rest, breathe, sit down, or even sleep. Then if it is ever unwanted, everything can be disassembled and moved to a new location or reclaimed, returning the site to a better place than it was before (see Figure 4.8). As Myllymäki (2021) imagines wooden arches rich in colour encircling the community kitchen, she also underlines that the proposed detailed design has to be reviewed in reality after acquiring available (locally reclaimed) materials, expertise from users and volunteers, and crafted and executed in a way that invites the community to be a part of the slow, eventful process, with the ability to empower people as described earlier in William Morris' vision.

Myllymäki (2021) discovered that through centring different living-beings with empathy in her work, her own 'internal architect values' shifted. Instead of controlling every detail and aspect of the project, she discovered that an architecture of empathy and repair

FIGURE 4.7 The wooden structure provides shadow and shelter for the inside, enabled by new and existing vegetation (Myllymäki, 2021).

FIGURE 4.8 The facade elements utilise existing fencing and reclaimed polyfibre, and create a rigid structure by steam bending and ancient wood joinery instead of nails or glue; it can be user-adapted, disassembled and moved (Myllymäki, 2021).

needs to encourage and enable people and other life-forms to adopt and adapt a place to its unique context that changes over time. This uncertainty indeed is a different kind of 'architecture' with a new aesthetic that might be viewed by the prevailing architecture

FIGURE 4.9 The communal kitchen is built with minimal construction as an open galley kitchen, in order to make everyone feel welcome to cook, chat over the counter, or stop for a cup of coffee (Myllymäki, 2021).

culture as 'messy', 'rough', 'incomplete', ' impermanent'. A new architectural thinking goes hand in hand with new ideas about what it looks like, as also suggested by Hill (2021), and how it is (re)presented and communicated. This is reflected in Myllymäki's (2021) illustrations, which are deliberately suggestive instead of directive, in order to encourage free interpretation, imagination and adaptation by local people and adapted to suit the locally available reclaimed resources. In the end, Myllymäki's make-shift inclusive intervention that is user-adaptable and repairable over time, might achieve a level of longevity and importance in people's lives hard to imagine with architect-controlled unadaptable buildings that might not meet users' needs and that might become prematurely obsolete.

Conclusion

This chapter highlights how architecture in the Capitalocene has been complicit in the exploitation of people, nature and other living forms for centuries. Architects' internal values have been both influenced and upheld by the prevailing profit-seeking societal system that normalised these architectural products based on exploitation and leading to embodied injustices.

We argue that the principles of radical inclusivity, where an approachable, unintimidating architecture can be used by different living-beings in different (adaptable) ways, along with biophilia's innate love towards diverse humans and all living creatures and topophilia's inclusion of the non-living natural world, might be a starting point in challenging and changing our internal values from exploitative to restorative.

Illustrated through the design process of a case study, we show how a restorative architecture that centres empathy, inclusivity and solidarity with people, nature and other living-beings helps repair past damages and rejects embodied injustices. Doing so requires a renewed thinking about our own role as architects and how and what architecture to create and (a)mend, and for whom. While the damaging impacts of our current architectural thinking are significant, the potential for renewed architectural thinking to contribute to a less exploitative and more just and inclusive society for all, is even more significant, and urgent.

KEY TAKEAWAYS FROM CHAPTER 4

- The current human-centric profit-seeking approach in design of the era of the Capitalocene has for centuries upheld an unjust system, exploiting both people and nature, but could be redirected towards an inclusive, biodiverse, restorative future with the principles of Biophilia, Topophilia and Radical Inclusivity.
- At present, architecture practice takes little notion in whom they are designing for, resulting in designs unfit for their purpose or the user, sequentially becoming obsolete prematurely.
- Empathy towards the other, whether human or non-human, is at the heart of Radical Inclusivity, that aims for adaptable, non-threatening spaces suitable for all different users.
- Biophilic (= 'love for nature') design centres around the fundamental rewilding of our cities as well as care for the nature surrounding them, reconnecting humans with nature once more.

Notes

1 Bio stands for 'life' and philia denotes fondness.
2 Travel to the site was not possible due to the global Coronavirus Pandemic.

Bibliography

Aalborg Charter. (1994). https://sustainablecities.eu/the-aalborg-charter/ (accessed 27 December 2021).

Beatley, T. (2020). *The emergence of biophilic design and planning, Re-envisioning cities and city life in Ecologies Design: Transforming Architecture, Landscape and Urbanism*, Editors: Maibritt Pedersen Zari, Peter Connolly & Mark Southcombe, Routledge, pp. 96–106.

Beery, T., Jönsson, K. I. & Elmberg, J. (2015). From environmental connectedness to sustainable futures: Topophilia and human affiliation with nature. *Sustainability*, 7, 8837–8854. https://doi.org/10.3390/su7078837

Beery, T., Stahl Olafsson, A., Gentin, S., Maurer, M., Stålhammar, S., Albert, C., Bieling, C., Buijs, A., Fagerholm, N., Garcia-Martin, M., Plieninger, T. & M. Raymond, C. (2023). Disconnection from nature: Expanding our understanding of human–nature relations. *People and Nature (Hoboken, N.J.)*, *5*(2), 470–488. https://doi.org/10.1002/pan3.10451

Blanco-Wells, G. (2021). Ecologies of repair: A post-human approach to other-than-human natures, *Front. Psychol.*, 8 April 2021, www.frontiersin.org/articles/10.3389/fpsyg.2021.633737/fulll https://doi.org/10.3389/fpsyg.2021.633737

Cole, A. (2015). *The Conscious Sub Syncretic System in Radical Inclusivity: Architecture and Urbanism*, edited by Krzysztof Nawratek, Alexander Horton-Howe, Benjamin Huggins & Matthew Oxley, Dpr Barcelona.

Crist, E. (2013). On the poverty of our nomenclature. *Environmental Humanities*, 3 (November 2013), 129–147.

Donovan, E. & Pelsmakers, S. (2019). Integrating sustainability in architectural design studio through blended learning. Conference Proceedings for Education, Design and Practice – Understanding skills in a Complex World, NY, June 2019, AMPS.

Fraser, M. (2015). Interview in *Radical Inclusivity: Architecture and Urbanism*, edited by Krzysztof Nawratek, Alexander Horton-Howe, Benjamin Huggins & Matthew Oxley, Dpr Barcelona.

Ghosh, A. (2021). *The Nutmeg's Curse: Parables for a Planet in Crisis*, John Murray Publishers.

Griffin, C. (2004). An introduction to biophilia and the built environment. *RMI Solutions*, 20(1), 7–11.

Hill, D. (2021). Slowdown city. *The Journal of Architecture*, 26(1), 67–72, DOI: 10.1080/13602365.2021.1883306,

Hajer, M. A. (1995). *The Politics of Environmental Discourse: Ecological Modernization and the Policy Process*. Clarendon Press.

Havik, K. (2014). *Urban Literacy: Reading and Writing Architecture*. NAI.

Havik, K. (2019). Passing the threshold of time: juxtaposing literary and architectural analyses of Amuri, Tampere (pp. 99–122), in Havik, K. (ed.) & Passinmäki, P. (ed.), Tampere, Finland: datutop. 144 p. (DATUTOP Occasional Papers; vol. 38).

Haraway, D. (2016). Tentacular Thinking: Anthropocene, Capitalocene, Chthulucene, Journal 75, September 2016 – e-flux Journal, 2016 (accessed 30 December 2021).

Mangold, W. (2010). Money-tecture... Or How Architecture Is Exploited by Capitalism, ACSA 99th annual meeting Conference proceedings, 'Where do you stand?', edited by Alberto Perez-Gomez, Annie Cormier & Annie Pedret.

Moore, J. (2019). Capitalocene and planetary justice, Who is responsible for the climate crisis?, Maize, www.maize.io/magazine/what-is-capitalocene/, (accessed 29 December 2021).

Myllymäki, M. (2021). *On How To Reimagine Lost Spaces Through The Concept Of A Table*. Master's thesis. Tampere University, Tampere. 2021

Nawratek, K. (2021). Hope in the Burning World (pp. 87–98), in Pelsmakers, S. & Newman, N., 2021. (Volume Guest Editors). RIBA *Design Studio, Volume 1: Architecture and the Climate Emergency: Everything Needs to Change*, RIBA Publishing.

Nawratek, K. & Nawratek, K. (2015). On the Frustrating Impossibility of Inclusive Architecture, (pp. 12–22) in *Radical Inclusivity: Architecture and Urbanism*, edited by Krzysztof Nawratek, Alexander Horton-Howe, Benjamin Huggins & Matthew Oxley, Dpr Barcelona.

Nisonen, E., Poutanen, J. & Pelsmakers, S. (2023). Responding to embodied injustices: introducing critical thinking and empathy in architecture education through design thinking, in *Architectural Thinking in a Climate Emergency*, edited by Brown, J., B. & Pelsmakers, S. Routledge.

Pelsmakers, S., Donovan, E., Moseng, K. & Eyebye, B. (2019). Developing architecture studio culture: peer-peer learning, Conference Proceedings for Education, Design and Practice – Understanding skills in a Complex World, NY, June 2019, AMPS.

Pelsmakers, S., Poutanen, J. & Saarimaa, S. (2020). (Hybrid) Architecture in and over time, in *Ecologies Design: Transforming Architecture, Landscape and Urbanism*, Editors: Maibritt Pedersen Zari, Peter Connolly & Mark Southcombe, Routledge.

Ryan, C. O., Browning, W. D., Clancy, J. O., Andrews, S. L. & Kallianpurkar, N. B.. (2014). 'Biophilic design patterns: Emerging nature-based parameters for health and well-being in the built environment.' *Archnet-IJAR: International Journal of Architectural Research*, 8(2), 62–76.

Söderlund, J. (2019). The emergence of a new social movement: Biophilic design. In: *The Emergence of Biophilic Design. Cities and Nature.* Springer, Cham. https://doi.org/10.1007/978-3-030-29813-5_1

Söderlund, J. & Newman, P. (2015). Biophilic architecture: a review of the rationale and outcomes. In AIMS *Environmental Science*, 2(4): 950–969. DOI: 10.3934/environsci.2015.4.950
Tabb, J. (2021). *Biophilic Urbanism, Designing Resilient Communities for the Future*. Routledge.
Till, J. (2020). Architecture after Architecture, in *Architects after Architecture Alternative Pathways for Practice*, edited by Harriet Harriss, Edited by Rory Hyde & Roberta Marcaccio, Routledge.
Tonkiss, F. (2015). Interview, in *Radical Inclusivity: Architecture and Urbanism*, edited by Krzysztof Nawratek, Alexander Horton-Howe, Benjamin Huggins & Matthew Oxley, Dpr Barcelona.
Tsing, A. (2015). *The Mushroom as the End of the World: On the Possibilities of Life in Capitalist Ruins*. Princeton University Press.
Tuan, Y. (1974) *Topophilia: A Study of Environmental Perception, Attitudes and Values*. Columbia University Press.
Tucker, E. (2021). A New Concept for Good Architecture: Embodied Justice – Metropolis, https://metropolismag.com/viewpoints/embodied-justice-2021/, (accessed 29 December 2021).
Ward, A. (1996). *The Suppression of the Social in Design Architecture as War in Reconstructing Architecture: Critical Discourses and Social Practices*, edited by Thomas A. Dutton & Lian Hurst Mann, University of Minnesota Press.
Wilson, Edward, O. (2016). *Half-Earth: Our Planet's Fight for Life*. New York: Liveright.
Vazquez, C. G. (2022). *Cities After Crisis: Reinventing Neighborhood Design from the Ground-Up* (1st edn). Routledge. https://doi.org/10.4324/9781003130857
Yusof, Y. M. & Kozlowski, M. (2015). Clients of Contemporary Urban Design: The Impact of Neoliberalism in *Radical Inclusivity: Architecture and Urbanism*, edited by Krzysztof Nawratek, Alexander Horton-Howe, Benjamin Huggins & Matthew Oxley, Dpr Barcelona.

5

ENACTED UTOPIAS

Learning from insurgent spatial practices

Dalia Milián Bernal, Mikko Kyrönviita, Elina Alatalo and Veera Turku

Shaping the city insurgently

In November 2021, we took part in a workshop organised within the framework of a seminar entitled Green City.[1] The workshop divided the group of participants into several teams and each team was asked to provide advice to one imaginary stakeholder who wanted "much more ambitious greening of the city… and *the city* to implement their commitment to increase and enhance nature everywhere". Stakeholders included a city gardener, a student, a politician, an urban planner, and an activist. Our team was given the task of advising the activist.

While we were able to provide some advice to our imaginary stakeholder, we did so with trepidation, aware that we were drawing on the knowledge we have gathered over the years about urban activism and their (often) insurgent spatial practices, from our research, teaching, personal experiences as well as the experiences of the many urban activists we have met in the process and who have shared their stories and knowledge with us – including some who are fighting to preserve the precarious green spaces still available in their cities.

It is with this in mind that we would like to return to the Green City workshop and advise all the aforementioned stakeholders that while they continue pressing *the city* to implement their commitment to greening the city, they also get informed, organised, armed (with gloves, boots, shovels, trowels, earth, seeds, and water), go to the streets, find a piece of land, and begin guerrilla gardening the city themselves. Alternatively, they should get some coloured ribbons and wrap them around trees to raise awareness about their importance, as done by the activists of the *Pro Eteläpuisto* (Figure 5.1), thereby temporarily halting the felling of approximately 300 old trees at an urban park in the city of Tampere in Finland (Mettinen, 2017). Or, in winter, give new life to Christmas trees and 'plant' them in an urban wasteland, as did the residents in Hochelaga-Maisonneuve district ("Résister et fleurir," n.d.), in Montreal (Canada), to stop the transformation of the wasteland into a container yard and show how it could become a park (Paré, 2022).[2] All the above are just a few examples of how to green and shape the city… *insurgently*.

DOI: 10.4324/9781003293903-6
This chapter has been made available under a CC-BY-NC-ND 4.0 license.

FIGURE 5.1 Photograph of activists of *Pro Eteläpuisto* fixing coloured ribbons around 300 old trees that would be cut down if a project of urban development, planned for the area, were to proceed, Tampere, Finland, 2016. Photo: Markku Rekola.

This chapter does not inquire into "things in space" – such as architectural projects – but into the "actual production of space" (Lefebvre, 1991), focusing on insurgent spatial practices – namely, spatial practices that occur outside, at the margins of, or against state-sanctioned and market-driven forms of urbanisation (Holston, 1995; Miraftab, 2009; Sandercock, 1998). We will illustrate how these practices can help shift architectural thinking in times of climate emergency and open emancipatory paths towards a spatial production that is not only more democratic but also more sustainable.

First, we draw on Henri Lefebvre's theory about the production of space to illustrate the way in which capitalism has become spatialised. We focus on the concepts of abstract and differential space. The former helps us visualise how space is produced under the current forms of capitalism and how this production contributes to and exacerbates the climate emergency. The latter, on the other hand, enables us to unveil the processes that challenge such production and the *ways* in which these forms of resistance become spatialised while remaining decoupled from the modes and relations of production inherent in capitalism. In this section, we will also emphasise the peculiar role of knowledge and how it has been deployed to further a mode of spatial production that is exploitative and violent towards all forms of life. In the third section, we trace several conceptualisations related to the spatialisation of insurgency followed by diverse cases from Latin America, Finland, and Canada that demonstrate *how* urban space can be produced with situated and experiential knowledge, care, and solidarity. We conclude by providing insights into how learning from insurgent spatial practices may open avenues for politically progressive architectural thinking with the potential to address the climate emergency.

The spatialisation of capitalism and its contestation

"If there is such a thing as a history of space," wrote Henri Lefebvre in 1974,[3] "then there is such a thing as a space characteristic of capitalism" (1991, p. 126). According to Lefebvre, (social) space is (socially) produced and each society, with its own modes and relations of production, produces its own space. "Capitalism," he tells us, has "produced abstract space" (1991, p. 53).

Abstract space

Abstract space is the contradictory space produced under current modes and relations of production, whereby space has become a concrete abstraction of the likes of money, commodities, capital and labour (1991, p. 27). Here, space is concrete in that it is "the material product of social relations" and abstract in that it is "a manifestation of relations" (Gottdiener, 1994, p. 130). Abstract space is the space of power – the power of money and the political state and elites – and, under capitalism, power is also manifested in and asserted through space (Brenner, 2009).

Architects and planners play a particular role in the manifestation and assertion of such power in space (Baeten, 2017; Harvey, 2013). To Lefebvre (1991), these disciplines represent the fusion of power and knowledge, resulting in an "authoritarian and brutal spatial practice" (1991, p. 308). Abstract space is instrumental and hierarchical and the spatial practice that produces it tends to erase difference and is inherently violent towards nature and the human body; not only does abstract space splinter the environment into formal and quantifiable fragments "made up of a series of zones defined by interdictions and bans" (1991, p. 35), it also fragments "the functions, elements, and moments of social practice" (1991, p. 52). Abstract space is the space envisioned by architects and planners (although, not only) constructed by abstract workers for abstract users employing violent tactics of erasure of difference and extraction of natural resources enabled via the power of the political state to further assert itself (see Baeten, 2017; Harvey, 2008).

It is increasingly clear that capitalism, an infrastructure that encapsulates not only an economic system but also "struggles over nature, social reproduction and public power" (Fraser, 2014, p. 56), plays a prominent role in the processes of urbanisation (Baeten, 2017; Brenner et al., 2012; Harvey, 2013; Lefebvre, 1991). Equally evident is that current modes and relations of production under capitalism are at fault for our current multiple yet interrelated environmental crises, including the climate emergency (Demaria et al., 2013; Klein, 2015; Monbiot, 2021), and that current forms of urbanisation exacerbate these multiple crises (Savini, 2021; Swyngedouw and Heynen, 2003). We would like to begin to address the proposition of this chapter by stating that, in a time of climate emergency, architectural thinking and practice must be decoupled from current capitalist modes and relations of production. Our aim hereafter is to illustrate and argue that other forms of spatial production are possible. As Lefebvre noted, abstract space "carries within itself the seeds of a new kind of space" – a "differential space" (1991, p. 52).

Differential space

Differential space is the space of those who have been excluded from the centrality of abstract space and of those who challenge the homogenisation and fragmentation inherent of it (Lefebvre, 1991). Differential space "puts an end to those localizations that shatter the integrity of the individual body, the social body, the corpus of human needs, and the corpus of knowledge" and its production restores the unity of social practices (1991, p. 52). Differential space is not attained by means of domination and property but through the appropriation of space. That is, through the occupation and transformation of space with our bodies and by means of spatial and embodied practices that help restore the relationship between body and space. This spatial practice cherishes socio-spatial relationships often built on solidarity and care. It generates collective knowledge and uses that knowledge to challenge inherited forms of spatial production and social organisation deployed through the policing of bodies in space. Building on Nast and Pile's (1998) argument that body/place relationships are political, we argue that such spatial practices are inherently insurgent; they challenge hegemonic power and illuminate emancipatory paths towards more sustainable and democratic spatial production.

Conceptualising the spatialisation of insurgency

Within the context of urban studies and planning, the notion of *insurgent* has been deployed to describe processes of urbanisation, forms of urbanity and citizenship, and planning practices that happen outside, against, or indeed, as a result of state-sanctioned "spatialised policies" and professionalised practices, including (but not limited to) urban planning, architecture, and urban policy (Holston, 1995; Karaliotas and Swyngedouw, 2019, p. 374; Miraftab, 2009; Sandercock, 1998).

Spaces of insurgent citizenship

In 1995, Anthropologist James Holston called for the (ethnographic) investigation into spaces of insurgent citizenship and employed the term *insurgent* to, on the one hand, "emphasize the opposition of these spaces of citizenship... to modernist spaces which physically dominate many cities today" and, on the other, "to emphasize an opposition to the modernist political project which absorbs citizenship into a plan of state-building" (Holston, 1995, p. 37). Holston understood modernist planning both as urban design and applied social science and critiqued modernity's "pervasive ideal... that the state... can change society and manage the social by imposing an alternative future embodied in plans" (Holston, 1995, p. 38). By utilising the term insurgent, he aimed to challenge the "modernist" idea that "the state is the only legitimate source of citizenship, rights, meanings, and practices" (Holston, 1995, p. 37) and called for a planning theory that is situated, ethnographic and which recognises that spaces of insurgent citizenship "introduce into the city new identities and practices that disturb established histories" (Holston, 1995, p. 44).

Insurgent planning historiography

Related to the disturbance of established histories, in a book of essays entitled *Making the Invisible Visible: A Multicultural Planning History*, Sandercock (1998) revealed how planning practices also take place outside, against, and often as a response to institutionalised forms of planning. She observed that, by emerging as a subfield of planning and not a subfield of history, planning history mostly "chronicled the rise of the profession, its institutionalization, and its achievements" without the "application of theories of power/knowledge/control to the domain of planning" (Sandercock, 1998, p. 3). Inasmuch, planning history disregarded the stories of oppressed and marginalised communities, such as women, people of colour, migrants, gays, and lesbians, the ways in which they experienced and/or were excluded from modernist planning, and how they, too, were "doing planning, contributing to city and community building, [and] researching urban problems" (Sandercock, 1998, p. 8). Through a series of essays written by different authors of diverse social backgrounds and fields of knowledge, Sandercock aimed to "revisit planning history and re-present it" and open to discussion the definition of planning (1998, p. 1). Inspired by Holston's notion of spaces of insurgent citizenship, Sandercock (1998) termed these stories and their practices as 'insurgent planning historiography' (1998, p. 2).

Insurgent planning

While Sandercock's insurgent planning historiography focused mostly on American peoples and practices unfolding in the United States, the concept of insurgent planning has been further developed and deployed to locate and understand spaces of insurgent citizenship and insurgent planning practices across the globe (Miraftab, 2018, 2009; Roy, 2005; Sletto, 2013). Miraftab (2018) views insurgent planning as radical and political planning practices unfolding "outside the state apparatus of inclusion... reflected predominantly on grassroots practices" and the practices of subordinate groups through which plans, policies, and spaces are also shaped (2018, p. 278). Building on Miraftab's concept, Huq (2020) argues that insurgent planning "is characterized by direct action, self-organization, autonomous politics, constant negotiation, and constant contestation" (2020, p. 379). When these actions "directly confront authorities, challenge, and destabilize the status quo", Miraftab (2018) refers to them as invented spaces (2018, p. 280). However, she also recognises that community-based informal groups and grassroots organisations are often subsumed and legitimised by donors, foundations, and governmental interventions, a practice she calls invited spaces. Miraftab (2018) notes that: "insurgent planning defies confinement and conformance of citizen's actions to liberal state and market norms, but also recognizes how citizens might use those norms to induce a rupture and create something new" (2018, p. 280).

Insurgent practices

An important characteristic of insurgent planning is that "the theorizing objects [...] are *practices* not individuals" (authors' emphasis, Miraftab, 2009, p. 281). Concordantly, García-Lamarca (2017) suggests that "insurgency can be effectively queried through practice – understood as open, temporarily unfolding, connected actions – to engage explicitly with more detailed socio-spatial and political processes around an emancipatory, democratic politics" (2017, p. 5). Thus, García-Lamarca suggests the notion of insurgent practices, which she defines as the "collective socio-spatial and political nexus of actions... that enact equality and disrupt the dominant production of space" (2017, p. 5).

Based on her study of and participation in housing recuperations in Spain, García-Lamarca (2017) argues that the theorisations of "insurgent citizenship and planning… lack profound engagement with spatiality" (2017, p. 6). To engage with spatiality, García-Lamarca (2017) deploys Henri Lefebvre's (1991) triad of space – spatial practice (the perceived space), representation of space (the conceived space), representational space (lived space) – and argues that insurgent practices prioritise representational space – lived space – "as use value, difference and social life are brought to the fore in and through the appropriation of space" (2017, p. 7).

Insurgent spatial practices

In a similar vein of thought but within the context of social movements in Latin America and deploying anarchist theory, geographer Lopes de Souza (2016) calls for the recognition of the "territorial work" of social movements (2016, p. 1297) and the importance of the social space within which these mobilisations emerge. For Lopes de Souza (2016) these social spaces include, "the built environment and its material substrate, […] processes of territorialization, de-territorialization, and re-territorialization, […] space as place (in the sense of perceived and lived space endowed with symbolic and emotional contents), [and] location" (2016, pp. 1300–1301). He argues that by engaging with space and spatiality, we can better understand how power is deployed in and through space, the "relative value (economic, political, strategic, and symbolic) of a specific space", as well as activists' agendas, demands, actions, resources, identity-building processes, and the way in which they resignify space (Lopes de Souza, 2016, p. 1301).

Lopes de Souza (2016) puts forth the term insurgent spatial practices, which may include: the appropriation of space "through or with the help physical presence"; the challenge of the "rights and territorial claims of private owners" and state-sanctioned "spatial rules by means of provocative signs"; the "re-functionalization and restructuration of material space" to adapt to new needs; "resignification of places" through culture, symbols and discourses; the construction of "alternative economic circuits"; and the creation of "spatial networks" (Lopes de Souza, 2016, pp. 1301–1302).

We build on all these perspectives and theorisations to engage with insurgent spatial practices. Unlike the utopias envisioned by the project of modernisation that aimed to manage the social by managing space (Holston, 1995), we see insurgent spaces as enacted utopias that materialise human needs, dreams, and desires in space where new and diverse identities, cultures, lifestyles, livelihoods, and multiple socio-spatial configurations may emerge.

The spatialisation of insurgency via appropriation

As noted above, the appropriation of (urban) space with the physical presence of bodies is an important characteristic of insurgent spatial practices. The presence of human bodies in urban space is still one of the most powerful tactics used by social movements to raise awareness, show political dissent and discontent, and achieve social change (for example, see Gersmann, 2011). Lopes de Souza (2016) refers to these appropriations of space as "territorialization in the strict sense" (2016, p. 1301); think of the classical street protest, or street blockades by groups such as Extinction Rebellion. Here, however, we will focus on forms of territorialisation and spatial appropriation whereby the existing (often abandoned) material space and urban infrastructures have been appropriated, refunctionalized, restructured, and resignified by the people who will use them (Lopes de Souza, 2016, p. 1301). Through appropriation, spaces are endowed with use-value, whereby the material substrate of urban space becomes an important resource in which alternative, social, and circular economies are practised; networks of solidarity and care are created; situated and experiential knowledge is generated and shared; new and diverse skills are developed, and new identities are formed. Often, these spaces may become breeding grounds for political action and social mobilisation critically needed in a time of climate emergency (IPCC, 2022).

In different geographical contexts, we have documented how abandoned urban spaces have been transformed by activists themselves into all sorts of cultural and working spaces utilising their bodies and corporeal physical work, a few tools, recycled and/or left-over materials and available resources (including material space), and their time (e.g., Alatalo et al., 2024a; Kyrönviita and Wallin, 2022; Milián Bernal, 2022, 2020). In contrast to big machinery utilised in construction sites to facilitate and speed up the construction process (e.g. cranes, excavators, etc.) as well as digital tools that allow for faster planning and design decisions to be visualised and made abstractly from somewhere else, by utilising their own bodies and energy to transform space, a scalar shift occurs, since our bodies represent a limit of what we can construct as well as how, how big, and how fast we can do so. This scalar shift is important to consider in architectural thinking in a time of climate emergency as it is what may enable a restoration of the relationship between body and space and allow time for nature to regenerate. In all the cases we have studied, people themselves take the space, clean it, prepare it to use for their needs. In this process, their bodies become their most valuable tool to work to transform appropriated space but also to gain situated knowledge; learning-by-doing but also by being present in space as opposed to engaging with abstractions of space through a computer screen.

In what follows, we present several case studies from our research in Latin America, Finland, and Canada. Our aim is to illustrate how insurgent spatial practices restore the relationship between body and space via appropriation (not domination) and how they offer insights into a production of space that is more democratic and sustainable.

Enacted utopias

Huerto Tlateloco, a centre for urban agriculture in Mexico, is a good place to begin with. It was created on an abandoned plot located in a central borough in Mexico City and is currently run by a small, non-profit organisation called *CultiCiudad S.A.* (Figure 5.2). Their aim is to generate and share knowledge about urban agriculture, to connect people with food and life cycles, and to provide new public use to otherwise abandoned urban spaces (CultiCiudad, n.d.; Milián Bernal, 2019). While *Huerto Tlatelolco* operates under a non-profit logic and mostly relies on volunteer work, the project has become a source of income and space to live for some people who take care of the space full time. Hence, the creation of this space is intrinsically linked to the livelihoods – and social reproduction more broadly – of the people who spend most of their day in this space. By working the

FIGURE 5.2 Photograph of the edible plants of *Huerto Tlatelolco*, Mexico City, Mexico, 2019. Photo: Dalia Milián Bernal.

soil with their hands and tools, planting the seeds, watering, and otherwise caring for the growth of the crops, the relationship between their bodies, their social practices, and the space they produce is rekindled. As Gabriela Vargas, one of the founders of the project, put it, *Huerto Tlatelolco* "transformed [her] life and makes [her] feel fulfilled" (Vargas in Fregoso Centeno and Moncayo, 2017).

In the context of Finland, activists have also found spaces at the edges and in the cracks of investment-heavy urban development that are free from conventional social norms and regulations (Franck and Stevens, 2006; Tonkiss, 2013). For example, there are several grassroots coworking spaces across different Finnish cities (for more details, see Alatalo et al., 2024a). Usually, this insurgent spatial practice is done by groups of people who locate spaces that have been left over from the rental market. These spaces are often in need of reparations, giving people the possibility to renovate the spaces themselves and to freely experiment with the kind of working spaces they want. These users often renovate the spaces themselves without prior knowledge of construction, renovation, or interior design. Thus, they create atmospheres that are different to the abstract and alienating styles circulating in design magazines (Arantes, 2019). Rather, they follow their instincts. In addition, many of these actors join self-organised coworking spaces due to their low cost, making it economically less risky to try out new forms of livelihoods. Here again, the production of space is linked to people's livelihoods, whereby appropriated and transformed spaces become their spaces of work: spaces of social practice. In addition to restoring the relationship between body and space, these grassroots coworking groups are also challenging established notions of what is considered work and a working space, as a boat, a theatre stage, or a shop can become a coworking space. The livelihood can be a collage of sources, some of them deriving from skills developed in these spaces (Alatalo et al., 2024a; Caperna et al., 2019).

Both *Huerto Tlatelolco* and the grassroots coworking initiatives have also become sites where knowledge is generated and shared. We have learned that insurgent spatial practices demand, deal with, and create situational and shared co-constructed knowledge. In the Mexican case, the actors organize workshops, training schools, and other events that help support other urban gardens and gardeners. Furthermore, the project forged collaborations with elementary schools that bring their pupils to learn about the process of growing an edible plant (and preparing a meal with it) (CultiCiudad, n.d.). In the coworking cases, the production of space is a process of learning by doing and it is worth noting how these 'design amateurs' avoid aesthetic memes of the design field and learn how to search for the materials, select the colours and compositions proper for their well-being (Caperna et al., 2019).

This was also the case of *Tikkutehdas DIY*, a skatepark built by a group of skaters who appropriated a space from a former matchstick factory in Tampere, Finland. The abandoned space offered endless possibilities for alternative uses and underground cultures to flourish and became a fertile ground for local skaters. The skaters cleaned the space and utilised materials from the surrounding area to create a skateable space. *Tikkutehdas DIY* was an ideal space for experimenting with building techniques and different materials that the skaters were not familiar with at first (Figure 5.3). Through labour and putting their bodies at work in communion with the space, the skaters learned about materials, the basics of skatepark building, and ways to work together for a common goal. *Tikkutehdas DIY*, which was demolished in May 2018, was a place for individual

FIGURE 5.3 The skaters in Tampere, Finland, appropriated an abandoned industrial space, started experimenting, and built an informal do-it-yourself skatepark. Photo: Niklas Pedersen.

and collective learning filled with ideals of decommodification, sharing, and caring (for more details, see Kyrönviita and Wallin, 2022).

Shared, situated and co-constructed knowledge is an often intangible characteristic of insurgent spatial practices but it can also be a tactic to deal with the lack of resources and (wo)manpower. *Parque Estrella* was an abandoned public park in San Juan (Puerto Rico, U.S.) that had become a public dumpster (Milián Bernal, 2022). It was later appropriated by the grassroots organization *La Maraña* and the neighbours living next to it and, together, it was transformed into a community park (Milián Bernal, 2022). All the work was done by volunteers and using only donated materials. Furthermore, they often lacked the knowledge to build what they wanted. Thus, they always used the opportunity to organize a workshop to learn how to do it themselves, but also generate and share knowledge in the process of the production of (differential) space. As one of the co-founders of *La Maraña*, told us: "If we were going to paint a wall, we did a painting workshop. If we were going to plough the earth, we did a ploughing workshop. If we were going to do an area for permaculture, well, we did a permaculture workshop" (S. Unanue, personal communication, February 21, 2018).

Cleaning the spaces, preparing them for use, learning by doing, and organising workshops take time. We have learned that the temporal scale of insurgent spatial practices contrasts with current capitalist accelerated forms of production: "time is money" goes the phrase. Slowness is, thus, a nuanced characteristic of insurgent spatial practices. For example, in a project to build a self-organised public sauna in the former industrial site of Hiedanranta in Tampere, Finland, negotiations took several years but it was constructed in a two-week timeframe due to the quick mobilisation and collaboration between different actors (Alatalo et al., 2024b). This relates to the issue of temporal scale: the moment of action might be sudden, as the right momentum and people with the relevant skills might come together. The sauna had no exact design when the construction started, hence the structures and solutions were negotiated on the go (Figure 5.4). Most of the participants had no previous skills in construction; part of the idea was to learn both the craft of building as well as agency in activating vacant places (Alatalo et al., 2024b; Havik and Arlandis, 2022).

The agility to build the sauna was made possible because people took on new and diverse roles, often outside their comfort zones. This meant, for example, mediating surprising quarrels, cooking for 30 people, or improvising information signs. As many things were designed at the same time as the building process, small groups were responsible for certain aspects of the construction, requiring trust amongst the diverse groups. This kind of action model and demand for self-organisation creates an interesting process of collective learning and is characteristic of insurgent spatial practices and projects (Alatalo et al., 2024b; Havik and Arlandis, 2022).

In Montreal, Canada, community activists transformed a former train shop into an autonomous community centre called *Bâtiment 7* (Kruzynski, 2018; Sévigny, 2021). Like the other cases discussed, *Bâtiment 7* provides an example of appropriation of vacant space and reutilising it to respond to the needs of local residents. Situated in the Pointe Sant-Charles, a suburb characterised by poverty, geographical isolation, and insufficient access to affordable and nutritious food options and basic services, *Bâtiment 7* has become home to a grocery store, co-op bicycle and car repair facilities, a metal workshop, a youth-run arcade, and a brew pub, to name a few. Before the COVID-19 pandemic,

FIGURE 5.4 View of the construction of the public sauna built on the former industrial complex of Hiedanranta in Tampere, Finland, 2019. Photo: Aleksi Martikainen.

there were 13 groups providing their services and hosting workshops where knowledge and skills were shared. The place continued to evolve when green and blue alleyways surrounding the building took shape, and new plans for a daycare and a food production hub evolved. Developed by citizens for citizens and governed horizontally by the collective "7 à Nous", *Bâtiment 7* has become like an alternative village, an enacted utopia, where one can spend the whole day.

However, a lot happened before *Bâtiment 7* opened its doors in 2018. When train shop number 7 became vacant, it was about to be a subject of very different development, driven by capitalist interests, disconnected from the neighbourhood, and the everyday life of its residents. Already in 2003, a group of community activists living in Pointe Saint-Charles planted a symbolic flag on the terrain that hosted old Canadian National Railway workshops. With this action, the community manifested their interest in transforming the immense train shop number 7 into a community centre. Soon after, the owner of the space, Canadian National Railway, transferred the lot to Groupe Mach, a real estate entity that aimed to bring Montréal Casino to the area. This started a protest movement and community organising, which kept the idea of a community centre alive for 14 years, and finally, in 2016, the neighbourhood associations took possession of the train shop. Also, the plans for a casino were finally buried (Bâtiment 7, 2022). Similarly to the case of the public sauna in Finland, the realisation of *Bâtiment 7* builds on the perseverance of the community groups and their ability to act when the possibility finally opened up.

FIGURE 5.5 Image of *Proyecto Aupa*, a temporary public space dedicated to children and their caregivers, built in a vacant plot in Tucumán, Argentina, 2017. Photo: Verónica Mansilla.

Duration is also a nuanced luxury characteristic of insurgent spatial practices and its spaces. Several cases we inquired into are temporary if not ephemeral in nature, even if not planned to be so. The process of the production of (differential) space may be slow, even in cases that are ephemeral. A case in point is *Proyecto Aupa*, a small public space for children and their caregivers that was planned to last ten days (Figure 5.5). It was built on one of the many vacant plots located in the city of Tucumán, Argentina, and the aim of the project was to raise awareness about the lack of public spaces, the vast number of vacant plots in the city, and to show ways in which these spaces could be repurposed for the recreation of children and their caregivers (1319. trecediecinueve, n.d.; Milián Bernal, 2020). Although the public space lasted less than ten days, it was two years in the making. During this time, the grassroots organisation, *1319.Trecediecinueve*, in collaboration with an engaged university teacher and her group of students, applied for funding and organised and took part in an array of workshops. These included an urban walk with children, elderly people, and their caregivers to understand their lives and challenges; a materials workshop to learn about their properties and select healthy and safe materials for the construction of urban furniture for children; another material workshop to learn different techniques to use felt; a workshop where they learned how children and adults experience space and how to

FIGURE 5.6 Compendium of photographs from the different workshops attended by *1319. Trecediecinueve*, the activist university teacher, and the students. Left: photograph of the group in a workshop where they learned different felt techniques. Above: photograph of a session during the workshop about children's and adults' experience of space. Below: The group in a session from the workshop about children's play. Photographs: Verónica Mansilla.

empathise; and a workshop where the aim was to re-learn how to play and be playful (Figure 6).

The reutilisation of abandoned or underused urban infrastructures might be one of the most obvious links to the sustainable urban development discourse (Kytölä et al., 2021). As already noted, the cases we engage with appropriate all sorts of abandoned urban spaces, from former industrial sites to abandoned vacant plots, from vacant private buildings to neglected public parks. Here, the material substrate of the city is a valuable resource put to new use.

In Nokia, Finland, a community of different actors transformed an abandoned factory building into a thriving complex that included working spaces, ateliers, a yoga hall, a stage, a restaurant, and different types of shared spaces (Figure 5.7). All of it was done slowly throughout the duration of nine years and using mostly recycled materials. The project was called *Tehdas 108* and here the flow of recycled materials was so abundant that they started repairing things, such as furniture and kitchen appliances they needed and then sold the overflow of things. In addition, they donated essential materials to other self-organised creative centres. Key people from *Tehdas*

FIGURE 5.7 Left: Former industrial building during the cleaning process, site of *Tehdas 108*, Nokia, Finland. Photo Andrei Fadeev. Right: View of an event inside *Tehdas 108*. Photo: Lauri Haavikko.

108 were recognised as locally important agents of eco-social sustainability transition by the Council of Tampere Region and their space was frequently visited by students interested in different levels of circularity that could be witnessed there (Pirkanmaan murrosagentit, 2021).

Another important aspect of this project was the way in which they collaborated and integrated youth, unemployed people, and people interested in doing internships. They offered a special community to join and to feel a sense of accomplishment by concretely doing something for the building. Nevertheless, after so many years, the *Tehdas 108* community was pushed out of the premises when these were sold to a private actor. The new owner did not understand the use-value of *Tehdas 108* nor the magnetism that their activities generated in the area.

Tehdas 108 illustrated that it is not only the built environment that is important to note here. Of interest is also the way that all sorts of recycled materials are mobilised and utilised. In some of the cases we follow, money is not even circulated (for example, *Parque Estrella*). Like *Tehdas 108*, often these differential spaces put in place practices of recycling and endow otherwise-considered waste or trash with exchange value. This also happens in *Huerto Roma Verde*, a land occupation of more than 6,000m^2 also in Mexico City that hosts another centre for urban agriculture and a whole array of activities

FIGURE 5.8 Photograph of the entrance to *Huerto Roma Verde*, a centre for urban agriculture located in a central borough in Mexico City, Mexico, 2019. Photo: Dalia Milián Bernal.

(Figure 5.8). In their case, the project has created an alliance with a plastic recycling company and visitors can bring their plastic and exchange it for seeds. On Saturdays, the space becomes a marketplace, where all sorts of social economies are hosted ("Huerto Roma Verde," n.d.).

Huerto Roma Verde is also a good case to return to the topic of situated, shared, co-constructed and experiential knowledge and how through this process of knowledge construction, knowledge goes from being intangible to concrete and tangible. As mentioned above, *Huerto Roma Verde* is a large land occupation at the heart of Mexico City. It began in 2012 and has since had to stand its ground against eviction orders. The land is owned by a public (governmental) entity. The place, however, has become an important part of the neighbourhood and public life. The knowledge they generated was built in conjunction with a variety of actors and in solidarity with one another. The knowledge and skills acquired through a long and slow process, as is the situation with all the cases mentioned above, became palpable in *Huerto Roma Verde* on 19 September 2017, when a catastrophic earthquake shook Mexico City. Several buildings collapsed during the tremor and many people needed shelter and food. *Huerto Roma Verde* became an important meeting point and centre where donations were gathered and distributed. Food was prepared on-site with the edible plants of the *huerto* and offered free of charge (Huerto Roma Verde, 2017). Tents were installed to provide shelter to those in need.

The building process and the new knowledge and skills developed were of concrete use to the skaters in Tampere as well, when the city of Tampere did not build a skatepark that had been planned by the city in collaboration with the skaters. The city's inability motivated the skaters to build a desperately needed new skatepark by themselves (Figure 5.9). This was a pivotal moment for the whole Tampere skateboarding scene. The skaters organised and formed an official association, *Pirkanmaan Kaarikoirat ry*, and started to promote skateboarding in different ways. This led to the process of planning and building a new skatepark with the skaters as experts possessing the knowledge and skills required in designing. They had acquired all this know-how by building an informal do-it-yourself skatepark in the yard of an abandoned matchstick factory (for more details, see Kyrönviita and Wallin, 2022). Since then, the skaters have learned to use their skills and knowledge in various contexts and contributed to urban development, culture, and even education (Kallio et al., 2021; Kyrönviita and Wallin, 2022).

Casa Taft 169, in San Juan, Puerto Rico, is another example of how knowledge is further mobilised and becomes an indispensable resource for political action. One of the main actors in the occupation of an abandoned house in Taft Street and its transformation into a civic centre, told us that after occupying the building they started to dig into its past, learning about its tax-debt, that the owner died, that there seemed to be no next-of-kin, and that the house had been declared a public nuisance. She also learned how this house was only one of thousands across the city and a result of an exodus of

FIGURE 5.9 Through building *Tikkutehdas DIY* skatepark, the skaters organised, formed an association *Pirkanmaan Kaarikoirat ry*, and gained the expertise to promote skateboarding more widely. Photo: Mikko Kyrönviita.

people moving from the island to the mainland United States, pushed off the island due to extreme austerity measures, part of a savage neoliberal system (Klein, 2018a, 2018b; Milián Bernal, 2022). It was through all of that slow but cumulative knowledge that the project gained energy around it and started a social movement to fight for the right to occupy and use buildings for community and social project that were declared a public nuisance (for detailed narration, see Milián Bernal, 2022). Knowledge, thus, is at the core of insurgent spatial practices and it is possibly the most powerful and valuable resource that is mobilised to catalyse systemic change.

Learning from insurgent spatial practices

In a time of climate emergency, architectural thinking and practice must be rethought. This links directly to the profession of architecture forming a "normative expert knowledge base within capitalist economies" which is established through education and professional practices leading to shared identities, perception of problems and solutions (Lorne, 2017, p. 274). In addition, architectural culture tends to emphasise aesthetics, style, form, and technique that often overshadow other aspects, such as the building process, and its consequences to social and ecological environments that, however, are inseparable from the practices of architecture (Awan et al., 2011; Hensel, 2012). Such inherited forms of knowledge and cultural conventions can hinder challenging the professional boundaries and the hegemonic power eventually preventing the much-needed change towards sustainability.

To think differently about architecture – that is, the production of space and its spatial practices – requires thinking differently about knowledge; how it is constructed and mobilised. As we have argued, architecture and planning are part of the processes through which power is manifested and asserted in space (Baeten, 2017; Harvey, 2013; Lefebvre, 1991). Under capitalism, accelerated by neoliberalism, architectural knowledge has often failed in ethical issues and holistic approaches to sustainability (Hensel, 2012; McNeill, 2006). Instead, it has been deployed in the production of abstract space to fulfil the uneven and unjust agendas of hegemonic power and the political elite which has contributed to the complex ecological, social, economic, and cultural crisis at hand (Lefebvre, 1991; Savini, 2021; Swyngedouw and Heynen, 2003).

At the same time, there is also a long history and plethora of contemporary practices of "alternative pathways" (Harriss et al., 2020) and "other ways of doing architecture" (Awan et al., 2011) that emphasise the need to thoroughly understand space through multidimensional, dynamic, and interdependent relationships of physical and material conditions and social and political processes. We see that such an engagement with space and spatiality is required more than ever to understand power relations and multiple values of specific spaces and to re-engage architectural thinking "with the politically progressive possibilities" (Lopes de Souza, 2016; Lorne, 2017, p. 282). Such possibilities lie in insurgent spatial practices. They challenge abstract space, "disturb established histories", and produce differential space (Holston, 1995, p. 44; Lefebvre, 1991). It is there where new practices and knowledges are created and mobilised for the construction of a better world.

Even though climate emergency, biodiversity loss, and human impact in the era of Anthropocene do not recognise any boundaries, the role of the "local" is not diminishing. Rather, such challenges are manifested locally and in interaction with a specific locality (Wittmayer et al., 2014). What is important from the perspective of learning is that insurgent spatial practices, such as the cases presented in this chapter, reassert "local knowledge claims and local identities against the simplifying and universalizing forces of global science, technology, and capital" (Martello and Jasanoff, 2004, p. 4). Insurgent spatial practices are solutions that meet the actual physical and social needs of local communities instead of privileging corporate interests or technological fixes and technocratic solutions (Bulkeley et al., 2015; Hensel, 2012). However, it is the uttermost challenge for any field, not just architecture, to be able to balance between different scales without falling into a "local trap" and not seeing the connectedness and impacts to larger scale systems, such as ecosystems (Jon, 2020, pp. 172–173). Still, learning from such situated knowledges (Haraway, 1988) opens avenues for "alternative politics of space" (Lorne, 2017, p. 282) and context-specific architectural concepts and "heterogeneous provisions for use and appropriation on a more finely graded scale" that actually contribute to sustainability (Hensel, 2012, p. 152).

Indeed, insurgent spatial practices are enacted utopias promoting various aspects of sustainability. People's needs, desires, and dreams are materialised and spatialised with people's own strengths while new skills and know-how are developed. Little available material resources are re-used and re-purposed via collaboration and acts of solidarity and care which, in turn, enhance social resources. Bearing in mind all the aspects mentioned above, insurgent spatial practices are ways of "staying with the trouble" as they are not projected in the future but responding to actual challenges here and now (Haraway, 2016, p. 1). Following Haraway (2016, p. 1), we would like architectural thinking to stay with the trouble and learn "to be truly present, not as a vanishing pivot between awful or Edenic pasts and apocalyptic or salvific futures, but as mortal critters entwined in myriad unfinished configurations of places, times, matters, meanings".

KEY TAKEAWAYS FROM CHAPTER 5

- By shifting the gaze from *objects* in space to the actual *production of space*, insurgent architectural and planning practices become visible everywhere and are released from the hostage of professional practitioners, conventions, the state, and the markets. This shift potentially alters the philosophical underpinnings of these disciplines, their objects and subjects of theorisation and knowledge, and, thus, also the way these are taught and practiced.
- The climate emergency necessitates that current architectural and planning practices as well as their philosophical underpinnings are altered and requires the formation of new forms of collaborations, whereby local people, their actions, knowledge, networks, and resources are taken seriously. We argue that insurgent spatial practices do this and hold the potential to address social, environmental, and climatic injustices and mitigate and adapt to the impact of climate change.
- Rather than integrating insurgent spatial practices into current mainstream architecture and planning, we call for the reconceptualisation of these practices and the formation of a different architecture and planning agent; one who is politically engaged for the construction of a just and sustainable world.

Funding

The research discussed in this article was funded by the Consejo Nacional de Ciencia y Tecnología (CONACyT) Mexico; the School of Architecture at Tampere University, and received support from the Strategic Research Council of the Academy of Finland under Grant 289691.

Notes

1 The seminar was organized by the research community of Sustainable Transformation of Urban Environments (STUE) at Tampere University, Finland.
2 We do not subscribe to the practice of felling trees for Christmas celebrations. Our interest with this example is the tactics used to 'green' the city.
3 Henri Lefebvre published *The Production of Space* in 1974. It wasn't until 1991 that it was translated into English.

Bibliography

1319.trecediecinueve, n.d. aUPA Space. Available at: https://trecediecinueve.wixsite.com/1319x/aupa-space03

Alatalo, E., Leino, H., & Jokinen, A. 2024a. Heterotopic diversity of coworking spaces: Providing adaptive capacity for cities. *Cities*, 104717.

Alatalo, E., Leino, H., Laine, M., & Turku V. 2024b. Heating up the sauna: analogue model unraveling the creativity of public participation. *Planning Theory* 23:1, 22–41.

Arantes, P.F., 2019. *The Rent of Form, Architecture and Labor in the Digital Age*. University of Minnesota Press.

Awan, Nishat, Schneider, Tatjana, & Till, Jeremy, 2011. *Spatial Agency: Other ways of doing architecture*. Routledge.

Baeten, G., 2017. Neoliberal Planning, in: Gunder, M., Madanipour, & A., Watson, V. (Eds.), *The Routledge Handbook of Planning Theory*. Taylor & Francis Group, pp. 105–117.

Bâtiment 7, 2022. Bâtiment 7. Available at: www.batiment7.org/fr/accueil-2/ (Accessed: 2 February 2022).

Brenner, N., 2009. What is critical urban theory? *City* 13, pp. 198–207. https://doi.org/10.1080/13604810902996466

Brenner, N., Marcuse, P., & Mayer, M., 2012. *Cities for People, not for Profit: Critical Urban Theory and the Right to the City*. Routledge.

Bulkeley, H., Broto, V., & Edwards, G., 2015. *An Urban Politics of Climate Change: Experimentation and the Governing of Socio-Technical Transitions*. Routledge. https://doi.org/10.4324/9781315763040

Caperna, A., Tracada, E., Minervino, G., Alatalo, E., & Cerreta, M., 2019. *Spatial and Economic Smart Strategies for the 21st-Century Metropolitan City of Naples*. Springer, pp. 665–755. https://doi.org/10.1007/978-981-10-8588-8_12

CultiCiudad, n.d. *CultiCiudad*. Available at: culticiudad.org (Accessed 14 August 2023).

Demaria, F., Schneider, F., Sekulova, F., & Martinez-Alier, J., 2013. What is degrowth? From an activist slogan to a social movement. *Environmental Values* 22, pp. 191–215. https://doi.org/10.3197/096327113X13581561725194

Franck, K., & Stevens, Q. (Eds.), 2006. *Loose Space*. Routledge. https://doi.org/10.4324/9780203799574

Fraser, N., 2014. Behind Marx's hidden abode: For an expanded conception of capitalism. *New Left Review* 86, pp. 55–72.

Fregoso Centeno, I.C., & Moncayo, E., 2017. *Cultiva Ciudad*. Mexico.

García-Lamarca, M., 2017. From occupying plazas to recuperating housing: Insurgent practices in Spain. *Int. J. Urban Reg. Res.* 41, pp. 37–53. https://doi.org/10.1111/1468-2427.12386 (Accessed: 9 November 2023).
Gersmann, H., 2011. Anti-nuclear protests in Germany – in pictures. *The Guardian.* www.theguardian.com/environment/gallery/2011/nov/28/anti-nuclear-protests-germany-in-pictures.
Gottdiener, M., 1994. *The Social Production of Urban Space.* University of Texas Press, Austin.
Haraway, D., 2016. *Staying with the Trouble: Making kin in the Chthulucene.* Duke University Press.
Haraway, D., 1988. Situated knowledges: The science question in feminism and the privilege of partial perspective. *Feminist Studies.* 14, pp. 575–599.
Harriss, H., Hyde, R., & Marcaccio, R. (Eds.), 2020. *Architects After Architecture: Alternative Pathways for Practice.* Routledge. https://doi.org/10.4324/9781003007753
Harvey, D., 2013. *Rebel Cities: From the Right to the City to the Urban Revolution.* Verso.
Harvey, D., 2008. The Right to the City. *New Left Review*, no. 53.
Havik, K., & Arlandis, A.A., 2022. Choreography as a tool to understand architectural situatedness: a mediating intervention at Hiedanranta industrial heritage site, Finland. *Journal of Architecture.* 27, pp. 539–557. https://doi.org/10.1080/13602365.2022.2122533 (Accessed 9 November 2023).
Hensel, M.U., 2012. Sustainability from a performance-oriented architecture perspective – alternative approaches to questions regarding the sustainability of the built environment: Sustainability from a performance-oriented architecture perspective. *Sustainable Development* 20, pp. 146–154. https://doi.org/10.1002/sd.1531 (Accessed 9 November 2023).
Holston, J., 1995. Spaces of insurgent citizenship. *Plan. Theory* 13, pp. 35–51.
Huerto Roma Verde, 2017. Huerto Roma Verde. Facebook.
Huerto Roma Verde, n.d. Available at: http://huertoromaverde.org/
Huq, E., 2020. Seeing the insurgent in transformative planning practices. *Planning Theory* 19, pp. 371–391. https://doi.org/10.1177/1473095219901290
IPCC, 2022. Summary for Policymakers, Climate Change 2022: Mitigation of Climate Change. Contribution of Working Group III to the Sixth Assessment Report of the Intergovernmental Panel on Climate Change. Cambridge University Press.
Jon, I., 2020. Scales of political action in the anthropocene. Gaia, networks, and cities as frontiers of doing earthly politics. *Global Society*, 34, pp. 163–185. https://doi.org/10.1080/13600826.2019.1676702 (Accessed 9 November 2023).
Kallio, K.P., Jokela, S., Kyrönviita, M., Laine, M., & Taylor, J., 2021. Skatescape in the making: Developing sustainable urban pedagogies through transdisciplinary education. *Sustainability* 13, p. 9561. https://doi.org/10.3390/su13179561 (Accessed 9 November 2023)
Karaliotas, L., & Swyngedouw, E., 2019. Exploring insurgent urban mobilizations: from urban social movements to urban political movements?, in Schwanen, T. and Kempen, R. V. (eds.) *Handbook of Urban Geography.* Edward Elgar Publishing, pp. 369–382. https://doi.org/10.4337/9781785364600.00037
Klein, N., 2018a. *The Battle for Paradise: Puerto Rico takes on the Disaster Capitalists.* Haymarket Books.
Klein, N., 2018b. There's Nothing Natural About Puerto Rico's Disaster. *The Intercept*, 21 September 2018. https://theintercept.com/2018/09/21/puerto-rico-hurricane-maria-disaster-capitalism/ (Accessed 9 November 2023).
Klein, N., 2015. *This Changes Everything: Capitalism vs. the Climate.* Penguin Books.
Kruzynski, A., 2018. L'autonomie collective en action: Du Centre Social Autogéré de Pointe-Saint-Charles au Bâtiment 71. *Nouvelles Pratiques Sociales* 29, pp. 139–158. https://doi.org/10.7202/1043397ar (Accessed 9 November 2023).

Kyrönviita, M., & Wallin, A., 2022. Building a DIY skatepark and doing politics hands-on. *City* 26, pp. 646–663. https://doi.org/10.1080/13604813.2022.2079879

Kytölä, K., Sawczuk, P., & Huuhka, S., 2021. Think Before You Build, in: Pelmakers, S. & Newman, N. (Eds.), *Design Studio Vol.1: Everything Needs to Change: Architecture and the Climate Emergency*. RIBA Publications, pp. 25–33.

Lefebvre, H., 1991. *The Production of Space*. Blackwell.

Lopes de Souza, M., 2016. Lessons from praxis: Autonomy and spatiality in contemporary Latin American social movements. *Antipode* 48, pp. 1292–1316.

Lorne, C., 2017. Spatial agency and practising architecture beyond buildings. *Social & Cultural Geography* 18, pp. 268–287. https://doi.org/10.1080/14649365.2016.1174282 (Accessed 9 November 2023).

Martello, M.L. & Jasanoff, S. (Eds.), 2004. *Earthly Politics: Local and Global in Environmental Governance*. MIT Press.

McNeill, D., 2006. Globalization and the ethics of architectural design. *City*, 10, pp. 49–58. https://doi.org/10.1080/13604810600594639 (Accessed 9 November 2023).

Mettinen, P., 2017. *Eteläpuiston hävitettävät puut merkattiin*. Vasen Kaista.

Milián Bernal, D., 2022. Narratives of appropriation: Abandoned spaces, entangled stories, and profound urban transformations. *Writingplace Journal for Architecture and Literature*. 6, pp. 69–89. https://doi.org/10.7480/writingplace.6.6356 (Accessed 9 November 2023).

Milián Bernal, D., 2020. *Temporary Uses of Vacant and Abandoned Urban Spaces in Latin America: An Exploration*. Nordic Association of Architectural Research, pp. 181–209

Milián Bernal, D., 2019. Re-imaginando el espacio público en los intersticios del vacío y el abandono en las ciudades lationamericanas, in: *2o Congreso Ibero-Americano de Historia Urbana*. Mexico City, pp. 2435–2448.

Miraftab, F., 2018. Insurgent Practices and Decolonization of Future(s), in Gunder, M., Madanipour, A., & Watson, V., *The Routledge Handbook of Planning Theory*. Routledge, pp. 276–288.

Miraftab, F., 2009. Insurgent Planning: Situating Radical Planning in the Global South. *Planning Theory*, 8, 32–50. https://doi.org/10.1177/1473095208099297

Monbiot, G., 2021. Capitalism is killing the planet – it's time to stop buying into our own destruction. *The Guardian*, 30 October. www.theguardian.com/environment/2021/oct/30/capitalism-is-killing-the-planet-its-time-to-stop-buying-into-our-own-destruction (Accessed 9 November 2023).

Nast, H. & Pile, S., 1998. *Places through the Body*. Routledge.

Paré, J., 2022. Ray-Mont Logistiques: une centaine de sapins de Noël plantée sur le site. *Métro*.

Pirkanmaan murrosagentit, 2021. Pirkanmaan tulevaisuustarina. Inspiraatiota ekologisesti ja sosiaalisesti kestävään kehittämiseen. Pirkanmaan Liitto. Available at: https://projects.tuni.fi/ekotransitio/julkaisut/pirkanmaan-tulevaisuustarina/ (Accessed 9 March 2022).

Résister et fleurir, n.d. Available at: https://resisteretfleurir.info/ (Accessed 29 March 2023).

Roy, A., 2005. Urban informality: Toward an epistemology of planning. *J. Am. Plann. Assoc*. 71, pp. 147–158. https://doi.org/10.1080/01944360508976689 (Accessed 9 November 2023).

Sandercock, L., 1998. *Making the Invisible Visible*. University of California Press. https://doi.org/10.1525/9780520918573

Savini, F., 2021. Towards an urban degrowth: Habitability, finity and polycentric autonomism. *Environment and Planning A: Economy and Space*, 53, pp. 1076–1095. https://doi.org/10.1177/0308518X20981391

Sévigny, M., 2021. La trajectoire incertaine du Projet Bâtiment 7. Revue Possibles 45.

Sletto, B., 2013. Insurgent planning and its interlocutors. *J. Plan. Educ. Res*. 33, pp. 228–240. https://doi.org/10.1177/0739456X12467375

Swyngedouw, E., & Heynen, N.C., 2003. Urban political ecology, justice and the politics of scale. *Antipode* 35, pp. 898–918.

Tonkiss, F., 2013. Austerity urbanism and the makeshift city. *City* 17, pp. 312–324. https://doi.org/10.1080/13604813.2013.795332

Wittmayer, J.M., Schäpke, N., van Steenbergen, F., & Omann, I., 2014. Making sense of sustainability transitions locally: how action research contributes to addressing societal challenges. *Critical Policy Studies*, 8, pp. 465–485. https://doi.org/10.1080/19460171.2014.957336

PART 2

Architectural pedagogies to address the climate emergency

Creaturely pedagogy, collaboration and communities of equity, empathy and interdisciplinarity

In the second part of the book, we turn our attention to the education of architects. The majority – but not all – of the authors in this book, were educated as architects, and the majority – but not all – both teach and research architecture today. We are bound together not only by a shared concern for the role of architecture and architects in the climate crisis, but also by our shared responsibility to *do something* about it in our daily practice as educators. The five chapters in Part 2 consider the role of education in nurturing sustainable and critically reflective architectural thinking. The chapters have been chosen because of the remarkable innovation they demonstrate, and the practical insight that they offer other teachers confronting the challenge of creating equitable learning environments. In the five chapters that follow, you will find appeals for paradigmatic change as well as examples of how that change can be enacted. In common between our authors are appeals to dramatically change not only the content but also the form of architectural education: to dismantle colonial and anthropocentric worldviews that are entrenched in the academy, and to nurture instead communities of equity and empathy. Architectural education is pedagogically complex, perhaps more so than the education of future practitioners of allied disciplines in the creative arts, built environment, design or engineering disciplines Students of architecture are expected to earn both an academic and a professional degree; whenever professional or accrediting bodies like the ARB, RIBA, ACSA or AIA propose additions to architecture curricula (and in most recent years, those additions are often to do with climate, energy, materials, health and safety), educators face a challenge to find sufficient teachers with appropriate knowledge, hours in the schedule and credits in the curriculum.

Many of us are lucky to be blessed with unparalleled opportunities for pedagogical innovation. The design studio, which tends to occupy the largest amount of time, space and academic credits in the curriculum, is a brilliant space for pedagogical experimentation, innovation and discovery. Recent research on design studio pedagogies (Jones et al., 2025) has found that what is taught in studio is characteristically uncertain, complex and messy. What is learned in studio is typically diverse, and the learning experience is highly

DOI: 10.4324/9781003293903-7
This chapter has been made available under a CC-BY-NC-ND 4.0 license.

personal, transformational, yet ultimately contingent on the needs of the student. Most importantly, studio produces and reproduces cultures and praxes. We must recognise that the greatest and most progressive attributes of architectural education's signature pedagogy (the studio) are intimately bound up with its worst and most conservative tendencies. Yet the studio is also the site *par excellence* for the perpetuation of a hidden curriculum that prejudices certain individuals and groups. These prejudices inculcate negative behaviours, attitudes and value systems including but not limited to racism, sexism and ableism, to name just a few (Brown, 2021). As Felicity Atekpe argues in Chapter 6, we need to recognise that our dominant educational paradigms emanate from Eurocentric cultures and colonial export that date back to the fifteenth century. She proposes, instead, a creaturely pedagogy that de-centres our species away from the anthropocentric and anthropogenic worldviews and adopts instead an outlook of diversity, adaptability and restoration – as also discussed in Chapter 4. Atekpe, along with others in the book, argues for a shift from a mindset of abundance to one of scarcity, focusing on re-use, retrofit and designing with limited resources. The concept of ease is introduced, i.e. 'taking the time to evaluate and think more deeply without panicking will always result in better and quicker solutions that might expediate the cycle of corrective mistakes that have characterised previous responses to existential issues of sustainability and the climate crisis'. An environment of ease encourages critical and reflective thinking, which is further explored in Chapter 7. Atekpe also refers to this in relation to the ever-expanding architecture curriculum and that we need to re-evaluate what and how we teach and assess, with more of a focus on process instead of final design outputs, and that in so doing equity and inclusivity are supported.

Advocates for design studio point to the opportunities it provides students to engage in deep thinking around design problems. But in Chapter 7, Essi Nisonen, Jenni Poutanen and Sofie Pelsmakers interrogate the problem with Problem-Based-Learning (PBL). For architects, they argue, it is not helpful to view the world as a set of problems to be solved. They argue, instead, to view problems not as single difficulties but rather as opportunities to ask what is problematic about a given situation. Design should consist equally as much of *discovering* problems as exploring and solving them. Furthermore, the goals of PBL are opposed by the persistent master-apprentice model that continues to shape many learning environments, preserving rather than critiquing the asymmetrical and top-down power relationships in education. Ultimately learning and teaching architecture is about mindsets and values, and not simply about delivering sustainability knowledge. The authors argue that design thinking can go some way to 'help integrate optimal learning methods and mindsets into the architecture curricula' and that students should learn early on the iterative logic and collaborative nature of the design process, crucial for developing critical thinking skills. The authors argue that, together with blended learning, where active face-to-face learning events follow up more passive online learning activities support low-hierarchy learning environments and place empathy, collaboration and communication at the heart of the design studio culture. That said, not every chapter in this part of the book focuses on design studio. In the five chapters that follow, you will find precedents, suggestions and inspiration for teachers in all specialisms and pedagogical contexts. In Chapter 8, David Kroll, Susan Galavan and Veronica Soebarto highlight how the discussion of sustainability in architectural education is often limited to specific courses, often peripheral lecture, seminar or workshop courses in architectural technology or environment. They argue, instead, that there is an urgent need for it to become a central component in architectural education, especially design studio. As per the other chapters in this section and the book, they acknowledge that this will require a shift in the content and ways we teach (e.g. a multi-stakeholder, collaborative environment) but that this could be achieved through 'capstone' design courses by teachers who are committed to creating change, as a starting point. As they write: 'While this may not change the whole curriculum at once, it conveys the message that sustainability is not an "add-on" to be taught in specialised supporting courses only, but it is at the core of the degree and provides the underlying foundations of all architectural design and production.' As part of this shift, they also argue that we should practise what we preach, from the materials that student models are made out of (and how they are disposed) to the spaces we teach in (that often do not embody best practice principles), aligning the message of our teaching with everyday actions. In Chapter 9, we see a detailed example from the Department of Architecture at the University of Pretoria, in Gauteng province, South Africa. Writing from within an institution that was historically privileged by the totalising racist framework of Apartheid, Jan Hugo, Anika van Aswegen, Carin Combrinck, Dayle Shand and Chrisna du Plessis show how their department has engaged critically with the challenges at hand. The authors demonstrate how, with concerted effort over a period of time, an entire architecture programme can be reimagined as a transformative educational journey. Such change requires leadership and persistence, but importantly it must permeate into every course and every part of a programme. Common to all courses and all educational activities must be a recognition that an architectural education is an ecosystem of interrelated non-linear and iterative learning processes, and that the role of the school and others in the built environment is crucial to address the climate crisis and 'embracing epistemic diversity and transdisciplinary processes is critical, as the solutions should address basic developmental needs, undertake appropriate climate change adaptation strategies, and contribute to the global carbon mitigation agenda.' As in Chapters 4 and 7, the authors argue for an empathetic approach for people and environment, acknowledging that this shift is not easy and includes 'disruptive moments' to ignite new thinking and individual and collective student development. It means focusing on processes instead of product-oriented outcomes and the architect as a facilitator, as also highlighted in Chapter 7. The School's shift focuses in undergraduate courses on i) *my place in the world*, ii) *learning from this world and place*, and iii) *a collective vision for this world*; and masters courses focus on the iv) *Finding my role as practitioner in the real world* and v) *change agent: my professional responsibility towards others in this world.* The authors also raise a big challenge: for students to become true agents of change, they need to be able to do this through practice, yet our industry is slow-moving and not yet ready for this 'new breed of professionals', and alumni may instead forge new paths. This also highlights how transformational change continues after a student's architecture education. As the authors write: 'As the need for transformational learning is significant, it presents an exciting opportunity to redefine architecture in service of communities as prompted by the climate emergency.'

In Chapter 10, Elizabeth Donovan provides examples and evidence of how architectural approaches have changed in response to paradigm shifts and an uncertain future, even if only in small-scale experiments. If we are to produce future architects who can stand in solidarity with human and non-human stakeholders, then we need to be able to adopt different pedagogical approaches that nurture critical thinking as the foundation of design education where multi-species are considered, architects pool resources with other disciplines and act as facilitators (as also highlighted in Chapters 7 and 9). Donovan emphasises the importance of learning methods rather than teaching methods at the forefront of this paradigm shift with a focus on students being active participants instead of passive recipients and moving towards student-centred learning instead of the master–apprentice model that so many architecture teachers and schools still hang on to (see also Chapters 4, 6 and 7). Architecture is a dynamic form of knowledge, and as highlighted by Donovan, this requires a shift in focus from the object at the end of the education to the process of learning (e.g., the process of understanding and applying and testing knowledge). Scenario-based design and learning (as discussed in more detail in Part 3, Chapter 15) can help with embracing the unknown and including dynamic aspects of the setting, functions, users and climatic environments, and can help with 'imagining stories and narratives which provide inspiration for uncertain futures'.

Common to all of us is the need for teachers and practitioner-teachers (i.e. the architects and external experts who join us in the academy) to recognise the urgent need to transform from *authorities* to *facilitators*. When today's students become tomorrow's architects, the design problems that they will face will truly be *wicked*. They alone won't have the answer, nor will they have any hope of finding a single right answer or linear path towards it. Atekpe calls for 'a new signature pedagogy [that] must put humanity back into its rightful place as equals among all living beings, as the mammals we are, and develop a creaturely pedagogy that pays particular and equal attention to all other species with the wellbeing of our planet at the core' (Atekpe, p. 110).

Common to all these chapters about architectural education is the position (just as it was in Part 1 of the book) that we need to support dramatic value changes. We need to diminish the idea of the architect as an autonomous and expert individual, smashing the stubborn vestiges of the master–apprentice studio culture. To quote Elizabeth Donovan in Chapter 10, 'embracing the unforeseen future and recognising architecture as a dynamic form of knowledge is essential for addressing the ever-evolving challenges brought about by the climate emergency' (p. 183). If such changes feel overwhelming, we can take encouragement from Felicity Atekpe in Chapter 6, who reminds us: 'seismic shifts start with our everyday practice' (p. 109).

Bibliography

Brown, J.B., 2021. Making the studio smaller. *Design and Technology Education: An International Journal*, 26(4), pp. 256–260.

Jones, D. Brown, J.B., Boling, E., Corazzo, J., Gray, C.M., & Lotz, N., 2025. *Studio Properties: A Field Guide to Design Education*, London: Bloomsbury Visual Arts.

6

EVERYDAY, EQUITY AND EASE

Felicity Atekpe

Introduction

According to Bill N. Lacy "the best architects design more than buildings: their work ranges from the minute to the massive, from postage stamp to metropolis" (Hedjuk et al., 1988). In the more mischievous words of Bittertang Farm (2022), "Architecture won't be built solely from crystallised and solidified materials, but in coordination with living matter from bacteria to plants and other bioengineered products." Architectural education embraces this wide spectrum of knowledge, underpinned by a requirement for multiple perspectives and ways of teaching, which enable students to be taught empirical information – drawing on conventions, legal and statutory regulations, precedents, literature and critical thinking on typologies, an awareness of new technologies, an understanding of issues around sustainability – as well as art and, finally, the impact their designs will have in terms of psychology and sociology. However, the premise of this unique education, where students are taught to act as polymaths able to move with agility between many different and conflicting requirements, is fraught with ambiguities, tensions and conflicting responsibilities. It is especially problematic in the increasingly litigious and commercialised world of higher education, with the complexities experienced during the pandemic and the expanding content of its curriculum. Firstly, to achieve the broadest aims of education effectively and with equity, the profession must reflect society and allow a diverse representation of the world, adopting a planetary mindset that is aware of all living and non living things in the biosphere. Secondly, these ethical issues have fundamental implications on how we teach design and the role of architecture and the architect.

Sustainability, in its broadest sense, is a threshold concept (Cousins, 2006). When taken seriously, sustainability will demand a fundamental shift in how architecture is taught. We are all too familiar with the word and therefore fail to question and interrogate it in order to create new definitions to take on its all-encompassing implications for the field. Achieving true sustainability lies in meeting current needs without conceding future ones and balancing economic prosperity, environmental care and social well-being.

DOI: 10.4324/9781003293903-8
This chapter has been made available under a CC-BY-NC-ND 4.0 license.

Therefore, its implications are all-encompassing and require us to frame architecture education itself as an ecology. Understanding it as a system based on the interrelation of topics and interests will allow us to really affect radical, essential shifts in how we teach the architectural curriculum. We need to understand, as a starting point, the students' environment and everyday life and experiences, and design an education that will then also help students make sense of and navigate the wider world within which they research, learn and design.

Looked at through Bronfenbrenner's ecological systems theory (Bronfenbrenner, 1974 and 2009), we should view student development as a complex system of relationships affected by multiple levels of their surrounding environments, from the settings of family and school to broad cultural values, laws and customs, dividing our environments into five different systems: the microsystem, the mesosystem, the exosystem, the macrosystem and the chronosystem.

The microsystem is a student's most immediate cultural environment and is the most influential level of the ecological systems theory because it is the set up where they have the most direct contact with factors influencing development (Guy-Evans, 2020). It is particularly important in architecture because, as educators, we are teaching students a way of thinking through design and applying an open exploration where there are no right or wrong solutions. As such, the student will need to clearly demonstrate their approach, application and reflection in their design as a process. From the outset, sustainability needs to be at the core.

To address these implications of sustainability, we need to move from a mentality of abundance to one of scarcity in every aspect of our thinking. There are several opposing ways of addressing this. At one end of the spectrum is Sumita Singa's (2013) approach that prioritises retrofit and re-use over new construction, identifying site and social ecologies and upending our agreed historical and cultural references to include more complex narratives. At the other end of the spectrum is the Independent Group that came out of the Second World War and described by David Robbins (1990) as "democratizing the implications of the aesthetics of plenty" and promised a modern world of plenty by challenging the dominant culture in order to make it more inclusive of popular culture. This gives another clue as to how we might go about developing a novel and effective pedagogic approach by looking at the implications for design from the other end of such a lens, in other words democratizing the implications of scarcity. These strategies might provide valuable clues and solutions that were not apparent when viewed from the usual perspective.

This chapter is written with an explicit focus on the situation in the UK, acknowledging that the precise situation will differ from country to country. However, the density of UK architectural education, which includes 60+ schools and efforts by the RIBA to export their standards worldwide through International Validation, seen by some as a form of neo-colonial practice, means that my analysis and observations will have validity for readers globally.

Everyday

> As individuals express their life, so they are. What they are, coincides with their production.
>
> *(Marx & Engels, 1977)*

Seismic shifts start with our everyday practice. To consider the inputs that should shape architectural education in relation to its signature pedagogies (Shulman, 2005), it is imperative to question and better understand some of the basic premises of architectural education in comparison to other built environment disciplines in their everyday teaching processes. Firstly, why do architectural students – from the start of their first year – learn about the design of buildings largely through extended project work undertaken in the collective space of a design studio, and taught by practising architects? Why are they assessed largely on the basis of the design they produce? Why do we, as tutors, collectively mark the work in the presence of students rather than applying the anonymity used in other courses?

In contrast, structural engineers are taught largely from lectures and short practical demonstrations in engineering workshops taught exclusively by academics and assessed through exams. Landscape architects are taught to respond to "temporal change, transformation, adaptation and succession" and focus beyond humans. According to James Corner, landscape is "a medium uniquely suited to the open-endedness, indeterminacy and change demanded by contemporary urban conditions" and a good model for understanding process rather than outcome (Corner, 2006). Landscape architecture is a helpful comparison, particularly if we look at the proposals made by the late Professor Françoise-Hélène Jourda (2009), a renowned practitioner and academic, that through the consideration of ecological concerns we are confronted by the culture and relationships between that culture, geography and climate. She went on to say that by protecting nature, we are advancing culture and need to make architecture primarily by looking at ecological design more descriptively, analytically, and experimentally in our studio teaching environments. Our approach should be concerned with nature and the interconnectedness of systems to suggest an architecture that is rooted in ecology and serves all human and non-human things in this world similar to landscape architecture's focus.

Secondly, this observation from landscape architecture prompts us to consider what it is about the nature of architectural thinking and problem-solving – from the initial concept, through drawing and mark-making – that allows us to introduce ideas of temporality and interconnected systems. We should ask ourselves: what core aspects of our teaching methods persist, even as they evolve over time? These enduring elements could be seen as threshold concepts (Cousins, 2006) within architectural education. The answers will help us to scrutinise and improve the design of the quotidian outputs through the role of drawing (Delaney and Gorman 2011) and mark making (Orr 2010) and their particular contribution and impact on students' understanding in terms of design, evaluation and assessment. These questions provide food for thought but are beyond the scope of this chapter.

We need to develop a *creaturely pedagogy* that re-positions our species and moves from an anthropogenic worldview, as part of this, the principles of equity and ease in architecture education are further unfolded below.

Equity

The conventional wisdom of the Tower of Babel story is that the collapse was a misfortune. That it was the distraction or the weight of many languages that precipitated the tower's failed architecture. That one monolithic language would have expedited the building, and heaven would have been reached. Whose heaven, she wonders? And

> what kind? Perhaps the achievement of Paradise was premature, a little hasty if no one could take the time to understand other languages, other views, other narratives. Had they, the heaven they imagined might have been found at their feet. Complicated, demanding, yes, but a view of heaven as life, not heaven as post-life.
>
> *(Morrison, 1993)*

Toni Morrison (1993), has argued that narratives are one of the principal ways in which we absorb knowledge. Language is also one of the first ways of signalling change and as such creating a new lexicon needs to be an important part of this. From the perspective of equity, we need to move beyond our everyday architectural language and develop a new lexicon for architecture and practice that acknowledges and critically reflects on European dominance of pedagogy, practice and theory. This is understood as an epoch of inequity and abundance couched in the narrative of Darwin's survival of the fittest (Darwin, 1859). A new signature pedagogy must put humanity back into its rightful place as equals among all living beings, as the mammals we are, and develop a creaturely pedagogy that pays particular and equal attention to all other species with the wellbeing of our planet at the core. This new decentralisation of the human being might start by looking at our current curricula and pedagogies to identify what is missing and thereby help us to fill in the gaps and reframe our current thinking. We should let go of both Gropius's theories and organisation of the Bauhaus from the 1920s, which have been the basis of most of the education in the twentieth century:

> True creative work can be done only by one man whose knowledge and mastery of the physical laws of statics, dynamics, optics, acoustics equip him to give life and shape to his inner vision.
>
> *(Gropius, 1975)*

Modern architecture transcends the Bauhaus model of the lone genius architect. Today, building design is a dynamic symphony of collaboration – involving urban planners, engineers, community stakeholders, environmental experts and end-users. This collective approach ensures more nuanced, contextually responsive architectural solutions that reflect diverse perspectives and collective creativity. Instead, we need to devise a new lexicon of words and approaches which better reflect our current century and its urgent implications. Words such as "concept" explained as "the definition of conceptual architecture, evoking the idea it is the creation of an instant form, informs us that the better the concept can be, the better the final building may be as well" (Curl, 2007) but also "inward and outward polarity" promoted by Gropius and the Bauhaus should also be put aside for new words that better articulate the needs, both of our students' and of our planetary society, of the twenty-first century and beyond. Architecture is also produced by an orchestra of collaborators rather than the Bauhaus's promotion of one man as the artist and originator of buildings.

This new pedagogy needs to address equity by acknowledging that we do not all start from the same place and then adjust the imbalances at the start of an architectural education. It needs to have a holistic view of design and critical thinking, to include more explicitly the diverse technical and cultural contexts in a more integrated manner, an embodied practice inclusive of the othered and non-human. This way it will eventually start to engage with evolving art and scientific thinking and more accurately reflect the implications of Darwinian theory that explains "it is not the strongest of the species that survives, but rather, that which is most adaptable to change" (Darwin, 1859). The theory was never just about the survival of the fittest, but about the survival of an entire ecosystem. Adaptability as opposed to strength is a vital characteristic of biodiversity. Architectural education, as a pedagogic pursuit that is both practical and intellectual, has to be more adaptable. In order to demonstrate adaptability to practical experimental "hands-on" learning, we should look at best-case scenarios as well as examples from other disciplines.

We need to also look at the microsystems created by the dominant "signature" pedagogy of the design studio in architectural training and its celebration of the exceptional and outcomes rather than the process. Whereas to attain good sustainable solutions the process is a very important part of the solution. We might start with a critical reappraisal of project-based learning, which forms the basis of architecture's signature pedagogy. Project-based learning is an educational approach that emphasises learning through hands-on projects and real-world experiences (Blumenfeld et al., 1991). Biggs & Tang (2003) argue that there needs to be a re-establishment of a relationship in pedagogy between the student, teacher and task at hand, within a context of evolved learning outcomes and assessment. Biggs called this concept "constructive alignment" in our thinking about climate and ecologies. Whilst it is often associated with less traditional academic knowledge transfer, it is important to recognise that project-based learning can be a highly effective method of acquiring and applying both academic and practice knowledge. Architectural education already challenges the assumption that project-based learning is only suitable for less academic knowledge transfers by understanding the diverse ways in which projects can be designed and implemented. However, by further enhancing our assessment tools as Biggs & Tang suggest, we can better evaluate and explicitly assess students' authentic evaluation and understanding of both their academic knowledge and critical ability to apply that knowledge in real world situation, vital in the tackling of the climate context. That way, they could be rewarded not only for their solutions but more explicitly for their thinking process. This holistic approach would enrich their academic understanding by showing the interconnectedness and contextualisation of their ideas and their solutions, giving them more agency in their work. This applies especially to the role of practice, professionalism and ethics in architectural education. There can be no absolute answers while the emphasis needs to change on how we communicate, think and reflect on that thinking. Therefore, only the framework of learning outcomes can be the accurate judge of a broad range of design solutions, thereby providing more parity within the system and reasonably eliminating subjectivity.

The university educational model of the United Kingdom also fails because it integrates the equivalent of Stage 2 of the Plan of Works at Parts 1 and 2, so synthesising clients' requirements up to concept design in Design and relies on work experience to apply other broad skills, and particularly Part 3 to apply key aspects of that learning in order to demonstrate ability in technical design, statutory requirements, contracts and running projects on site. Otherwise, there is only a requirement for students to have knowledge, understanding and skills during their studies beyond concept design. This leaves "constructive alignment" (Biggs & Tang, 2003) vulnerable to the vagaries of the workplace. Namely, how the critical aspects of architectural education are understood and applied

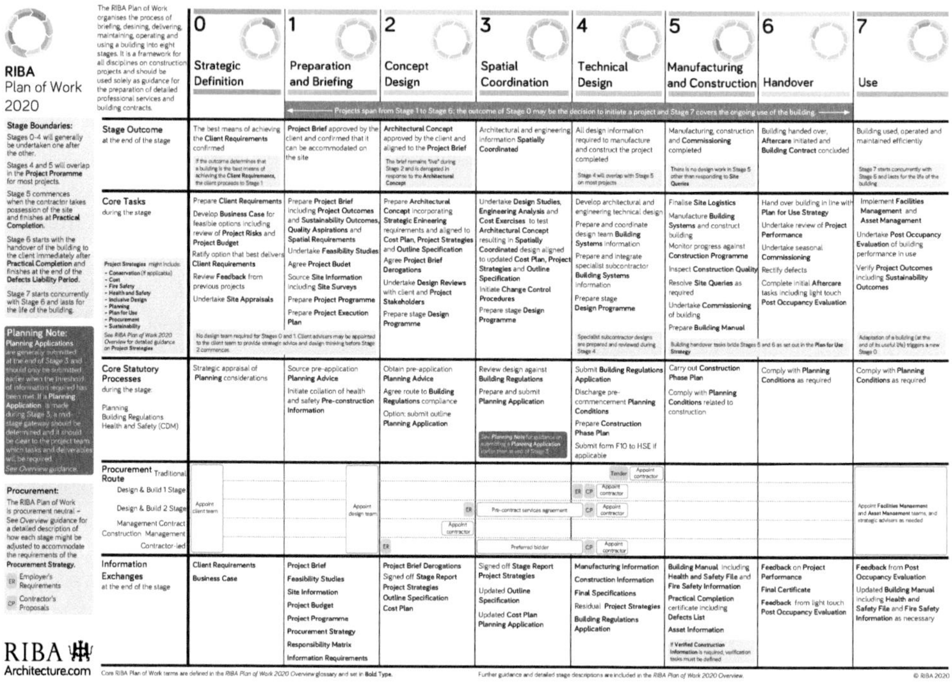

FIGURE 6.1 RIBA Plan of Work

within a professional setting. Whatever the creative solution; statutory requirements, legal obligations, construction and building technologies must be met to a standard that does not endanger lives and still caters for the more esoteric aspects of living as human beings in this shared world, such as delight, joy, caring and belonging. The implication of this proposal is for a more adaptable and equitable model that incorporates explicitly meaningful real-world connections whilst still allowing time for experiential learning. It should also be adapted to provide better and more intentional opportunities for team building and focus on student-driven learning and reflection. There is an assumption that too much of the real world will hinder or curtail students' creativity, but all our experiences show that it is in meeting the real world that real innovation occurs.

Equity is implemented through the adoption of an inclusive curriculum. An inclusive curriculum is based on the concept of Universal Design for Learning (UDL) (CAST, 2018) and is intended to improve the experience, skills and attainment of all students, including those in the protected characteristic groups of the Equalities Act 2010 in order to value diversity and provide equal opportunity for all (Advance HE). It should allow for a diversity of content, materials, ideas and methods of assessment, and involves purposefully integrating perspectives that expand and enhance the canon, both within individual papers and across the whole education. According to Armstrong, the characteristic of the inclusive curriculum has many facets, including attention, equality, ease, appreciation, encouragement, feelings, informative, difference and place (Armstrong, 2008).

Equity in education is also a social policy issue, and we should expand the capacity for public policy with a specific emphasis on the built environment and its impact on communities within and beyond cities. By adopting a restorative approach and breaking outdated binaries such as urban and rural, we can more effectively address global challenges that are acutely felt at the local/microsystem level, such as climate change, biodiversity loss, landscape, elderly care and housing, and create restorative practices designed to mend what has been broken. According to Paul Spicker, author of an introductory textbook on social policy (Spicker, 2014), education has been particularly significant as an instrument of social policy in the sense not only of policies for welfare but also as policies intended to deal with the structure of society. Professor Dame Janet Finch, in *Education as Social Policy* (1984) refers to the use of education as a vehicle for other types of social policy. For example, education provides a convenient basis for policy for children because of its universal coverage, the acceptance of responsibility for children's welfare and because it is easy to justify welfare measures in educational terms. A restorative approach also advocates a proactive rather than a reactive response to public policy in a rapidly changing world, allowing us to more effectively address issues related to health, human rights, climate, urbanism, diversity, equity and inclusion in resource-poor settings, both locally and globally. Restorative practices introduce a way of working with conflict that puts the focus on repairing and mending harm that has been done. It is an approach to conflict resolution that includes all parties involved. This approach to equity has been described as radical architectural pedagogies by Froud and Harriss (2015) that should allow for a warp and weft delivery of teaching practices that embed climate and ecological literacy (Blessinger, 2018) into every part of everyday course delivery. Every aspect of the teaching should look at how to frame society, territory, political and economic implications as a way to provide catalysts to create the places of our dreams (Kolb, 1984).

If we apply this to our training, with Bronfenbrenner's theory in mind, there is work to be done in architecture education that is beyond the academic setting and moves into a societal level as an initial step. Change can only come through addressing vertical and horizontal inequalities in architecture school. Vertical and horizontal inequalities are concepts used in the study of social and economic disparities and concern different aspects of inequality within societies. Vertical inequalities relate to the distribution of income between social groups and horizontal inequalities to the distribution of opportunities and constraints within social groups with a common identity. Vertical inequalities refer to the disparities in the distribution of income, wealth or resources between different social groups. These groups can be categorised based on various factors such as income level, social class, ethnicity, race, gender or other socio-economic characteristics. Vertical inequalities highlight the gap in access to economic resources and opportunities between different strata of society. Addressing vertical inequalities often involves policies and measures aimed at reducing income disparities and promoting economic equity. Horizontal inequalities, on the other hand, concern disparities in the distribution of opportunities, constraints and privileges within social groups that share a common identity. This common identity can be based on factors such as ethnicity, religion, nationality or other social affiliations. Unlike vertical inequalities, horizontal inequalities focus on how certain groups face differential treatment or exclusion within their own communities, even when those communities may be relatively homogeneous in terms of their shared identity. Addressing horizontal inequalities typically involves measures to promote inclusion, social cohesion and equal opportunities within these identity-based groups (Sen, 1973). It is important because equitable policies work for everyone and how we treat others and the othered directly reflects our level of compassion and care as a civil society.

Ease

> Space is substance. Cézanne painted and modelled space. Giacometti sculpted by "taking the fat off space". Mallarmé conceived poems with absences as well as words. Ralph Richardson asserted that acting lay in pauses… Isaac Stern described music as "that little bit between each note – silences which give the form"… The Japanese have a word (ma) for this interval which gives shape to the whole. In the West we have neither word nor term. A serious omission.
>
> *(Fletcher, 2001)*

The concept of "ease" is paramount because it facilitates the other qualities identified as essential in terms of confronting the anxiety of learners and learning. Instead of filling up the curriculum with ever-more content that results in a superficial understanding of the issues, we need to decide which are the critical skills essential to the learning process of our students and the scope of content that it should encompass. We should think of it in terms of lengthening as well as shortening the time it takes to qualify as an architect, to cultivate ease in attainment of this training. It is important because without ease there is no innovation (The Inclusive Curriculum, 2020). We can develop frameworks that combine perceptions, strategies and values (Kleiman, 2009) and adapt a student-centred approach, allowing students to bring to the classroom a knowledge of the achievements of their cultures, incorporating this awareness into the curriculum through co-creation (Petty, 2009). Effective learners acquire knowledge, understanding and skillsets in a bewildering number of ways, each revealing distinctive strengths and a lack of alternative styles (Kolb,1984).

The term understanding is a key part of architectural pedagogy, relating to the amount of knowledge that students have acquired throughout their education, putting more informed options in place to achieve a defined result. Yet, understanding is a product of time, and students will vary in the length an experience takes to acquire a relevant amount of information to be able to comment on the subject matter by means of their respective designs.

To facilitate ease, we need to pay greater attention not only to what learners are interested in but also to how they learn in order to design new curricula. Students who are anxious because they are not doing very well are generally reluctant to abandon their existing methods for completely new untried alternatives. This is true even if these methods are clearly not working for them. The attention of architectural education needs to also re-focus on the problem of our existing assessment methods, which introduce too much subjectivity and underlying and problematic cultural biases into the process. This is evident in the way students from underrepresented groups do much worse in architecture due to assessors having different personal or cultural characteristics and lived experiences. It is a fact, for example, that in the UK black students fare worse in assessment compared with their white peers, and this is accentuated for black females and/or those from poor backgrounds, while assessment panels are still mostly white, middle class and male (Orr, 2010ab). So far, peer-to-peer and auto reviews (which are students reviewing their own work either individually or amongst themselves and providing formal feedback) have been an invaluable way to address these prejudices and highlight the importance of this to create a culturally responsive, course context (Angélil & Hebel, 2008). We must think of new ways of assessment or at the very least, adopt better, tried and tested solutions from design-teaching pedagogy.

Returning to the need for a creaturely pedagogy, to try and design new ethical interspecies relations we should begin with ease because it affords attentiveness: a "creaturely" attentiveness to the material ways that human and non-human lives are entangled and mutually vulnerable in the everyday world. Attentiveness foregrounds the everyday as a crucial site for transforming human–non-human relations in the era of the post-Anthropocene or Symbiocene. But also, the ambivalent hum of everyday life which Lili Pickett-Palmer (2019), a film and screen academic at Cambridge University, has described as encompassing boredom and emptiness, experiences of domestic and institutional violence, and intense moments of joy. In her case she is looking at trans cinema as a medium for encountering, imaging and affectively engaging with non-normative forms of relationality but in our case it can be about putting the relationality at the heart of architectural education.

To incorporate a "creaturely" attentiveness in our education, there is potential for architects to draw insights from the field of social anthropology, especially in terms of ethnographical fieldwork and the critical examination of complicity and ethics. By incorporating these elements into architectural practice, it is possible to create more inclusive and socially responsible designs that address the needs of diverse communities. This is because social anthropology's signature pedagogy of ethnographical fieldwork involves immersive research and participant observation, which could serve as a valuable approach for architects. Architects often design spaces and buildings for diverse communities, and understanding the cultural, social and human aspects of these spaces can be crucial in creating inclusive, culturally sensitive and functional designs that cater for a broader range of people. The recognition of complicity in settler colonial violence and saviourism within social anthropology's history is an essential step towards addressing such issues. Similarly, the field of architecture should also engage in critical self-reflection and acknowledge its own historical and contemporary challenges, such as issues related to colonialism, gentrification and displacement. By confronting these issues, architects can work towards more ethical and inclusive design practices. Cross-disciplinary collaborations between social anthropologists, architects and other friends could be beneficial. These collaborations can lead to innovative solutions and a more comprehensive understanding of the complex social, cultural and spatial factors that influence the built environment. Like social anthropology's evolving ethical standards, architects can develop and adhere to ethical guidelines that prioritise community well being, cultural preservation and environmental sustainability. Architecture and anthropology have always been concerned about issues at the intersection of behaviour and the built space of human societies and cultures differentiated by trying to understand what being human means and providing conditions for being human and responding to the human condition (Askland HH et al., 2014). Anda-Ioana Sfintes (2019) expands on the idea of giving new meaning to fundamental concepts and what we should be aiming for as "an architecture that accommodates the ever-changing social needs; of socio-spatial interactions as interactions fostered by design; of the boundary as a dynamic relation between spaces and – more than that – between their social components; of the in-between as the most socially, emotionally, experiential and transformative contemporary space through the creativity and the performances it facilitates; of (anthropological) architecture as an equilibrium between architectural perception and anthropological field research."

Ease can be cultivated by confronting the anxiety of learners and challenging project-based learning, our signature pedagogy. First-hand experience is a central and invaluable component of architectural education, and it aligns with the principles of experiential learning as described by David A. Kolb (1984). His model posits that learning occurs through a continuous cycle of concrete experiences, reflective observation, abstract conceptualisation and active experimentation. In the context of architecture and addressing various ecological concerns (rural, urban and social), first-hand experience can be a driver for an exploration of a new, radical alternative to turning practice into learning and architectural education can become a dynamic and transformative process that immerses students in the complexity of rural, urban and social ecologies. This approach encourages critical thinking, problem-solving and the development of innovative, contextually relevant architectural solutions. Furthermore, embracing first-hand experience and experiential learning can lead to a more profound understanding of the social and environmental impacts of architecture, promoting sustainable and inclusive design practices in both urban and rural settings. It encourages architects to be socially responsible and environmentally conscious, addressing the complexities and challenges of today's diverse ecologies. Students' creative projects are so closely linked to their own identity and life experience that it is no coincidence that students of architecture suffer the most psychological problems during their studies (McClean et al., 2020). This way of looking at their education by introducing "ease" should better support and encourage students to explore the potential of the academic freedom they are provided with, whilst inhabiting a safe environment that allows them to explore, experiment and even fail as part of a growth process.

Similarly, although architectural academics are amongst the most hardworking and committed teachers, we should also re-think the balance between teaching practitioners and academics, and their role in architectural education. Teaching practitioners who are the core of the tutors involved in design teaching in schools of architecture are generally from practice. They often have little understanding of the pedagogy of teaching. They are also afforded the least training, preparation, oversight, staff development, familiarity with institutional processes and values but have the largest influence in the student experience through studio teaching. Conversely, full-time academics often have the least experience of practice and are at a disadvantage in the immediate application of their ideas and concepts.

Finally, the statistics are alarming and give a concrete reason why looking at ease will benefit the profession in the long term. According to the *Architect's Journal*, only 15–18% of students who begin architectural education qualify as architects in the UK (Anon, 2022). These statistics get much worse when we look at the number of women or students from global majority backgrounds who go on to qualify (Campbell-Stephens, 2021). This is an enormous attrition rate in a profession in which an overwhelming number of prospective students when interviewed for admissions, told me that architecture has been a childhood dream. Further sobering evidence can be found in the RIBA Clients' Satisfaction Survey (2021): registered architects score highly in design, particularly for private domestic and commercial clients where the satisfaction is generally above 70% and highest for developing and interpreting the brief. Less so with contractor-clients where the score is around 50%. However, when it comes to satisfaction with their ability to provide the wrap-around services, this drops by 20% across the board. A better integration of professional practice in architectural education, especially in the context of the ARB and RIBA Part 1 & 2 criteria, highlights a common concern in architectural curricula where design studio minimises professional practice and the structure of modules tend to give it less weighting signalling to students that they are less important. However, achieving the right balance between design studio work and professional practice-related aspects is crucial for producing well-rounded architects. Addressing these observations by providing "ease" as a way of achieving balance in education is critical to producing well-prepared architects who can navigate the complexities of the field. Addressing the weighting and structure of modules and providing ease in transitioning between design and practice-focused elements, can lead to more comprehensive and effective architectural education.

The idea of ease may seem in direct conflict with the urgency of the climate agenda and requirements, but I believe they are not only closer than they first appear, but also entirely compatible. Taking the time to evaluate and think more deeply without panicking will always result in better and quicker solutions that might expediate the cycle of corrective mistakes that have characterised previous responses to existential issues of sustainability and the climate crisis. What we require in terms of effectively addressing the climate and ecological crisis we face is a personal transformation in our architectural educational environments to take on board their all-encompassing requirements. When education is presented as personal transformation it becomes more difficult, indeed probably impossible, to commodify. Transformation is not consumed, it is undergone (Land et al., 2016).

Conclusion

> In the end we will conserve only what we love, we will love only what we understand, and we will understand only what we are taught.
>
> *(Baba Dioum, 1968)*

Essential shifts are required in architectural education to address the climate and ecological crises to address sustainability and explain how it is currently failing some people, many other creatures, and nearly all of nature. Architecture's pluralist aims, which means that curricula are forever being added to and not distilled, but also in its equity, not just with those who are easy to identify as under-represented. Equitable and socially just policies for the built environment and beyond lie at the intersection of horizontal and vertical inequalities and place, compelling us to replace our short-term reactive approaches with long-term proactive, healing, restorative approaches. This can be done by focusing on and influencing policy towards looking at championing change through addressing emerging trends and challenges, technological advancements and evolving socio-political landscapes within and across the built environment and architectural education.

Signature pedagogies for architecture and other built environment and social science subjects are useful in interrogating architecture's general degree stance, apropos of the breadth of knowledge students are expected to acquire. Focusing on what is relevant to achieve its overarching aims, whilst addressing the need for ease or a pause in architectural education should expand to include sustainability in its everyday practice. If sustainability is taken as the threshold concept, it will also help address diversity and be more focused on professional skillsets, which are also shifting in the face of the societal crises faced. This is not easily implemented in an increasingly complex world of information and knowledge overload. Therefore, reflection on the usefulness and relevance of curricula and programmes and what needs to be taken out of architectural education in order to implement these fundamental shifts is required. In order to provide more tenable solutions for most humans, non-humans and the biosphere, we need to also develop a creaturely pedagogy that moves from an anthropogenic worldview forged by dominator practices emanating from Eurocentric cultures and their colonial export since the fifteenth century, and reinforces diversity, adaptability and restorative practices in our curricula. Finally, we need to find new equitable assessment methods to allow everyone a better progression at our institutions, irrespective of their background.

Higher education institutions are notoriously resistant to change, despite the overwhelming evidence of best practices and the need to change signature pedagogies (Shulman, 2005). This is particularly acute in architecture. We should perhaps not try to master everything but instead widen horizons and learn processes to respond to the change needed, also attracting the widest possible range of students to our courses. We should aim to give students tools to navigate the world by gaining a methodology to be able to make good equitable decisions. In the context of the global climate and ecological emergency, architecture has a critical role in playing out and exploring future scenarios, testing the effects of changing environmental processes upon life and design whilst using design projects as a tool to communicate and represent these often-intangible global threats to new audiences and in new ways. This is important not because they benefit global majority students, professionals and publics, but because these emergencies affect us all. Restorative practices benefit us all – from the air we all breathe, through the communities we all form part of, to the world we all inhabit.

The world has changed, and there has at last been an official acknowledgement of this within the profession in terms of the introduction of the ARB's Tomorrow's Architects outcomes and behaviours model of education as well as the RIBA's The Way Ahead. However, to maintain the momentum for transformation, it is crucial for academics, professional institutions and their members to actively embrace systemic change. As Ibrahim Thiaw, Executive Secretary of the United Nations Convention to Combat Desertification (UNCCD 11.05.22), emphasized in a 2022 address on drought, addressing deep-rooted failures requires more than marginal adjustments – it demands a fundamental shift towards sustainable solutions. This perspective is just as relevant to architecture and its teaching, reinforcing the need for bold, transformative action within the profession.

KEY TAKEAWAYS FROM CHAPTER 6

- Sustainability must become a central tenet of architectural education. This requires a shift from a mindset of abundance to one of scarcity, focusing on re-use, retrofit and designing with limited resources to encourage more ecologically responsible thinking.
- Architectural education must embrace equity. This includes developing a curriculum that recognises diverse cultural perspectives, challenges Eurocentric dominance and incorporates localised environmental knowledge for holistic ecological literacy.
- The emphasis should be on design as a continuous, adaptive process rather than a final product. Project-based learning and experiential teaching models help students to appreciate sustainable practices by engaging with real-world ecological challenges and the complexity of environmental systems.
- Architectural education must create an environment of ease. Ease can allow students to discard any internal sense of urgency to create the best possible conditions for critical and reflective thinking.

Bibliography

AdvanceHE (nd) Inclusive curriculum. AdvanceHE: www.advance-he.ac.uk/inclusive-curriculum [Accessed 10 December 2022]

Angélil, M. & Hebel, D. (2008) *Deviations: Designing Architecture – A manual*, Actar/ Eth Zurich, Birkhäuser Architecture.

Anon (2021) 'The Secret Architect: The school takes action on a 70% student drop-out rate' *Architects Journal*, 30 July. www.architectsjournal.co.uk/news/opinion/the-secret-architect-the-school-takes-action-on-a-70-student-drop-out-rate [Accessed 10 December 2022]

Armstrong, F. (2008) "Inclusive Education" in Richards, G. & Armstrong, F. (eds) (2008) *Key Issues for Teaching Assistants: Working in diverse and inclusive classrooms*, Routledge.

Askland, H.H., Awad, R., Chambers, J. & Chapman, M. (2014) Anthropological quests in architecture: Pursuing the human subject. *International Journal of Architectural Research*, 8(3) (November 2014), pp. 284–295.

Biggs, J. & Tang C.S. (2003) *Teaching for Quality Learning at University: What the student does*, The Society for Research into Higher Education and Open University Press, 2nd edn, Chapters 1&2.

Bittertang Farm (2022) http://bittertang.com/mainpage/index.html [online] [accessed 14 May 2022].

Blessinger, P., (2018) Rethinking higher Education in the service of humanity. [online] University World News. Available at: www.universityworldnews.com/post.php?story=20180711094726636 [accessed 25 February 2020].

Blumenfeld P. C., Soloway E., Marx R. W., Krajcik J. S., Guzdial M. & Palincsar A. (1991) Motivating project-based learning: Sustaining the doing, supporting the learning. *Educational Psychologist*, 26, pp. 369–398.

Bronfenbrenner, U. (1974) Developmental research, public policy, and the ecology of childhood. *Child Development*, 45, 1–5.

Bronfenbrenner, U. (2009) *The Ecology of Human Development: Experiments by Nature and Design*. Harvard University Press.

Campbell-Stephens, R.M. (2021) *Educational Leadership and the Global Majority: Decolonising Narratives*. Palgrave Macmillan.

CAST (2018). Universal Design for Learning Guidelines version 2.2. Retrieved from https://udlguidelines.cast.org/?utm_source=castsite&lutm_medium=web&utm_campaign=none&utm_content=aboutudl [Accessed 12 February 2021].

Colomina, B., Galán, I.G., Kotsioris, E. & Meister A., (2022) *Radical Pedagogies*. The MIT Press.

Corner, J. (2006) Terra Fluxus in Waldheim, C. (ed.) *The Landscape Urbanism Reader*. Princeton Architectural Press, pp. 20–33.

Cousins, G. (2006) Introduction to Threshold Concepts, Planet no.17 (Document).

Curl, J.S. (2007) *A Dictionary of Architecture and Landscape Architecture* (2nd edn). Oxford University Press, p. 192.

Darwin, C. (1859). *On the Origin of Species by means of Natural Selection, or Preservation of Favoured Races in the Struggle for Life*. John Murray.

Dioum, B. (1968) Paper presented at the general assembly of the International Union for the Conservation of Nature & Natural Resources, New Delhi, Seattle Public Library Archive.

Finch, J. (1984) *Education as Social Policy*. Longman.

Fletcher, A. (2001) *The Art of Looking Sideways*. Phaidon Press.

Froud, D. & Harriss, H. (2015) *Radical Pedagogies: Architectural Education and the British Tradition*, RIBA Publishing p. xii

Gropius, W. (1975) The Theory and Organisation of the Bauhaus, in Herbert Bayer, Walter Gropius and Isa Gropius, eds., *Bauhaus 1919–1928*, Museum of Modern Art, p. 22.

Guy-Evans, O. (2020). Bronfenbrenner's ecological systems theory. *Simply Psychology*.

Hedjuk, J., Lewis, D. & Diller, E. (1988) *Education of An Architect: Irwin S. Chanin School of Architecture of the Cooper Union*. Rizzoli, pp. 7–9.

Jourda, F.H. (2009) *Petit manuel de la conception durable* (Little Manual of Sustainable Design), Archibooks + Sautereau Editeur

Kleiman, P. (2009) Design for learning: a guide to the principles of good curriculum design. [e-book] PALATINE (HEA). Available at: http://cielassociates.co.uk/wp-content/uploads/2014/08/KLEIMAN2009DesignforLearning.pdf [Accessed 24 February 2022].

Kolb, D.A. (1984) *Experiential Learning*. Prentice-Hall.

Land, R., Meyer, J. H. & Flanagan, M. T. (Eds.). (2016) *Threshold Concepts in Practice*. Springer.

Lappin, S. (Ed.); Delaney, M., Gorman, A. (2011) *Studio Craft & Technique*. The School of Architecture, University College Dublin.

Marx, K. & Engels, F. (1977) *The German Ideology*, C.J. Arthur (Ed.), Lawrence & Wishart. p. 42.

McClean, D., Holgate, P., Bloice, L. & Murray, I. (2020) Mental Health in UK Architecture Education: An analysis of contemporary student wellbeing. An Initial Study. Robert Gordon University, UK, [RIBA Research Grant (17 February 2020)].

Morrison, T. (1993) *The Nobel Lecture in Literature*. Chatto & Windus.

Orr, S. (2010) The Role of Professional Judgement in Assessment: A Journey from Accuracy to Zing. Inaugural Professorial Lecture. York St John University. January 2010. chrome-extension://efaidnbmnnnibpcajpcglclefindmkaj/https://eprints.soton.ac.uk/382757/1/final-pdf.pdf [Accessed 20 June 2025]

Orr, S. (2010) *Making Marks: Assessment in Art and Design*, York St John University. Summer 2010 chrome-extension://efaidnbmnnnibpcajpcglclefindmkaj/http://arts.brighton.ac.uk/__data/assets/pdf_file/0004/65308/Susan-Orr-article-Issue-10-pp-9-13.pdf [Accessed 20 June 2025]

Petty, G. (2009) *Evidence Based Teaching – A Practical Approach*. Nelson Thornes.

Pickett-Palmer, L. (2019) A creaturely cinema of the everyday: Maggie in Wonderland (Sweden, 2008), Slices of Everyday Life Conference, Cambridge, Cinemuse Space.

Robbins, D. (1990) *The Independent Group: Post-war Britain and the Aesthetics of Plenty*. The MIT Press.

Sen, A.K. (1973) *On Economic Inequality*. Clarendon Press.

Sfintes, A.I. (2019) Architecture and Anthropology. Working in between Concepts, IOP Conference Series: Material Science and Engineering, 471(7), 072027.

Shulman, L.S. (2005). *Signature Pedagogies in the Professions*. Daedalus, pp. 52–59.

Singa, S. (2013) *Architecture for Rapid Change and Scarce Resources*. Routledge.
Spicker, P. (2014) *Social Policy: Theory and Practice*. Policy Press
The Equalities Act 2010: www.legislation.gov.uk/ukpga/2010/15/contents [Accessed 10 December 2022]
The inclusive curriculum (2020) (Webinar)[online] Available at: www.timetothink.com/thinking-environment/the-ten-components/ [Accessed 10 October 2021].
UNCCD Publication (11 May 2022). Drought in Numbers 2022: restoration for readiness and resilience [online] Available at: www.unccd.int/resources/publications/drought-numbers [Accessed 13 May 2022]

7

RESPONDING TO EMBODIED INJUSTICES

Introducing critical thinking and empathy in architectural education

Essi Nisonen, Jenni Poutanen and Sofie Pelsmakers

Climate emergency and radical rethinking

Architecture has been complicit in the climate and biodiversity crises, forcing architects to finally examine their extractive and exploitative design and decision-making processes. The impacts of architects' actions extend far beyond these dual crises, creating and sustaining other social and environmental consequences, which can all be referred to as the "embodied injustices" of architecture (Tucker, 2021).

For decades, architects' design decisions and material specifications have justified the exploitation and destruction of ecosystems and the extraction of natural resources, reinforcing the separation of humans and nature. There has been a profound lack of empathy and consideration towards e.g., the health and wellbeing of disadvantaged and vulnerable communities affected at the sites of production as well as the projects' impacts on increasing climate injustices, child-labour use or other human rights abuses. By ignoring societal challenges and denying the impact of projects' embodied injustices, architects allow themselves to comfortably continue to design to their intersubjective "architect" values. These values and canonical references, material cultures and stubborn aesthetical values are what got us in this situation to begin with, through "the spatialisation of an out-of-control capitalism" (Till, 2021).

We cannot build our way out of these embodied injustices and emergencies, because building is what has caused them in the first place. The climate emergency reframes our value systems: from what we build, for whom, to how we build it, and if we should be building at all. It requires us to bring empathy, critical thinking and a sense of broader responsibility into the design process, our values and our roles as architects.

The foundation for this change must be laid in architecture education, where (professional) values, cultures and norms are established that one later builds on in practice. To engage students in the active promotion of sustainability both in their work and communities, students need to be taught with an immersive focus on their values and behaviour, not solely delivering sustainability knowledge and positive attitudes (Pappas et al., 2012; Wodika & Middleton, 2020). However, architectural education typically

DOI: 10.4324/9781003293903-9
This chapter has been made available under a CC-BY-NC-ND 4.0 license.

does not respond well to such a vast, seemingly unsolvable net of problems that extend beyond the physical environment to social and societal structures. Finding a frame or an approach for holistic sustainable architecture[1] has become increasingly difficult as often in design curricula sustainability is still treated as a distinct topic separate from the design process or even its desired outcome. Treating sustainability as a phenomenon instead of a way of thinking has resulted in many students gaining a narrow understanding of sustainability aspects and their interdependence (Donovan & Pelsmakers, 2019, p. 220). Clearly, teaching sustainability knowledge is not enough: a radical reimagination of how we work and what we do has to be borne out of completely different internal values and new design cultures, implying a new architecture (Hill, 2021; Atekpe, 2025. Previous attempts have shown that we cannot achieve a complete rethink of what architecture is about by using the same methods, attitudes and values that we have instilled over many decades: architects should be trained to be flexible and holistic thinkers rather than linear spatial problem solvers (Deutsch, 2020; Nisonen, 2022).

This chapter investigates how these challenges might be met in architecture education and therefore by extension in practice. First, the status quo is explored through *Problem-based Learning* as a typical existing architecture education pedagogy, delving further into the setting of *Design Problems*, the *Design Studio* and typically associated problems with this. Next, *Design Thinking* theory is unfolded and analysed as a potential new approach for architecture education. It must be noted that while there are exceptions to the general ways of teaching architecture, and that avoid the pitfalls and risks with it, this chapter focuses on the prevailing ways of teaching in most schools around the world.

Exploring the status quo

Problem-based learning

Problem-Based Learning (PBL) is one of the most prominent pedagogies in architecture education. Literature describing PBL as a theoretical approach exists, but there is very limited literature available on the actual relevancy and effectiveness of its *implementation* in architecture education (Bridges, 2006). This section gives an overview of the theory of PBL and its typical application in architecture.

PBL was developed to tackle the established top-down power positions in education: teachers were acting as gatekeepers for "justified knowledge", serving it to the students. This practice ignored the student as an individual and rather saw them as a passive subject absorbing information (Margetson, 1991, pp. 47–48). In contrast, PBL was based on the idea that the starting point of education is the curiosity of the learner (Drinan, 1991). In the 1990s PBL further evolved through its actual implementation, e.g. PBL was developed to allow for students to become active learners who can internalise a desired, complex body of knowledge through creatively studying "a real-world problem" in different scales and contexts, together with others. Students would no longer solely learn to acquire knowledge but to also gain capability and agency (Drinan, 1991; Boud & Feletti, 1991).

In 1991 Drinan explored the limits of problem-based learning by listing the desired outcomes of PBL (Table 7.1) from its original purposes (1–5) to ones that emerged later through its development (6–9). A less successful PBL curriculum often encompasses the points 1–5, where the goal is for students to gain relevant information and organise it to

TABLE 7.1 Synthesising the original and developed purposes of PBL (after Drinan, 1991)

Original purpose of PBL: students to gain relevant information and organise it to make decisions and to independently learn more	1. developing the ability to make decisions (solve problems) 2. raising awareness of the complexity of real-world issues 3. acquisition of, or exposure to, a body of knowledge 4. motivation for learning through use of professionally relevant material 5. developing the capacity for self-directed learning
Developed purpose of PBL: the necessity of reflection and conceptualisation	6. developing the ability to extend learning beyond the presented situations (problems) into new ones 7. generating the desire and ability to think deeply and holistically 8. generating an enthusiasm for learning from and utilising all life's experiences in personal, professional and community development 9. encouraging a search beyond one's own preconceptions, so becoming ultimately innovative and critical with respect to self and one's profession

make decisions and to independently learn more. This typically excludes the necessity of reflection and conceptualisation (points 6–9), leading to a lack of comprehension of the depth and breadth of the "problem" and does not motivate to challenge or reframe it (Drinan, 1991; Engel, 1991).

Design problems

From early on, many authors have recognised the limits of the term "problem" in PBL, which can mislead one into thinking that there is something negative to solve, something to react to, rather than an opportunity to explore (Drinan, 1991; Bridges, 2006). A "problem" should not refer to *a single difficulty*, but rather to *what is problematic about a situation*, and represent a broader phenomenon with a complex context (Margetson, 1991).

When evaluating the applicability of PBL in architecture education, it is important to note that design problems are often described as "wicked" (e.g. Buchanan, 1992; Earle & Leyva-de la Hiz, 2020), meaning that they cannot be comprehensively described, nor do they have a beginning or an end. They have multiple possible solutions, and the success of these alternatives usually cannot be objectively evaluated. Often these different solutions prompt new questions and new problems to solve, making each design problem unique (Lawson 2019).

Attempts to formulate a PBL curriculum in architecture often tend to tackle either the problem itself (reshaping the design problems into cases), or the integration of the curriculum to better support the problem-solving process (e.g., establishing a shared theme as an umbrella encompassing all individual courses in a year) (Maitland, 1991; Bridges, 2006). When PBL is used to reformulate design problems, the risk is to diminish architectural thinking into a diagnostic process of cause and result, which overlooks the complexity of design problems. When the attention is on the integration of content and the creation of a desired body of knowledge, the threat is that the design approaches are steered into a pre-defined, uncritically derived solution. It has also been noted that designers tend to approach problem-solving by growing a better understanding of problems through trying and retrying different solutions to them (Dorst, 2011; Lawson, 2019). This contrasts the original purpose behind PBL, where problems were solved primarily through theoretical research (Boud & Feletti, 1991).

The design studio

In architecture, *the design studio* is typically used to assist a student to solve a brief for a building or urban plan (i.e. the problem). The creative studio work is usually underpinned by individual design tutorials and passive learning (supporting lectures). At the end of the design process, the solution is presented by the student and critiqued by a jury of tutors and practitioners. This culture often contributes to creating a mastery syndrome, where the teacher acts as the master in a public critique with little room for collaborative and empathetic dialogue with the students (El-Latif et al., 2020).

Moreover, the way problems are typically presented in the design studio is by already describing a solution (the design brief), which is hardly ever the right answer to the initial problem (what is the *brief itself* reacting to?) (Lawson, 2019). This culture of problem-posing mimics the dominant situation in practice, where the briefing process is usually one-sided; the designer often becomes a part of the briefing process in late stages, or after its completion (Paton & Dorst, 2011). Design should, however, consist equally as much of *discovering* problems as exploring them; solely solving the problem is often far less essential than reframing and questioning it (Drinan, 1991; Lawson, 2019; Deutsch, 2020; Nisonen, 2022).

If the brief is solely created by the teachers, mimicking real life, and if students are unable to question the starting point of their designs, they are pushed to follow a fairly linear process towards a certain type of desired solution (Melles, et al., 2015). A pre-defined design brief generally prevents students from forming a deep understanding of the problem context and prohibits them from getting personally motivated in the design process (Paton & Dorst, 2011). While Maitland (1991) refers to the architectural studio as a "problem-based learning feature", the commonly used master–apprentice-based model in architecture education actually promotes a culture of learning that PBL initially tried to tackle.

While the design studio and master–apprentice culture provide a long-established method of transmitting tacit knowledge (i.e., intangible professional know-how) to new generations (Nonaka & Takeuchi, 1995), embodied injustices sit at the heart of the culture and the educational methods we use. The system contains several fragilities even though the design studio's benefit is in creating a community of learners and peer support (Deamer, 2020; Pelsmakers et al., 2020).

Firstly, the current system is based on the idea (or ideal) that the teachers are distinguished designers, masters who know it all, i.e., the very "gatekeepers" PBL originally aimed to address. The tacit knowledge base of the practitioner-teachers is simultaneously highly valuable, but a threat if the teachers as individuals are not "flexible and holistic thinkers" (Drinan, 1991; Deutsch, 2020) who actively upskill themselves and reflect on values, norms, content and teaching methods, and adapt their thinking to meet the challenges of the climate emergency.

Secondly, teachers' tacit knowledge base renews slowly, solely based on their own interests, rendering both teachers and students prisoners of their own (unnoticed) inflexibilities and biases (Salama, 2021; Nisonen et al., 2024. Furthermore, teachers coming from practice rarely have pedagogical training, and often teach as they were taught without critically reflecting on their practices (Hill, 2017; Donovan & Pelsmakers, 2019; Nisonen, 2022).

Thirdly, if the design studio is based on the guidance of an individual teacher, the students miss the contamination of different ideas and approaches, which are the incentives for critical reflection that is derived from the input of several teachers (Pelsmakers et al., 2021; Nisonen, 2022; Nisonen et al., 2024).

Finally, the field of architecture typically promotes individualistic designer capabilities, a star architect syndrome that the studio culture tends to emphasise. This misleads one to think that complex design problems could be solved by one individual alone. It also fails to mimic reality, where complex design issues require not only research and development but are also tackled in multi-disciplinary teams (McDonnell, 2018). Together, all the points unfolded above may lead to pedagogically unsound situations.

However, the design studio as an educational environment does not *have* to be the problem. This, in turn, requires a deliberate reset of our values and praxis (e.g. Donovan et al., 2021, Nawratek, 2021; Nisonen, 2022; O'Dwyer et al., 2023).

In summary, PBL was originally a solution to a problem. Teachers were gatekeepers of knowledge, and students were not encouraged to discover. In allowing the discovery of solutions, a new issue emerged: teachers often became gatekeepers of problems, and students were typically prohibited from seeing their origins. At present, the key problems stem from the typical top-down problem-posing (of the design tasks) and the lack of integrated and holistic sustainable architecture knowledge. Given how architecture education has focused on design as an individual effort, it lacks empathy with others and common goals to strive for. In the climate emergency, conventional architecture education needs a new paradigm where we reposition ourselves as teachers and architects from authorities to facilitators.

Towards a new paradigm

Sustainability is an archetype of a wicked problem: unformulated, confusing and complex, with no "right" way to solve it (Earle & Leyva-de la Hiz, 2020). Wicked problems are best addressed if the design process shows empathy for the social and physical context of the problem (Gasparini, 2015, p. 6). Most importantly, it is difficult to tackle wicked problems with rational-analytic, linear problem-solving methods of PBL.

To help students become holistic and transformative thinkers in the era of the climate emergency, the iterative problem-solving approaches of Design Thinking might provide tools for both, creating a framework for creating knowledge, as well as evaluating and transforming one's own values and approaches (Earle & Leyva-de la Hiz, 2020). In education, rather than restraining students into absorbing information and repeating predetermined practices, the goal of Design Thinking is to engage the student in a process of '*active experimentation, concrete experience, reflective observation and abstract conceptualization*' (Earle & Leyva-de la Hiz, 2020; Beckman 2020) and to develop a deep, empathic understanding of user needs and the problem context (Beckman, 2020). Contrary to PBL, Design Thinking suggests that students actively participate in the *problem posing*, not solely in the search of a "solution". For learning about embodied injustices, the holistic nature of sustainability and its integration into design, this empathic positioning and evaluation of one's own approaches, values and attitudes, is crucial.

A designer needs to understand their design process in order to change it (Deutch, 2020; Lawson, 2019). In architecture education more specifically, the understanding of design processes is passed on primarily as tacit knowledge through design activities (likely originating from the idea of "design can only be learned through designing"), rather than being explicitly taught or analysed (Lawson, 2019; Maitland 1991; Schön 1987 in Donovan & Pelsmakers, 2020; Deutsch, 2020; Nisonen, 2022). The way Design Thinking has unveiled the recurring stages and activities of a design process has helped demystify the creative problem-solving process, "name" issues and allow the use of "targeted" activities to help release tensions in specific parts of the design process. This is further explored in this section.

Background: characteristics of Design Thinking

Literature reviews analysing the nature of Design Thinking (Hassi & Laakso, 2011; Micheli, et al., 2019) identified key systematically repeated characteristics in its implementations and research. For this research, the characteristics were formulated under two (overlapping and interacting) categories: "empathy, collaboration and communication" and "critical thinking" (Table 7.2). These categories are used to explore the applicability of Design Thinking in architecture education.

The following section further unfolds the two categories and their connection to learning and teaching architecture. The role and meaning of "divergent and convergent approaches" and their connection with blended learning strategies will also be explored.

TABLE 7.2 Summarising the connection of Design Thinking characteristics and the defined categories (Based on: Hassi & Laakso 2011; Micheli et al. 2019; Beckman, 2020).

Category	*Aim*	*related Design Thinking characteristics*
Empathy, collaboration and communication	– empowerment/sense of collective/community through connecting narratives and logics – setting aims and value base of a design project and positioning oneself as a designer – using (visual) communication to encourage dialogue and inspire to design against the status quo – representing and transforming knowledge through visuals and prototypes – seeking validation	– user centeredness & involvement – collaborative (interdisciplinary) work style – creativity and innovation – thinking by doing – iteration and experimentation – visualising
Critical thinking	– putting aside personal prejudices and biases – forming appreciation of diversity of perspectives – forming deep understanding of problem context – generating ways to address or "solve" the problem in chosen frame (= concepts) – refining, combining and sorting concepts – self-reflection on own practises and solutions	– experimental and explorative mindset – future-orientation and optimistic mindset – blended analysis and intuition – tolerance of uncertainty and failure – holistic view – abductive reasoning – iteration

Empathy, collaboration and communication

The way we establish and nurture our interpersonal skills – human relations and our ability to work cooperatively – determines our actions towards the environment. (Pappas et al., 2012; Kondrad et al., 2021). Learning from others, building and maintaining trust and being able to tackle conflicting interests are paramount skills for sustainability.

In Design Thinking, *a collaborative work style* promotes a culture of utilising interpersonal and interdisciplinary knowledge and its diverse perspectives for tackling wicked problems (Hassi & Laakso, 2011; Micheli et al., 2019). Merging these perspectives "in a meaningful and novel way" is achieved through *interaction*, rather than working alone (Hassi & Laakso, 2011). Collaboration and interaction also strengthen designers' ability to be truly empathetic: when based on individual "low-level mindreading", what one thinks is empathy can just in fact reflect one's own beliefs, motivations and judgements (Heylighen & Dong, 2019, p. 116).

In Design Thinking, the iterative problem-solving process is formed around user-centeredness *and user involvement* to empathise with the physical and emotional expectations, needs and behaviours of users, human and non-human alike (Hassi & Laakso, 2011; Micheli et al., 2019). Design Thinking uses tools like personas and journey mapping to form a deep understanding of users' interactions with design; the use of these tools is based on analyses of user data.

Design Thinking's approach to creativity distinguishes it from the traditionally more individualistic architectural design process. Contrary to seeing creativity as a characteristic or skill of the designer, Design Thinking considers *creativity and innovation* as *shared experiences* that motivate engagement in the problem-posing and problem-solving processes. The ability to visualise is considered to be a tool for communicating intangible concepts and ideas. The informality of e.g. sketching and mind-mapping is considered to lower the threshold for sharing ideas (Hassi & Laakso, 2011; Micheli, et al., 2019).

In Design Thinking, establishing a "shared language" as well as shared aims and values is considered to happen through communication and is seen to enable validation, empowerment and a sense of community (Beckman, 2020). The creation of this shared language is also paramount for the process of transforming tacit knowledge into explicit knowledge (Nonaka & Takeuchi, 1995). To achieve this, the design process should involve stages explicitly dedicated to shared experiences and collective knowledge creation, like brainstorming sessions, peer discussions and workshops. In architecture education, introducing a culture of teacher-facilitated, active peer-learning might help tackle the existing top-down power structures enforced by the master–apprentice model. (Drew & Mackie, 2011; Sgambi et al., 2019; Mahmoud et al., 2020; Nisonen, 2022)

Approaches to critical thinking

A holistic approach to sustainability helps to unveil the complex interdependence of social and societal structures and allows students to critically position themselves within them (Argento et al., 2020). In design, this translates to adopting a profound understanding of the interdependence of the problem's aspects and a deep understanding of its context, i.e. gaining a holistic view of the problem (Micheli et al., 2019, p. 135). This understanding should expand from functional needs to emotional, social and cultural needs (Deutsch, 2020).

Design Thinking aims for one to form an appreciation of diversity, and to provide means for self-reflection and withstanding uncertainty. Putting aside personal prejudices and being able to celebrate this multitude of different, often contradictory, alternatives is also a crucial skill in approaching wicked problems (Hassi & Laakso, 2011; Leifer & Steinert, 2011; Micheli et al., 2019; Earle & Leyva-de la Hiz, 2020).

A culture of being guided and positively challenged by constraints and obstacles, as well as having a desire to improve an existing situation, requires *future orientation*, *optimism* and an *experimental and explorative mindset* (Hassi & Laakso, 2011). In a learning environment, transformative behaviour and tolerance of uncertainty can be fostered by providing students a safe (physical and mental) environment that allows for experimentation (Leifer & Steiner, 2011; Elsbach & Stigliani, 2018).

Problem-posing (and solving)

A "problem space" refers to the mental representation of a problem and all the possible paths to solving it. In Design Thinking, the problem space is considered to exist parallel with the solution space (see Figure 7.1). Design problems and design solutions evolve separately in these two spaces but affect and interact with each other constantly. Cycles of iteration happen both within the spaces and between the spaces, and this process eventually narrows down the solution space, resulting in a holistic understanding of the problem and its context and allowing for "final" design approaches to emerge (Du et al., 2012; Nisonen, 2022).

Thinking by doing refers to this process of knowledge creation through a movement between the problem space and solution space. This movement encompasses experience, conceptualisation, observation and experimentation (Figure 7.1). It originates

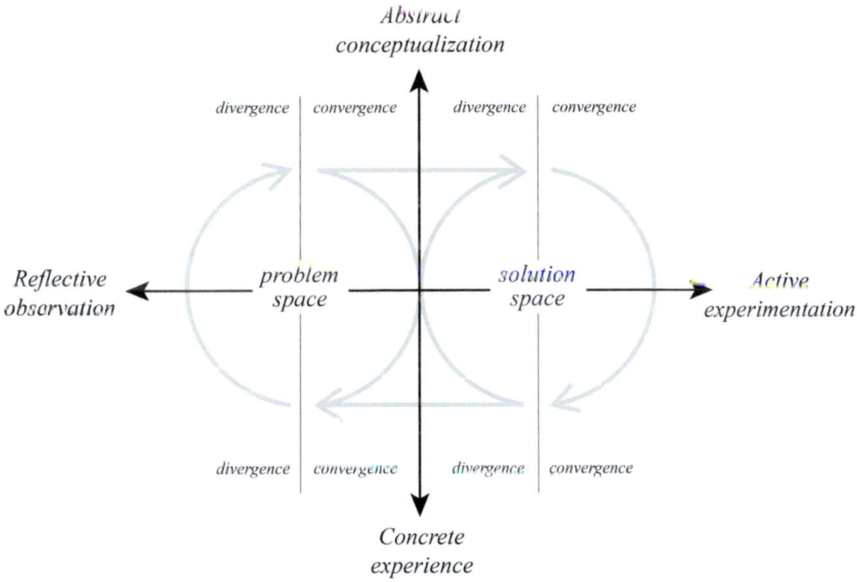

FIGURE 7.1 The Design Thinking process, adapted from Beckman, 2020.

from the logic of experiential learning, where two approaches for "grasping" experience and "transforming" experience are juxtaposed (Kolb, 2014; Beckman, 2020). In DT, the evolution of design problems and solutions happens through a combination of divergent and convergent approaches. Divergence and convergence have a dual meaning; moving further out and zooming closer into a problem, but also moving between personal perceptions and a shared understanding (Leifer & Steinert, 2011; Dorst, 2011).

Diverging, or distancing oneself from the problem, generates and acknowledges alternative approaches to a problem (Hassi & Laakso, 2011). Diverging helps the designer to choose a viewpoint from which the issue is explored, i.e. how they initially frame the problem. During framing the designer empathises and positions themselves in relation to the problem (not when they solve it) (Dorst, 2011). To establish a frame, questions like "what are the facts?", "what do they imply?", "what is taken for granted?" and "is there something that prevents us from seeing other options?" should be addressed through research and observation (Beckman, 2020). Answering these questions is the basis of critical thinking; the designer should recognise patterns and relationships that may be conflicting and paradoxical and through this, also become aware of their own prejudices and biases (Hassi & Laakso, 2011).

Converging brings the various problem approaches together through establishing a dialogue: different viewpoints are evaluated against each other (and integrated) to create a (new) common understanding, i.e. reframing the problem. (Dorst & Cross, 2001; Dorst, 2011). A cycle of divergence and convergence typically results in the emergence of a design approach. Through the analysis and critical evaluation of the design approach in the solution space, new perspectives to the initial problem emerge, and a new cycle of divergence (research for additional information) and convergence (synthesising the information and creating a new possible design approach) begins, often moving back to the problem space (revisiting and refining the "problem" the design approach is trying to solve) (Du et al., 2012; Lawson, 2019).

It is recognised that generating out-of-the-box ideas during the diverging process is easy, but people have a tendency to revert back to more average ideas during the converging process due to fear of mistakes (Beckman 2020; Mahmoud et al., 2020). One possible mechanism leading to this could be in the oversimplification of complicated matters both in the problem space (oversimplifying insights that emerge from research) and solution space (over-exaggerating the effects of design approaches) (Beckman, 2020), and the realisation that ideas might have been too far-fetched, prompting some to lose motivation in their ideas, thinking they have to start over (Mellon et al., 2015; Lawson, 2019). This kind of thinking is especially common for students early in their studies or less-experienced designers (Mahmoud et al., 2020). Familiarising students early on with the iterative logic and collaborative nature of the design process might help overcome these failures or moments of despair and help students stay engaged in the problem-solving process (Leifer & Steinert, 2011; Du et al., 2012), which is also paramount for the development of critical thinking skills (Deutsch, 2020; Rodríguez Aboytes & Barth, 2020).

In learning and teaching, explicit iterative movement between the problem space and solution space would create a different culture for establishing a desired body of knowledge; problem posing is based on shared experiences and co-creation, rather than being posed top-down (Dorst & Cross, 2001; McDonnell, 2018). Successfully supporting this iterative process of holistic knowledge creation calls for a matching pedagogical approach.

Blended learning (BL) could be considered a suitable option, as its logic is based on optimising the use of different teaching strategies to best support different learning modes (i.e. experience, conceptualisation, observation and experimentation), typically combining face-to-face learning events with online-learning (see e.g. Nisonen, 2022). To help with diverging from the problem, and to individually perceive and process information, students should be provided with general background knowledge through lectures, reading and discussions. In BL, this is often done through flipped classroom methods, i.e. pre-recorded lectures and other online resources (Donovan & Pelsmakers, 2019; Perbandt et al., 2021). To help converge and create a new understanding of the problem and its context, active learning methods (Drew & Mackie, 2011; Sgambi et al., 2019; Nisonen, 2022) embedded in DT practices, e.g., discussions, workshops and brainstorming sessions (McDonnell, 2018), should be organised along individual guidance sessions. Discussions and shared experiences help with the application of theoretical knowledge, i.e. transforming the abstract ideas formed in the problem space to concrete design approaches in the solution space (Donovan & Pelsmakers, 2019; Nisonen, 2022). The core of this process is what DT refers to as integrative thinking; the possibility to face several opposing ideas simultaneously, and instead of choosing between them, produce a creative resolution which releases the "tension" and uses the strengths of the several ideas (Hassi & Laakso, 2011; Micheli et al., 2019).

Instead of the linear status quo practises in the design studio, we argue that establishing connections between the iterative movement between the problem space and the solution space and different blended learning strategies can provide fruitful insights for the future of architecture education: finding a balance between co-creation, dialogue and individual reflection is key for supporting students' process of becoming holistic and transformative thinkers understanding both themselves and the surrounding society.

Conclusion

Architects need to achieve a transformative change in their values and design processes to support society's urgent transformation. This requires discovering and integrating new approaches for holistic sustainable architecture and implementing an explorative and experimental mindset that helps us unveil, critically evaluate and radically rethink architecture.

Problem-based learning was originally adopted to allow for students to become active learners who can grasp complex issues through creatively studying a "real world" problem in different scales and contexts, together with others. Architecture education often claims to be following PBL, but we argue that the predominant master-apprentice model and problem-posing culture opposes the initial goal of PBL, which was to tackle the established top-down power positions in education.

With decreasing resources and increasing knowledge demands the traditional design studio needs a rethink (Hill, 2017; Donovan & Pelsmakers, 2019). We suggest that

Design Thinking could be used for rooting a deeper appreciation for iteration and reflection to the design studio: together with blended learning strategies, design thinking can help integrate optimal learning methods and mindsets into the architecture curricula. *Establishing a culture of participatory, holistic problem-posing* will help students to critically position themselves in the design process and get motivated in transformational thinking against the status quo. To support the development of students' interpersonal skills, low-hierarchy learning environments encouraging co-creation and dialogue should be established and nurtured. *Empathy, collaboration and communication* can be enforced through active, (face-to-face) learning events where students come together with teachers acting as facilitators, not authors of the collective knowledge creation. These dialogic learning events could also potentially be used for updating the tacit knowledge base of teaching staff.

Students cannot be encouraged to be innovative and transformative without the teaching staff promoting the same behaviour. It is vital to encourage teachers to try new approaches; often the idea of thinking outside the box can feel intimidating or futile at first, prohibiting new approaches from being tested, but when implemented, new approaches are mostly greeted with excitement and engagement by students (Beckman, 2020).

An understanding of design as a process-forward, collective and holistic activity should encompass the whole curriculum, both in the attitudes of educators and the structure and goal of the design studio. Architects should 'redefine' themselves from problem solvers to holistic thinkers and facilitators of transformation.

KEY TAKEAWAYS FROM CHAPTER 7

- The role of practitioner-teachers (and by extension, architects) needs to transform *from authorities to facilitators*: design problems are wicked and do not have one right answer or a right, linear path towards it. Instead, learners create knowledge and define (design) problems and their solutions from unique viewpoints. This needs to be acknowledged through the teaching and learning cultures of the design studio.
- Empathy, collaboration and communication need to lie at the heart of the design studio culture. *Iterative processes cannot be controlled* through top-down cultures or power structures between teachers and learners but should instead be *facilitated* through low-hierarchy and dialogic teaching practices and peer-to-peer cultures that acknowledge the diversity of learners and the complexity of (solving) wicked problems.
- Design should consist equally as much of *discovering* problems as exploring and solving them. Learners need to be given the possibility to question the starting points of their designs. Allowing the learners to form an in-depth understanding of the design problem and its context is vital for holistic sustainable architecture and develops learners' critical and reflective thinking skills.
- Different phases of the problem-posing and problem-solving processes require different teaching and learning activities to best support them. Blended learning strategies can support the iterative design process by balancing online-learning with face-to-face activities (see e.g., Nisonen, 2022).

Note

1 Holistic sustainable architecture comprehensively considers high standards in all aspects related to sustainability in architecture, and includes " the quadruple climate change challenges (i.e., climate change mitigation, adaptation, restorative design and climate justice)" (Pelsmakers et al., 2022).

Bibliography

Atekpe, F., 2025. Chapter 6, "Everyday, equity and ease", in this volume.

Argento, D., Einarson, D., Mårtensson, L., Persson, C., Wendin, K. & Westergren, A., 2020. Integrating sustainability in higher education: a Swedish case. *International Journal of Sustainability in Higher Education*, 6(21), pp. 1131–1147.

Beckman, S. L., 2020. To frame or reframe: Where might design thinking research go next. *California Management Review*, 2(62), pp. 144–162.

Boud, D. & Feletti, G., 1991. What is problem-based learning?. In: *The Challenge of Problem Based Learning*. St. Martin's Press.

Bridges, A., 2006. A Critical Review of Problem Based Learning in Architectural Education. Proceedings of the 24th Conference on Education in Computer Aided Architectural Design in Europe, pp. 182–189.

Buchanan, R., 1992. Wicked problems in design thinking. *Design Issues*, 8(2), 5–21.

Deamer, P., 2020. Design pedagogy: The new architectural studio and its consequences. *Architecture_MPS* 18(1), 2. DOI: https://doi.org/10.14324/111.444.amps.2020v18i1.002.

Deutsch, R., 2020. *Think like an Architect: How to develop critical, creative and collaborative problem-solving skills*. RIBA Publishing.

Donovan, E. & Pelsmakers, S., 2019. Integrating Sustainability in architectural design studio through blended learning. Conference Proceedings for Education, Design and Practice – Understanding skills in a Complex World, NY, June 2019, AMPS.

Donovan, E., Kozminska, U., Ruiz Allen, N., & Hilberth, T., 2021. Architectural Learning for a Sustainable Future (pp. 77–86), in Pelsmakers, S. & Newman, N. (Volume Guest Editors), RIBA *Design Studio, Volume 1:* Architecture and the Climate Emergency: Everything Needs to Change, RIBA Publishing.

Dorst, K., 2011. The core of "design thinking" and its application. *Design Studies*, Issue 32, pp. 521–532.

Dorst, K. & Cross, N., 2001. Creativity in the design process: co-evolution of problem-solution. *Design Studies*, 22(5), pp. 425–437.

Drew, V. & Mackie, L., 2011. Extending the constructs of active learning: implications for teachers' pedagogy and practice. *The Curriculum Journal*, 4(22), pp. 451–467.

Drinan, J., 1991. The limits of problem-based learning. In: *The Challenge of Problem-Based Learning*. St. Martin's Press.

Du, J., Shikai, J. & Jihong, L., 2012. Creating shared design thinking process for collaborative design. *Journal of Network and Computer Applications*, 35(1), pp. 111–120.

Earle, A. G. & Leyva-de la Hiz, D. I., 2020. The wicked problem of teaching about wicked problems: Design Thinking and emerging technologies in sustainability education. *Management Learning*, 52(5), pp. 581–603.

Engel, C. E., 1991. Not just a method but a way of learning. In: *The Challenge of Problem Based Learning*. St. Martin's Press, pp. 23–33.

El-Latif, M.A., Al-Hagla, K.S. & Hasan, A., 2020. Overview on the criticism process in architecture pedagogy, *Alexandria Engineering Journal*, 59(2), pp. 753–762, https://doi.org/10.1016/j.aej.2020.01.019.

Elsbach, K. D. & Stigliani, I., 2018. Design thinking and organizational culture: A review and framework for future research. *Journal of Management*, 44(6).

Gasparini, Andrea A. (2015). Perspective and Use of Empathy in Design Thinking. International Conferences on Advances in Computer-Human Interactions ACHI. ISSN 2308–4138. pp. 49–54.

Hassi, L. & Laakso, M., 2011. Making sense of design thinking. IDBM papers vol. 1. Helsinki: IDBM program, Aalto University, pp. 50–63.

Hill, D., 2021. Slowdown city. *The Journal of Architecture*, 26(1), 67–72, DOI: 10.1080/13602365.2021.1883306,

Hill, G. A., 2017. The "Tutorless" Design Studio: A Radical Experiment in Blended Learning. *Journal of Problem Based Learning in Higher Education*, 5(1), pp. 111–125.

Kolb, D., 2014. *Experiential Learning: Experience as the Source of Learning and Development* (2nd edn). Pearson.

Kondrad, T., Wiek, A. & Barth, M., 2021. Learning process for interpersonal competence development in project-based sustainability courses – insights from a comparative international study. *International Journal of Sustainability in Higher Education*, 22(3), pp. 513–560.

Lawson, B., 2019. *The Design Student's Journey*. Oxon: Routledge.

Leifer, L.J. & Steinert, M., 2011. Dancing with ambiguity: Causality behavior, design thinking, and triple-loop-learning. *Information Knowledge and Systems Management*, 10(1), pp. 151–173.

Mahmoud, N. E., Kamel, S. M. & Hamza, T. S., 2020. The relationship between tolerance of ambiguity and creativity in architectural design studio. *Creativity Studies*, 13(1), pp. 179–198, 2020.

Maitland, B., 1991. Problem-based learning for an architecture degree. In: *The Challenge of Problem-Based Learning*. St. Martin's Press.

Margetson, D., 1991. Why is problem-based learning a challenge?. In: *The Challenge of Problem-Based Learning*. St. Martin's Press.

McDonnell, J., 2018. Design roulette: A close examination of collaborative decision-making in design from the perspective of framing. *Design Studies*, 57, pp. 75–92.

Melles, G., et al., 2015. Problem finding through Design Thinking in education. *Innovations in higher education teaching and learning*, 3, pp. 191–209.

Micheli, P. et al., 2019. Doing design thinking: Conceptual review, synthesis and research agenda. *J Prod Innov Manag*, 36(2), pp. 124–148.

Nawratek, K. Hope in the Burning World (pp. 87–98), in Pelsmakers, S. & Newman, N., 2021. (Volume Guest Editors). RIBA *Design Studio, Volume 1: Architecture and the Climate Emergency: Everything Needs to Change*, RIBA Publishing.

Nisonen, E., 2022. Architecture as sustainability – educating the transformative practitioner. Master's Thesis, Faculty of the Built Environment, Tampere University.

Nisonen, E., Pelsmakers, S. & Kallio, K. P., 2024. (A)mending Architecture Education in the Climate Emergency: a focus on values and cultures. *Architectural Research in Finland*, *8*(1), 239–264. https://doi.org/10.37457/arf.146880

Nonaka, I. & Takeuchi, H., 1995. *The Knowledge-creating Company: How Japanese companies create the dynamics of innovation*. Oxford University Press. p. 304.

O'dwyer, S., Geoghegan, E., Nisonen, E., Castano-de la Rosa, R., Pelsmakers, S, Lykouras, I., Donovan, E., Alvise Bragadin, M., Morganti, C. & Coraglia, U. M., 2023. Architectural education: methods for integrating climate change design (CCD) in the curriculum. AMPS Conference, April 2022, A Focus on Pedagogy: Teaching, Learning and Research in the Modern Academy. AMPS Proceedings Series 28.2.

Pappas, E. et al., 2012. Using Bloom's Taxonomy to teach sustainability in multiple contexts. *Journal of Cleaner Production*, 48 (2013), pp. 54–64.

Paton, B. & Dorst, K., (2011). Briefing and Reframing. Proceedings of the 8th Design Thinking Research Symposium (DTRS8) Sydney, 19–20 October 2010. Edited by Dorst, K., Stewart, S., Staudinger, I., Paton, B. & Dong, A.

Pelsmakers, S. Donovan, E., Moseng, K. & Eyebye. B., 2020. Developing architecture studio culture: peer-peer learning, Conference Proceedings for Education, Design and Practice – Understanding skills in a Complex World, NY, June 2019, AMPS.

Pelsmakers, S., Donovan, E., Hoggard, A. & Kozminska, U., 2022. *Designing for the Climate Emergency – a Guide for Architecture Students*. RIBA Publishing.
Pelsmakers, S., Donovan, E., Poutanen, J., Hoggard, A., Sandström, N., & Friedman, K. 2021. *Architecture Design Studio in Uncertain Times: Digital Teaching and Learning Implications*. Scroope, (30), pp. 47–72.
Perbandt, D., Heinelt, M-S., Bacelar-Nicolau, P., Mapar, M. & Caeiro, S. S., 2021. Towards effective e-learning on sustainability: a case study-course on participatory processes in environmental politics. *International Journal of Sustainability in Higher Education*, 22 (4), pp. 801–832.
Rodríguez Aboytes, J. G. & Barth, M., 2020. Transformative learning in the field of sustainability: a systematic literature review (1999–2019). *International Journal of Sustainability in Higher Education*, 21(5), pp. 993–1013.
Salama, A. M. (2021). *Transformative Pedagogy in Architecture and Urbanism*.
Routledge.
Sgambi, L., Kubiak, L., Basso, N. & Garavaglia, E., 2019. Active learning for the promotion of students' creativity and critical thinking. *Archnet-IJAR: International Journal of Architectural Research*, 13(2), pp. 386–407.
Till, J., 2021. Architecture after Architecture, in *Architects after Architecture – Alternative Pathways for Practice*, edited by Harriet Harriss, Rory Hyde, Roberta Marcaccio, Routledge.
Tucker, E. A New Concept for Good Architecture: Embodied Justice – Metropolis, https://metropolismag.com/viewpoints/embodied-justice-2021/, [accessed 29th December 2021].
Wodika, A. B. & Middleton, W. K., 2020. Climate change advocacy: exploring links between student empowerment and civic engagement. *International Journal of Sustainability in Higher Education*, 21(6), pp. 1209–1231.

8

CHANGING CULTURES

Sustainability in architectural education and design teaching

David Kroll, Susan Galavan and Veronica Soebarto

Sustainability in architectural programmes

Architectural schools or programmes in many universities have included sustainability as part of their mission and marketing, often backed up by research activity and expertise in that area. Professional registration and accreditation bodies such as the Architects Registration Board (ARB) in the UK are also actively revising their assessment criteria to give sustainability a more central role in architectural education (Architects Registration Board, 2022). These are certainly steps in the right direction that are likely to lead to positive changes. However, it is all too easy to call for more top-down regulation and to defer responsibility, rather than consider how changes can be made by those on the ground.

Currently, the teaching of sustainability in architecture programmes is often confined to specialised courses and is too easily dropped or forgotten in other parts of the curriculum, such as the core design courses (Oliveira, Marco & Gething, 2018). How can we move to a more comprehensive engagement with sustainability in architectural curricula?

A few years ago, one of the authors taught a workshop on designing a sustainable house at a local high school near Adelaide. The children were creative, excited, engaged and asked good questions. At the end of the workshop, one of them asked what they should now do with all the materials that they used and the models they built during the workshop. The student seemed surprised that they would now have to throw all the material away. The overflowing bins after the workshop is an image that is still in our head. It was fun and probably instructive for the students to make the physical models. However, the large amount of waste afterwards did not align with the message to the students. The workshop was about sustainable design but left a pile of hard-to-recycle waste of paper, mixed with cellular tape, modelling clay and sticks. Perhaps much of our teaching and practice in architecture at university deals with that same dilemma. Our curricula include courses that teach students how to design sustainable buildings, essentially trying to change what *others* should do (builders, developers etc.), while not examining our own behaviours and practices. Simply teaching more sustainability courses with better

DOI: 10.4324/9781003293903-10
This chapter has been made available under a CC-BY-NC-ND 4.0 license.

FIGURES 8.1 AND 8.2 Models created during a workshop to design a sustainable house at a high school in Greater Adelaide, Australia. Photographs by David Kroll.

content is unlikely to change this dynamic. To make fundamental and lasting improvements, we need to also examine our teaching and what we do. We need to walk the talk.

What are the issues?

The available literature on the challenges of integrating sustainability in architectural education is comprehensive and it is worth summarising some of the issues that have been identified by researchers. A 2020 study of students' perception of sustainability in architectural education showed that they generally see sustainability as a key part of their studies (Boarin, Martinez-Molina & Juan-Ferruses, 2020). However, students also noted that, while sustainability is part of the curriculum, the specific skills and knowledge that are taught are often not made explicit enough and are therefore difficult to apply. In another survey, many alumni fail to attain basic literacy in building science, envelopes and systems, making them insufficiently equipped with the skills to engage in environmentally sustainable design in the workplace. In addition, a lack of exposure to renovation projects during their studies meant that graduates were not well prepared for renovation and reuse work, another key component of current architectural practice (Grant, 2020). The contradiction between message and practice is another concern that was raised. One of the students commented on the inconsistency between the willingness to discuss sustainability and the personal commitment of staff and students, suggesting that they are 'not willing to do a simple thing like turn off the computer or lights after leaving the room, or walk 5 min instead of driving' (Boarin, Martinez-Molina & Juan-Ferruses, 2018). The comment highlights how important it is for the content and message of sustainability in teaching to be supported by concrete action.

With a focus on building energy use-related content, the perceptions of educators and students in architecture have been examined, for example, by Oliveira et al. (2018). Overall, the UK-based study showed that building performance simulation has become widely accessible to architectural students and designers in part due to user-friendly software tools. However, one of the issues raised by students was a discrepancy and lack of integration between what is seen as creative design teaching and more technical content. While energy- related content is important in the research in many schools, it could be perceived by students as an add-on, rather than a central component of the design agenda. This same division was perceived between staff responsible for technical and non-technical content. Only in recent years have efforts been made to move sustainability to a more central position within the studio environment.

Donovan and Pelsmakers (2019) have successfully embedded sustainability in design thinking at the Aarhus School of Architecture in Denmark, by experimenting with blended learning, a hybrid teaching method that combines e-learning with conventional face-to-face interaction (see also Chapter 7 and 10). Another example is the state-wide initiative in the Republic of Ireland where the Building Change project aims to transform education across the six architecture schools. The primary vehicles for change are the Design Studio pilots, empowering both staff and students to collectively re-imagine the curriculum through a climate action lens (Building Change, 2025). The results thus far are very promising, enabling radical shifts in thinking on material selection, performance and designing for disassembly in design studio projects (see Figure 8.3). However, others have suggested doing away with the traditional architectural design studio environment altogether, considering it an outdated format that is ill-equipped to meet current societal challenges.

FIGURE 8.3 Model of 'Gateway' building by third-year architecture student Aidan Doherty investigating low carbon approaches to design at the Atlantic Technological University Sligo, Ireland as part of the Building Change project. (Photo: Aidan Doherty)

For example, Deamer argues that we should design 'in lieu of objects, scenarios for people to interact' in collaboration with experts from the construction industry (Deamer, 2020). There have also been calls for architects to take account of the procurement process: who builds, with what materials, from where and using what means to consider the broader environmental impact of built forms.

Due to the multi-stakeholder nature of architecture, curriculum development cannot be done in isolation (Iyer-Raniga and Tony Dalton, 2017). Iyer-Raniga et al. point out that '(...) educational institutions need to work in tandem with several other institutional agencies to drive, complement and support changes for low carbon futures; including government, industry practitioners and peak industry bodies' (p. 1). The Building Change project mentioned above engages with over 40 partners, including 24 enterprise partners, the Royal Institute of the Architects of Ireland (RIAI) and community groups and government. The re- training of architects goes hand in hand with that of the wider construction industry and there have been a number of calls for a major reform of construction sector skills training to meet zero-carbon policy objectives in the UK (Killip, 2020). For architecture schools, it is important to keep in mind all the parties involved in this transition and not give students an illusion of the 'lone rider' architect that saves the day, but rather to develop a better understanding of the agency and influence of architectural professionals within broader stakeholder networks. These aspects of working in a multi-stakeholder environment can be integrated into the curriculum, for example, through industry collaborations and involvement. The hierarchies of the universities themselves are part of this wider stakeholder network.

Many universities are now making concerted and systematic efforts to reduce their carbon footprints. Architecture schools can be part of this bigger picture and can make an important contribution.

Changing cultures

In a 2007 study on sustainability literacy in construction education, Murray and Cotgrave highlighted the need to develop a clear understanding of sustainability (Murray & Cotgrave, 2007). The authors recognised the important role of professional bodies in changing priorities in higher education, advocating for '(...) institutions to make small, steady and incremental changes to their curricula to provide them with the time they need to develop the expertise, awareness and motivation of their own staff' (p. 20). Sixteen years on, it is clear that this very tentative form of incremental change is not going to be enough, or fast enough to meet the challenges of today's climate emergency where extreme weather events are taking lives at an increasing pace. Radical culture change is urgently required, what Nisonen and Pelsmakers call 'a shift of architects' core values and attitudes – even what it means to be an architect' (Nisonen & Pelsmakers, 2022). What does such a culture change mean for architectural education and the profession? According to Jeremy Till, there needs to be a move away from the self-absorbed obsessions with aesthetics and taste, the overreliance on technology as an escape route from the climate crisis and the view of the profession as separate from external dependencies (Till, 2023). Till calls for a radical shift in thinking from a culture of power and superiority to one of collective dialogue, allowing room for critical reflection and enabling a more democratic and inclusive approach to share societal goals. Putting our own house in order is surely a necessary part of creating this broader change.

Changing existing cultures in architecture schools is difficult because of path dependency and they are not alone in this challenge. The research field of 'sustainability transitions' analyses and evaluates such socio-technical transitions at various scales, and the case studies show that these transitions, despite the apparent obstacles, can and do happen (Markard, Raven & Truffer, 2012). The complex interconnected processes involved take place at several levels and are often top-down, as well as bottom-up. One of the key ingredients to these kinds of transformations is a small network of actors that support and facilitate the changes (Geels, 2011). Critical mass theory asserts that change begins with a small group that then draws in the participation of others and eventually becomes large enough to fundamentally affect the organisation as a whole (Marwell & Oliver, 1993). A more cautionary perspective on transition management and sustainability transitions questions with good reason whether genuine change is possible within a system that is so dependent on high carbon emissions (Shove & Walker, 2007). This dependency is also evident in higher education. Tertiary educational buildings themselves are significant producers of carbon emissions. Lecture theatres, for example, generally tend to be windowless for practical reasons, which means high electricity use for lighting, ventilation, heating and cooling, even when the temperature outside is comfortable. Another example is laboratories or design studios with a controlled indoor environment, which need to be accessible around the clock. This culture of dependency on high carbon emissions is difficult to change but doing nothing or business as usual is not an option in the face of impending climate change disaster.

A radical rethink of architectural education and starting new schools from scratch is one option. However, a step-by-step transformational approach seems equally important in order to change existing curricula and institutions that better prepare students as sustainable architects and professionals. The authors are interested in this transformational approach, as we work within already established institutions, and we want to take an active role in the broader sustainability transition of architectural education. To move from analysis paralysis to practical steps of changing cultures, the following points and suggestions from our own ongoing journey are hopefully useful to others and are intended as conversation starters. We recognise that transforming architectural education to become more sustainable in its content and practices is an ongoing process and conversation.

Walking the talk

Leading experts on creating and sustaining culture change emphasise the need for the message and values to be aligned with and supported by actions (Taylor, 2005; McHale, 2020). Architecture programmes, which teach ways to make our future built environment more sustainable, should be able to demonstrate these values by taking steps in the right direction themselves. Misalignment between message and action was a point of criticism from students in the studies reviewed above (Boarin, Martınez-Molina & Ferruses, 2020). If we cannot make any changes within our own immediate surroundings, how can we claim to be able to change our broader built environment? And how can that message be believable to students? There may already be architecture programmes that successfully 'walk the talk'. However, examples of university programmes in architecture or construction that advertise how they are run with a low or positive carbon footprint seem to be much more difficult to find online than architecture courses that teach sustainable design.

Walking the talk evidently includes broader aspects of sustainability, discussed in other chapters, and requires a holistic approach, which includes a reassessment of how the messages that are part of the teaching content, are also practised by staff and students. If collaborative working is a shift that we argue is needed in the skillset of architects, for example, then teachers also need to set projects up that way and collaborate themselves. Walking the talk then means more inclusive and democratic teaching processes, for example, with suitable spaces. This chapter considers resources that can help with some of the environmental issues but there are evidently many aspects that are not covered here but that also need to be addressed to practise what we preach. We will need to carefully consider our own environmental footprint and carbon impact to make the vision of a more sustainable built environment believable within our immediate surroundings.

Practical steps, which can be taken and are often at least partly within our control, include changing our behaviour and the resources we use in teaching, research and administration. We can choose to find sustainable means for travelling to work and to avoid or reduce unnecessary travel (for talks or study trips, for example). We can also avoid or reduce unnecessary plastic waste during the day, for example for takeaway coffee or lunch. For our teaching, we can change the resources used for model making and printing. For model making, for example, we can provide guidelines to encourage students to reuse or to use materials that are made from natural and renewable resources, such as cardboard or wood, and to use more eco-friendly, biodegradable glues. The amount and type of printing are another consideration for change. At times, it may be important for students to print work for pedagogical reasons. However, the overall amount of printing can be reduced and replaced by digital presentations. Furthermore, when needed, other environmental considerations can be part of the course deliverables, such as the use of matt or recycled paper when printing is required. Staff stationery and printing can equally be reviewed in terms of improving the environmental footprint. In addition to considering the life cycle of materials, the right bins need to be available for staff and students for any waste to be recycled. This may seem obvious, but it is surprising how many design studio end-of-semester cleanouts seem to simply end up in non-recyclable waste bins.

Beyond improving our environmental imprint on a personal and programme level, there may be ways to link these to other levels in the institutional hierarchies. Many tertiary institutions have already or will implement plans to reduce their carbon footprint (Monash University, 2024). Architectural programmes could be part of this broader sustainability plan at the institution level, which can include carbon footprint tracking. Furthermore, we can use these broader strategies to lobby for a more sustainable operation of our facilities, such as motion sensors to be installed for lighting and air conditioning, so that they switch off when students are not there. Finally, communicating the message to stakeholders is essential for creating lasting culture change and for it to be supported by students and staff.

Bigger picture and visions

To achieve lasting changes in architectural education, goals are important to set out expectations and the following resources may be useful in that regard. The UN Sustainable Development Goals (SDGs) are controversial because they can be seen as simply supporting the kind of continued economic growth and development that has led to the climate emergency in the first place (Swain, 2018). Green growth so far has failed to deliver on lowering carbon emissions and it seems necessary to ultimately consider how architectural professionals can contribute to degrowth and restoration instead (Gunderson & Yun, 2017). Such bigger-picture goals that have international backing can still be a helpful starting point to inform architectural education. The most obvious goal related to architecture is number 11, 'Make cities and human settlements inclusive, safe, resilient and sustainable' (United Nations, 2024). The other goals also directly and indirectly relate to architecture and our interventions in the natural and built environment have a significant impact in achieving these. The Union Internationale des Architectes (UIA) provides *An Architecture Guide to the UN 17 Sustainable Development Goals* in two volumes with useful example projects that have contributed to achieving these goals (International Union of Architects, 2023).

Another resource more specifically conceived for an architectural audience that can be used to guide curriculum content is the Copenhagen Declaration. UIA launched the declaration at the UN Climate Change Conference in Copenhagen in 2009 by initiating its Sustainable by Design strategy to reduce, or reverse, the negative impact of the built environment on the global climate. The strategy itself is only a one-page document but comprises the key goals that underpin sustainable design (International Union of Architects, 2009). The declaration emphasises essential components including: engaging multiple stakeholders; high- performance design; respecting heritage, culture and social values; life cycle analysis and management; reducing carbon imprints; healthy materials; improving quality of life, as well as advancing economic well-being and providing opportunities for community engagement and empowerment. The strategies outlined in the document can be taken as a starting point for principles to be integrated across courses.

The Copenhagen Declaration is, of course, not the only such document that can be helpful to outline the underlying principles and values. Another relevant declaration to drive sustainable design is the Hannover Principles, which was the basis for the design guidelines for the Expo 2000 World's Fair (William McDonough & Partners, 1992). More recently and widely covered by the media, architects signed the Architects Declare Climate & Biodiversity Emergency, which began in 2019 with a declaration signed by the 17 UK recipients of the Stirling Prize that year. Since then, the declaration has expanded as Built Environment Declares to other occupations and has had more than 7,000 signatories across 28 different countries (Architects Declare, nd). The 11 points of Architects Declare, such as designing beyond the standard of net zero carbon in use, overlap and align with parts of the Copenhagen Declaration and UN Sustainability Goals. While many practices have signed up to Architects Declare, any architecture department that supports these principles, would surely need to use these to guide their courses and integrate them as a core component of the curriculum too. One criticism of Architects Declare has been that numerous practices readily declared their commitment but the translation into concrete actions still seems to be less apparent (Hocknell, 2023). A similar criticism about discrepancies between words and action could be directed at architectural education. The Architects Climate Action Network (ACAN) aims to address this need for concrete action and, as one of their initiatives, publishes a useful education toolkit with tangible and actionable suggestions to implement into academic teaching (ACAN, nd). Their proposed nine key points provide detailed recommendations on how to initiate cultural changes, revise the curriculum content, better support student agency and lead by practice.

Starting with our own courses

Existing institutional cultures are difficult to change (see also the path dependency mentioned above) and key decisions may at times be outside of our control. A university-wide waste policy or building operations, for example, may be difficult to change overnight by individual teaching staff. However, there are always steps that can be taken to change the things that we do have control over. One of these could be to begin with our own courses and to make sure that sustainability goals inform the course content at every level, from teaching materials to deliverables and lectures. Once we consider courses holistically by using these principles as the core content and values, other more detailed aspects can follow. If the underlying goal is to consider the lifecycle of materials, for example, then that surely would have to be part of the material specification in the course deliverables for a design studio. Another example is the difficulty of integrating performance evaluation into the design studio content. As evidenced in research studies mentioned above, this is often difficult to achieve within the allotted time available for design development. However, if it is not possible within the timeframe of the studio, performance evaluations could be conducted after the completion of the design studio in conjunction with other courses. Students still get the experience of how performance simulations can be used to inform design. In practice, the simulations may not always be completed by the designers themselves, but having first-hand experience provides a good grounding about how design can be adjusted to improve performance.

If there is broader support from staff in the architecture programme, another practical step may be to start by building up the sustainability content for core courses first, such as capstone design studios at the undergraduate and postgraduate levels. While this may not change the whole curriculum at once, it conveys the message that sustainability is

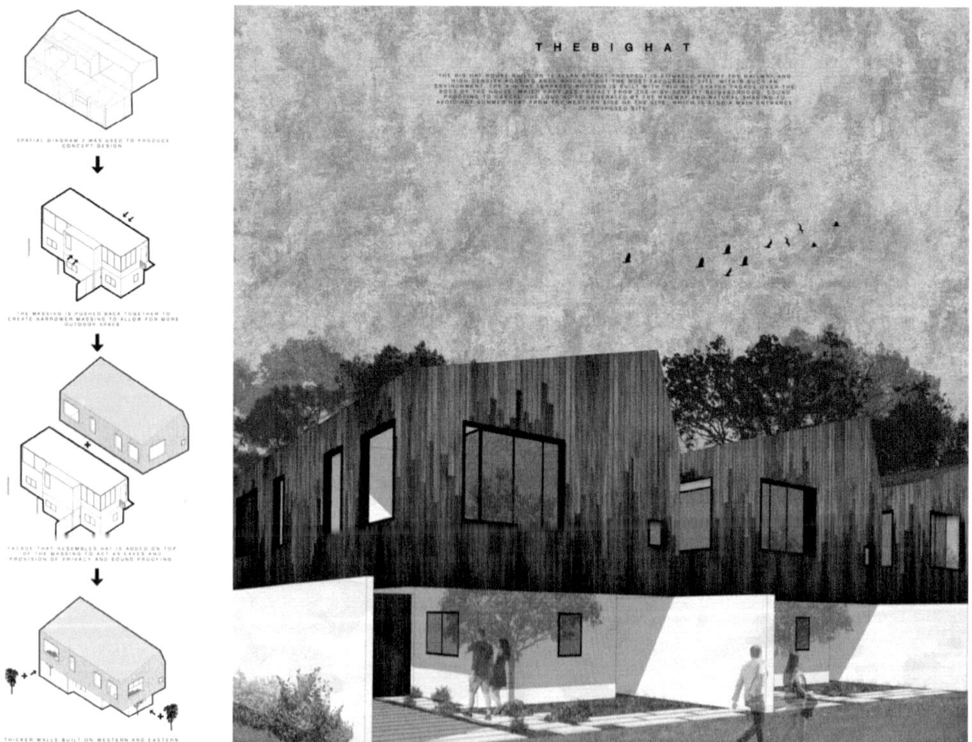

FIGURE 8.4 House design proposal by Gloria Goyo Choi with a creative solution to reduce overheating, from an undergraduate design studio at the University of Adelaide.

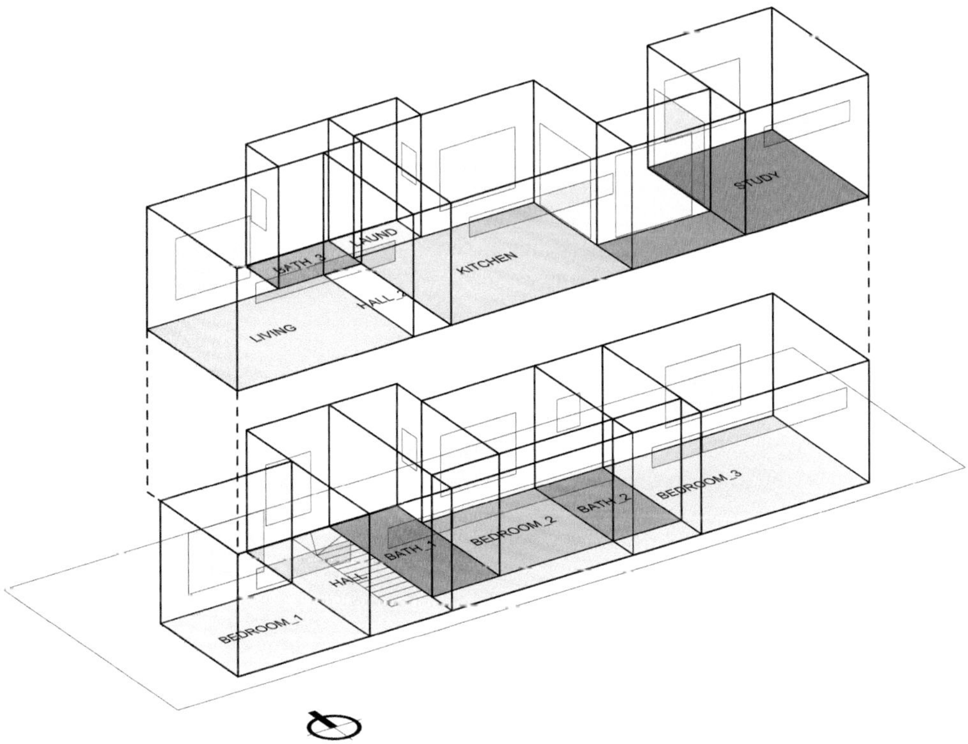

FIGURE 8.5 shows an image from a thermal performance study of the same design using Ladybug tools with Rhino/Grasshopper (Juliana Croffi).

not an 'add-on' to be taught in specialised supporting courses only, but it is at the core of the degree and provides the underlying foundations of all architectural design and production.

Critical mass and managing change

Sustainability transformations take place at multiple levels, from bottom-up initiatives of a small group of actors to top-down changes implemented through regulations, and each of these plays a role. Throughout, there are likely to be inevitable obstacles and resistance to changing the status quo, which may come from inside or outside the organisation, as well as from our own ingrained habits and behavioural patterns. A key component to lasting change is that it is carried by its stakeholders and becomes an ingrained part of the culture (McHale, 2020). Apart from changes to our habits, one way to initiate change would be to work with other members of the department who are interested and keen. Incremental changes by a small group may not immediately show large-scale results but can build the momentum and critical mass that leads to an overall transformation of the programme's culture.

While obstacles are unavoidable, it is also worth keeping the broader support and allies in mind. At least since *Architects Declare*, sustainable architecture is no longer a minority movement, but the direction of future architectural design and practice. Any architecture programmes making changes now are likely to be on the right side of history, and those that are not, on the wrong side. The local architects' professional body, practices, and associations, for example, may be potential partners and already strong advocates. An example is the introduction in 2019 of the Leadership in Sustainability Prize category to the annual Architecture Awards in Australia. Discussions are now underway to make sustainability a more integral part of the selection criteria for all awards, and the award of the RIBA Royal Gold Medal in 2023 to Yasmeen Lari for her low-carbon architecture in Pakistan is a welcome change in this regard (Ravenscoft, 2023). The architecture programmes that transform their own footprint and teaching are part of this broader industry change and send the right message.

Beyond environmental sustainability

The UN Sustainable Development Goals and Copenhagen Declaration emphasise that sustainability is not only environmental, as often presented in architectural education. Sustainability of our built environment can only be achieved when social values, economic growth and equity, including for disadvantaged communities, are addressed at the same time (Williamson, 2003). While architects are typically not in control of all these factors, our design decisions have implications for and influence our environmental, social and economic future. A holistic perspective on sustainability in the built environment should therefore be taken as a starting point for a department to teach sustainability across the entire architectural curriculum. Such an interdisciplinary perspective can help with changing the focus from merely creating more efficient solutions within the same growth model that has led to the climate emergency, to instead approach development more holistically as part of its broader ecology (Shove, 2017).

FIGURE 8.6 'North Adelaide Complex' project by Zihe Chen for 'Studio Architecture', a core Masters design studio at the University of Adelaide.

For example, a design studio can be integrated with a course on architectural technologies that runs in parallel and students consider aspects such as building structure, energy use, thermal performance, the lifecycle of materials and potential for adaptive reuse to inform design decisions and to aim for an overall net zero or net positive development.

Courses in humanities subjects such as history and theory, for example, while normally part of architectural curricula, are not necessarily taught and discussed in the context of sustainability. Much of our current understanding of architecture seems based on paradigms that place it in opposition to its environment rather than considering how it can enhance and contribute to it (Webster, 2011). To this end, Calder and Bremner argue that the climate emergency and a building energy-use perspective fundamentally change our view of architectural history, which in turn also shapes expectations for our future built environment (Calder and Bremner, 2021). How, after all, can we design the future if we do not know how we got here? Since the emergence of the environmental humanities over a decade ago, there have been growing efforts to make carbon more central to historical narratives and this research could inform and guide lecture content in this regard (Hochhäusl et al., 2018).

Another example is structural design courses. With 60% of a building's embodied carbon emanating from its structure, these modules also need to embed climate action within their programmes (Arnold et al., 2020). Sustainability can no longer be seen as important only for technical courses nor can it be perceived as solely about environmental sustainability and building energy use, which only addresses part of the issue. In this way, all staff on an architecture programme would be having the same conversation, but from different perspectives. As Nisonen and Pelsmakers argue, sustainability should be 'the starting point, the process and the outcome of all our actions' (Nisonen & Pelsmakers, 2022). A thorough and holistic examination of teaching practices and content throughout the curriculum is needed to make sustainability the starting point. It means questioning how we think about architecture, our lifestyles, influences and roles, as well as its production and materials.

Conclusion

For architectural education to address the climate emergency, one of the hardest things is not only to change how and what we design but to also change ourselves and our industry. To go beyond solutions and fixes for others, our behaviour, habits and professional choices need to be part of that change. So ingrained is our path dependency, that the task looks nearly impossible but also seems to be the only way to create the change we seek to be part of.

Future schools of architecture surely need to be zero carbon themselves, like the buildings we want to design. That begs the question of what such a school of architecture that walks the talk could look like.

An example of architectural education underpinned by sustainability in its teaching and practices is the Centre for Alternative Technology (CAT) in Wales (CAT, 2023). Since its beginnings in the 1970s, CAT has strived to create architecture that achieves a lower carbon footprint. The curriculum of its M.Arch Sustainable Architecture seems to be led throughout by this consideration in the design, technical, social, professional as well as history and theory course content. The architecture of the school itself is also driven by these zero-carbon ambitions and students take a hands-on approach to the creation of more sustainable architecture. Even if we work in institutions that do not currently function in that way, we can take steps in the right direction by rethinking the approach to our teaching and practices from the ground up to make sustainability a starting point and core component. We hope that the considerations and the resources mentioned here provide food for thought and action. They have for us.

KEY TAKEAWAYS FROM CHAPTER 8

- Addressing the climate emergency requires a radical cultural shift in architecture, moving beyond aesthetics and technology reliance to embrace a more collaborative, democratic approach.
- The discussion of sustainability in architectural education is often limited to specific courses. There is an urgent need for it to become a central component in architectural education, including core design courses as well.
- Architecture programmes should embody sustainability principles in their operations, aligning the message with actions. Simple changes, such as reducing model-making waste or choosing eco-friendly materials, make the sustainability message tangible and credible for students.
- Architects operate within a multi-stakeholder environment, and curriculum reforms should reflect this by including collaborations with industry, government and communities.

Bibliography

ACAN. 'Education Toolkit and Educators Workshops'. Accessed 4 June 2023.

Architects Declare, nd. *UK Architects Declare Climate and Biodiversity Emergency.* Accessed 3 November 2024. www.architectsdeclare.com/about

Architects Registration Board. 'Modernising Architectural Education and Training'. Accessed 11 January 2022. https://arb.org.uk/modernising-architectural-education-and- training/

Arnold, Will, Mike Cook, Duncan Cox, Orlando Gibbons, and John Orr. 'Setting Carbon Targets: An Introduction to the Proposed SCORS Rating Scheme'. *The Structural Engineer* 98, no. 10 (1 October 2020): 8–12. https://doi.org/10.56330/SQDI8782

Boarin, Paola, Antonio Martinez-Molina, and Ignacio Juan-Ferruses. 'Understanding Students' Perception of Sustainability in Architecture Education: A Comparison among Universities in Three Different Continents'. *Journal of Cleaner Production*, 248 (1 March 2020): 119237. https://doi.org/10.1016/j.jclepro.2019.119237.

Building Change, nd. *Building Change: Designing a Resilient Future through Architectural Education.* Accessed 24 January 2025. https://buildingchange.ie/

Calder, Barnabas, and G. A. Bremner. 'Buildings and Energy: Architectural History in the Climate Emergency'. *The Journal of Architecture* 26, no. 2 (2021): 79–115.

Monash University, 2024. 'Campus Sustainability', accessed 3 November 2024, www.monash.edu/campus-sustainability

Centre for Alternative Technology. 'Centre for Alternative Technology'. Accessed 4 June 2023. https://cat.org.uk/.

Deamer, Peggy. 'Design pedagogy: The new architectural studio and its consequences', *Architecture_MPS* 18 (1 September 2020). https://doi.org/10.14324/111.444.amps.2020v18i1.002.

Hocknell, C., 2023 'The Hollowness of Architects Declare Should Serve as a Warning to the Industry', *Dezeen*, 26 April. www.dezeen.com/2023/04/26/architects-declare- sustainability-chris-hocknell-opinion/

Donovan, Elizabeth & Sofie Pelsmakers. 'Integrating sustainability in design studio through blended learning: education, design and practice'. Edited by Ellyn Lester.

Education, Design and Practice, AMPS Proceedings Series, 2019, 211–220.

Geels, Frank W. 'The multi-level perspective on sustainability transitions: Responses to seven criticisms'. *Environmental Innovation and Societal Transitions* 1, no. 1 (2011): 24–40.

Grant, Elizabeth J. 'Mainstreaming Environmental Education for Architects: The Need for Basic Literacies' 1, no. 1 (4 September 2020): 538–349. https://doi.org/10.5334/bc.41.

Gunderson, Ryan & Sun-Jin Yun. 'South Korean Green Growth and the Jevons Paradox: An assessment with democratic and degrowth policy recommendations'. *Journal of Cleaner Production* 144 (15 February 2017): 239–247. https://doi.org/10.1016/j.jclepro.2017.01.006.

Hochhäusl, Sophie, Torsten Lange, Ross Exo Adams, Daniel A. Barber, Aleksandr Bierig, Kenny Cupers & Isabelle Doucet, et al. 'Architecture and the Environment'. *Architectural Histories* 6, no. 1 (14 December 2018). https://doi.org/10.5334/ah.259.

International Union of Architects. 'Architecture Guide to the UN 17 Sustainable Development Goals'. Accessed 2 June 2023. www.uia- architectes.org/en/resource/architecture-guide-to-the-un-17-sustainable-development-goals- english/.

International Union of Architects. 'Copenhagen Declaration', 7 December 2009.

www.nzia.co.nz/media/1054/uia-copenhagen-declaration.pdf.

Iyer-Raniga, Usha & Tony Dalton. 'Challenges in aligning the architecture profession in indonesia for climate change and sustainability'. *Procedia Engineering*, International High- Performance Built Environment Conference – A Sustainable Built Environment Conference 2016 Series (SBE16), iHBE 2016, 180 (1 January 2017): 1733–1743. https://doi.org/10.1016/j.proeng.2017.04.336.

Killip, Gavin. 'A Reform Agenda for UK construction education and practice'. *Buildings and Cities* 1, no. 1 (21 August 2020): 525–537. https://doi.org/10.5334/bc.43.

Markard, Jochen, Rob Raven & Bernhard Truffer. 'Sustainability transitions: An emerging field of research and its prospects'. *Research Policy*, Special Section on Sustainability Transitions, 41, no. 6 (1 July 2012): 955–967. https://doi.org/10.1016/j.respol.2012.02.013.

Marwell, Gerald & Pamela Oliver. *The Critical Mass in Collective Action*. Cambridge University Press, 1993.

McHale, Siobhan. *The Insider's Guide to Culture Change: Creating a Workplace ThatDelivers, Grows, and Adapts*. HarperCollins Leadership, 2020.

Murray, Paul E. & Alison J. Cotgrave. 'Sustainability literacy: The future paradigm for construction education?' *Structural Survey*, 2007.

Nisonen, Essi & Sofie Pelsmakers. 'Architectural Education in the Climate Emergency'. *Architectural Review*, 15 September 2022. www.architectural- review.com/essays/architectural-education-in-the-climate-emergency.

Oliveira, Sonja, Elena Marco& Bill Gething. 'Towards an energy "literate" architecture graduate? UK educators" and students' evaluation'. *Architectural Engineering and Design Management* 14, no. 4 (2018): 317–329.

Ravenscroft, T., 2023. 'Yasmeen Lari Wins 2023 RIBA Royal Gold Medal', *Dezeen*, 26 April 2023. www.dezeen.com/2023/04/27/yasmeen-lari-riba-royal-gold-medal-2023/

Shove, Elizabeth. 'What Is Wrong with Energy Efficiency?' *Building Research & Information*, 29 August 2017. www.tandfonline.com/doi/abs/10.1080/09613218.2017.1361746.

Shove, Elizabeth & Gordon Walker. 'Caution! Transitions ahead: Politics, practice, and sustainable transition management'. *Environment and Planning A: Economy and Space* 39, no. 4 (1 April 2007): 763–770. https://doi.org/10.1068/a39310.

Swain, Ranjula Bali. 'A critical analysis of the sustainable development goals'. *Handbook of Sustainability Science and Research*, 2018, 341–355.

Taylor, Carolyn. *Walking the Talk: Building a Culture for Success*. Random House Business Books, 2005.

Till, J. 'Architecture criticism against the climate clock'. *The Architectural Review*, (April 2023), pp. 6–10.

United Nations. 'THE 17 GOALS | Sustainable Development'. 2024. Accessed 3 November 2024. https://sdgs.un.org/goals.

Webster, Helena. *Bourdieu for Architects*. Routledge, 2011.

Williamson, Terry, Antony Radford & Helen Bennetts. *Understanding sustainable architecture*, Spon Press, 2003.

9

PREPARING ARCHITECTS TO SERVE AS CHANGE AGENTS

Jan Hugo, Anika van Aswegen, Carin Combrinck, Dayle Shand and Chrisna du Plessis

Introduction

Globally the climate change crisis is deepening, with the increased occurrence of environmental and social emergencies affecting multiple communities. The World Bank (2021) reports that $215 billion will be needed to adapt South African cities to current and expected climate change impacts. While this highlights the demand for response measures, it also emphasises the extent of resources needed to undertake these measures. However, the rapidly changing nature of sub-Saharan cities means that how these actions translate and manifest in our cities is not a simple matter. Sub-Saharan cities are characterised by the proliferation of poorly serviced auto-constructed (slum) developments (UN-Habitat, 2014). These urban conditions, largely due to rural–urban migration, will only worsen as a result of climate change-induced migration, while climate change-driven impacts will exacerbate the already vulnerable conditions in these cities. This necessitates flexible development processes and spatial fluidity that engages these rapid emergent processes in our growing cities. Ultimately these trends call for alternative developmental solutions that challenge the current status quo in our built environment (Combrinck & Venter, 2020). Aligning contextual relevance with processes that foster climate awareness is therefore critical to enable appropriate climate action.

While being inundated by the knowledge of the crisis and its effects, the spatial design professions often spurn responsibility to act, thereby limiting the capacity of those entering the professions to develop and participate in climate change solutions (Stupar et al., 2017). If Southern African built environment professionals are to address the climate crisis in these complex informal contexts, embracing epistemic diversity and transdisciplinary processes is critical, as the solutions should address basic developmental needs, undertake appropriate climate change adaptation strategies, and contribute to the global carbon mitigation agenda.

While education plays an important role in achieving adequate climate awareness, Hornsey et al. (2016) argue that identity is an important factor affecting climate awareness and action. Effectively educating climate awareness requires holistic approaches as

DOI: 10.4324/9781003293903-11
This chapter has been made available under a CC-BY-NC-ND 4.0 license.

it is not only dependent on disseminating climate knowledge, but also requires instilling a moral responsibility towards climate change action (Killip, 2020). This calls for the education of architects who can address the climate emergency by assimilating not only technical knowledge, but also collaborative approaches that instil cognitive diversity, transactive memory, open-mindedness, and alternative practices beyond utilitarian capabilities (Leal Filho et al., 2021; Srivasta, 2020; Stevenson & Kwok, 2020). Thus, fostering individuals who are change agents requires transformational education pathways, which translates into adjusting learning content, learning outcomes and our notion of success within the pedagogic environment, finally placing emphasis on societal transformation (UNESCO, 2020).

Transformational learning is key to leverage the impact that young practitioners can have in the built environment. Amongst others it requires a shift in how individual students understand their own roles and identities and how they relate with the "other" and the world (O'Sullivan, 2002). It "involves experiencing a deep, structural shift in the basic premises of thought, feelings, and actions" (O'Sullivan, 2002, 11).

As contribution to the education for climate awareness discourse, we present an educational framework used by the University of Pretoria's (UP) Architecture Department, which as an educational ecosystem promotes changing the cognitive, social, emotional and behavioural dimensions of learning, as proposed by the Education for Sustainable Development (ESD) Roadmap (UNESCO, 2020). A curriculum structured along intentional stages of learning punctuated by disruptive dilemmas (Mezirow, 1997), and cognitive threshold concepts (Meyer & Land, 2003), inculcates transformational learning pathways to turn climate awareness, knowledge, and design values into actionable responses. It engenders a design practice that critically questions our role as designers in the built environment and engages with the climate crisis through structured transformational processes. Our modalities of teaching and methods of learning are considered critical to leverage current built environment practice as a catalyst of change in our cities and promote climate action (Grant, 2020; Mataya et al., 2020; Nyka, 2019; Stupar et al., 2017).

In this chapter we deliberate on the current educational ecosystem and transformational learning embedded within the curriculum and how this change has taken place in the UP Architecture Department. Key pedagogical techniques and modalities informing the programme structure are discussed, and finally, the chapter reflects on the outcomes from this pedagogical procedure and how the transition to the built environment industry becomes the final transformational process.

Transformational learning

The UNESCO ESD Roadmap (2020:18) calls for 'big transformation', both as individual action and reorganising societal structures for a sustainable future. This is considered as the foundation for the transformative learning approach advocated at the UP Department of Architecture. Transformative learning theory is understood as "the process of effecting change in a frame of reference", and this is informed by one's assumptions and influences the resultant actions (Mezirow, 1997: 5). Perspectives and preconceived ideas can restrict one's understanding of the world and ability to identify complex issues embedded in design challenges. However, by reflecting on our own biases, we may develop an awareness of other perspectives, including that of other people, species, or environmental entities such as our Savanna biomes.

Transformative learning's two dimensions, habits of mind (fixed "set of codes"), and points of view (flexible value judgements and attitudes) (Mezirow, 1997: 5–6), explain how students experience and process a 'disorienting dilemma' (Mezirow, 1978) or unfamiliar situation. When a new experience presents itself, unless they can shift their perspective to expand their understanding, their default position normally reverts to a place of comfort that is tried and tested.

Traditional learning theory focuses on cognitive and ethical considerations of student development as sequential stages or positions of growth (Perry, 1970). Other views include development as a series of perspectives towards 'ways of knowing' (Belenky et al., 1986). A deeper investigation into 'procedural knowledge' reveals the relevance of 'separate knowing' and 'connected knowing' (Belenky et al., 1986:100) in the context of sustainability thinking. Separate knowing relates to the traditional view of the expert (architect as analytical thinker that evaluates and reasons), compared to connected knowing's empathetic and subjective engagement to gain deeper insight (Belenky et al., 1986: 104–105, 112–113). Connected practices allow designers to move beyond their own biases and foster a 'social and emotional learning dimension', highlighting empathy and compassion to balance cognitive learning approaches (UNESCO, 2020: 17). A balance between the two perspectives is needed to foster a holistic approach that addresses the climate change challenges affecting communities.

We argue that if conventional architectural practices include an empathetic approach, a renewed awareness of people and the environment becomes entwined in understanding of architecture students. This capacity to shift perspectives does not come easily, as students are confronted with disruptive moments as part of their learning, which are difficult to navigate and negotiate. In such instances, threshold concepts, i.e. those concepts that once understood, change the way a person thinks about a topic (Meyer & Land, 2003), are experienced as 'conceptual gateways' or a 'liminal space' of development (Meyer & Land, 2005: 373, 375). Threshold concepts are seen as transformative, irreversible, integrative, troublesome or bounded (Meyer & Land, 2005: 373–374) according to the level of study and the nature of the disruption.

These unsettling experiences provide the opportunity for self-reflection and meaning-making of the content and process leading to this moment. Mezirow (1997) refers to two types of critical reflection of and on assumptions: objective reframing (interactions with others and actions taken by the student) and subjective reframing (own assumptions). Critical reflective practices furthermore consider process and understanding to include 'reflection in action' by focusing on an activity, 'reflection on action' as a review of a past event (Schön, 1983) and 'reflection for action' as projecting the impact of activities into the future (Thompson & Thompson, 2008:16).

The learning environment of architecture students provides opportunities for layered and nuanced design investigations. Project engagement requires them to explore a variety of ways that enable the surfacing of issues that are not always architectural.

In such instances, the problem arises of how or whether they should entertain that in their projects. This emphasises the reality that climate change is a multi-faceted problem that requires a holistic approach to future change (Leal Filho et al., 2021). Triple-loop learning offers the context for decision making, by focusing on 'how do we decide what is right? (Medema et al., 2014: 27), versus what or why interventions are needed. This reaches beyond the understanding of the phenomena and highlights the assumptions and contextual understanding of the larger issues involved in the design project. Berke (in Medema et al., 2014: 27) outlines the different levels of "multi-loop social learning processes": macro (governance and societal structural conditions), meso (actor networks) and micro (participating stakeholders). The larger environmental context in which architectural projects are situated, and the heterogeneous impact of climate change affecting the biophysical, socio-cultural and socio-ecological of our cities calls for a multi-scalar investigation across these levels to make sense of the complexity.

The current educational model of our department has an ecosystemic, multi-scalar focus and embraces transformative practices across different years of study. Disruptive moments are specifically determined to ignite new thinking and to allow student development on an individual, but also collective level.

Educational Ecosystem

While the notion of a transformation journey implies a start and finish, it is important to acknowledge that personal transformative growth is not necessarily always similar nor linear. Rather, it is a life-long iterative and continuous process. Didau (n.d.) proposes a non-linear/non-hierarchical pedagogical taxonomy that can be related to an 'interactive dialogic approach' in architectural education (Salama, 2015, 6). While Table 9.1 unpacks the main themes with disruptive dilemmas, threshold concepts and the evolving position of the student as change agent in a linear format, Figure 9.1 is a closer representation (being non-linear/non-hierarchical) of the implementation and pedagogic practice of the department's educational ecosystem as an integrated learning process with a widening focus of responsibility.

The various threshold concepts are introduced following a linear format in Table 9.1, but are often reused and referred back to in subsequent modules, with knowledge deepening as it builds on the learning of previous years (Figure 9.1).

Broadly, the undergraduate and postgraduate programmes have different foci. The undergraduate programme focuses on the student as an individual functioning in, and contributing to, a larger community and system, introducing main themes to highlight i) *the individual as not separate from the world*, ii) *us learning from the world*, and finally, iii) *cross-disciplinary collaboration creating alternative visions for this world*. The Postgraduate programmes focus on the iv) *individual as professional* and v) *change agent*. Table 9.1 further indicates the expansion of the student's locus of influence and role as change agents across the five years of study, considering and expanding individual ethical positions as they progress with their studies. The emphasis shifts from the individual to including others and finally, confronting the identified design problem as situated within both the individual's ethical values within the department's ethos, as well as the ambit

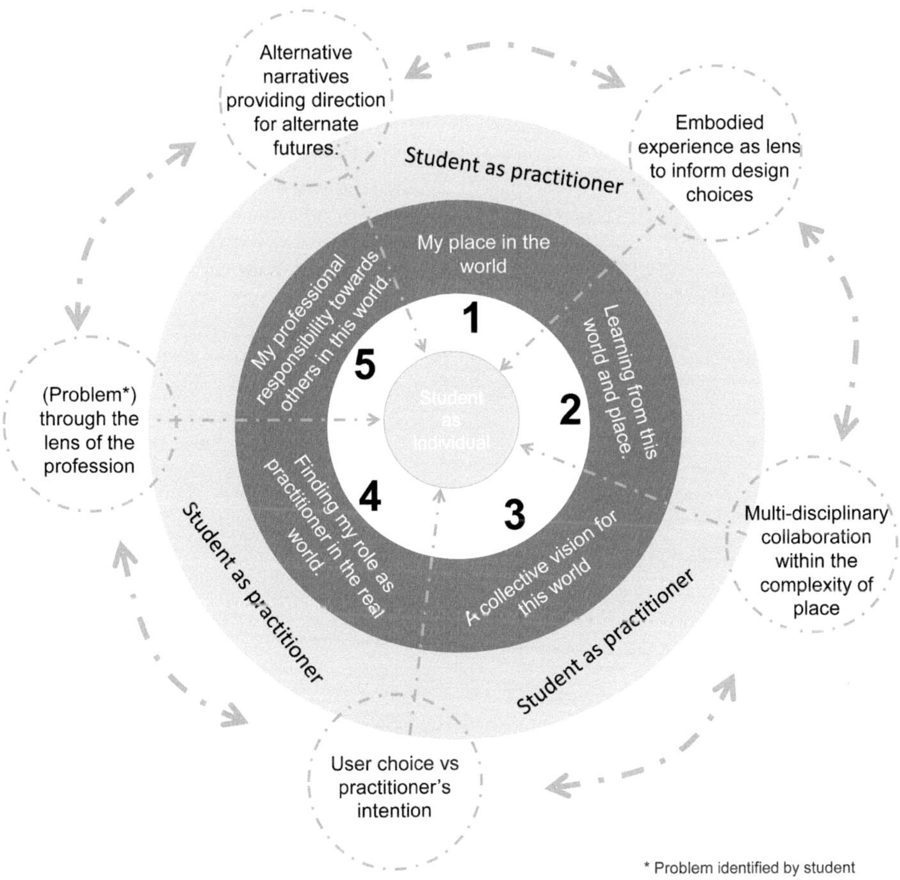

FIGURE 9.1 The education ecosystem as an integrated learning process.

of the architectural profession and its mandate. Building on the embedded community engagement model developed in the department (Combrinck & Venter, 2020), the education model progresses from limited engagement (starting with awareness of end users and philanthropic actions) towards embedded collaboration to develop user-centric design strategies in the students' final year. At this point the students are equipped to engage as professionals and change agents in community settings.

Students are introduced in the first year to the values and principles of the ecological worldview that underpins the department's ethos. This worldview is reinforced in subsequent years, both in the theory modules and design studios.

Transitioning themes of climate awareness education

The five phases and their implementation outlined in Table 9.1 are now discussed in more detail by focusing on the particularities of the disruptive dilemmas embedded within each theme as experienced per year.

TABLE 9.1 Detailed unpacking of the educational ecosystem (Positionality based on Combrinck & Venter, 2020).

	Year	*Themes/ Disruptive dilemmas*	*Threshold concepts*	*Positionality as change agent*
Student as individual	1	My place in the world / Alternative narratives providing direction for alternate futures.	• World views and values as boundaries for behaviour and responses to change. • Living in nested systems of increasing complexity. • Being part of the web of life. • Inter-being and holism. • Regenerative thinking	Towards awareness. • Realising my impact on and responsibility for other people and the environment
	2	Learning from this world and place / Embodied experience as a lens to inform design choices.	• Introducing and applying ecosystemic knowledge and skills in the built environment. • Building capability for climate action. • Long-term planning and life cycle thinking.	Towards philanthropy • Exploring generosity through design for communities and the environment.
	3	A collective vision for this world / Multi-disciplinary collaboration within the complexity of place.	• Collaboration with other spatial design disciplines towards self-efficacy. • Engaging with society and the system for appropriate adaptation. • Articulating future visions for alternative urban environments. • Critically engaging with monitoring and evaluating design proposals.	Towards activism • Supporting action for positive change through design decisions for people and the environment
Student as practitioner and change agent	4	Finding my role as practitioner in the real world / *User choice versus practitioner's intention.*	• Confronting current and future climate change impacts by participating in local communities and projects. • Exploring the alternative roles of the profession to find deep-seated solutions. • Developing appropriate adaptation solutions by including user perspectives or using normative dissociation. • Using empathy building and relational understanding for user-centred adaption. • Critical reflective processes to review application successes.	Towards solidarity • Understanding through empathy and reciprocity between people, communities and the environment.

(Continued)

TABLE 9.1 (Continued)

Year	*Themes/ Disruptive dilemmas*	*Threshold concepts*	*Positionality as change agent*
5	My professional responsibility towards others in this world / The (crisis / theory / context*) through the lens of the profession. *Student-led choice	• Negotiating real-world problems and solutions through the practitioner's ethical values. • Confronting the changing roles of the profession within a changing world and climate. • Exploring alternative processes towards new solutions that address the climate crisis. • Conflict between the conventional professional outcomes and the needs contextual deep-seated climate change adaptation solutions present.	Towards collaboration • Participating with communities to establish partnerships through mutual relationships.

Theme 1: My place in the world

Before any technical knowledge and principles are instilled, the programme aims to open designers cognitively and emotionally to transformative processes. Drawing on approaches such as Appreciative Enquiry (Bushe, 2013) the purpose of the programme is to build the cognitive capacity, emotional resilience and habits of mind that would enable students to effectively engage with the complexity and uncertainty of life in the twenty-first century, including climate change challenges. The process starts by positioning the climate (and broader ecological) crisis as a shared crisis that requires collective responsibility. This is used as a disorienting dilemma (Mezirow, 1978), with a series of short tasks and critical self-reflections forcing students to examine both their own vulnerability and their complicity in the crisis and how this relates to their own assumptions, worldviews and value systems. There is a deliberate focus on personal values and worldviews that establishes the premise for the long-term transformational process which continues beyond the formal education period.

Students are deliberately made uncomfortable through discussions, lectures and tasks which challenge them from three directions: a) the existential threat posed by the Anthropocene; b) interrogating a worldview based on separation, duality, reduction, certainty and control by introducing different perspectives drawn from indigenous knowledge systems, traditions such as Buddhism and Hinduism, and the sciences of quantum physics and ecology; and c) the conflicting and competing perspectives on sustainability and how it should be achieved. Following Alrøe and Kristensen (2003, 69), this process gradually encourages students to question their assumptions about their place in the world by extending the individual self through identification with other individual selves (e.g. through sympathy, empathy, or disinterested benevolence), thus cultivating a feeling of

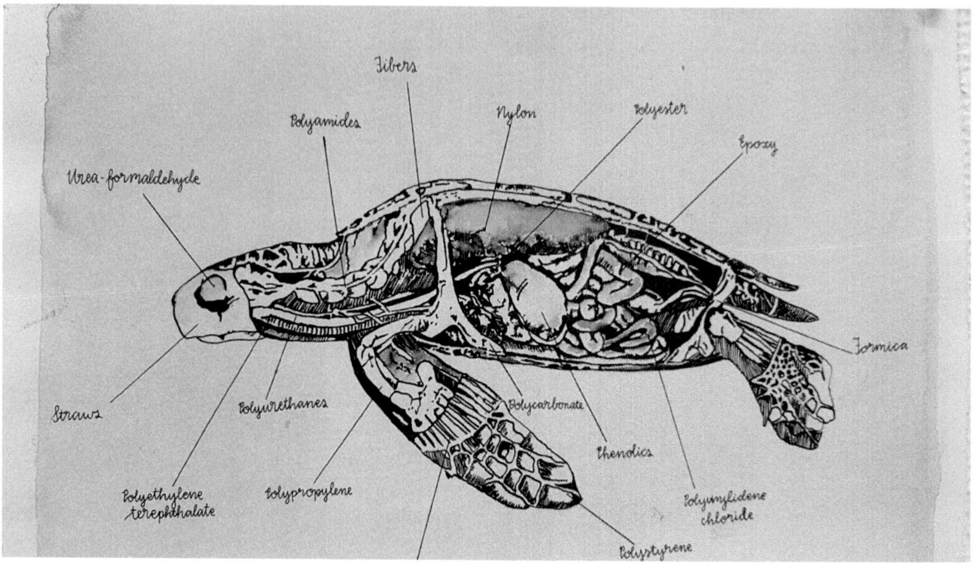

FIGURE 9.2 A biological study of contamination.

Source: Robyn Wolmarans, 2019).

responsibility, and then a systemic extension that expands the boundaries of the self so that "there is no sharp boundary between self and non-self" (i.e. transpersonal awareness). This culminates in a cathartic ritual or artwork (Figure 9.2) in which students can express feelings of fear, grief and anger, before they are led into the next phase.

The second phase of the programme follows the Appreciative Inquiry stages of Discover, Dream and Design (Bushe, 2013), during which students explore the world through their 'new eyes' to see opportunities for intervention. They are introduced to concepts such as systems thinking, social-ecological systems, biomimicry, biophilia and regenerative design thinking, and invited to first envision a new story for their world and then re-design their own immediate environments to match this vision.

An important difference in this phase is that it does not focus on sustainability, but rather on regenerative design practices and transformative resilience. This prompts the students to take responsibility for climate action in envisioning the crisis as an opportunity to develop transformative solutions for diverse ills (Du Plessis, 2022; Folke et al., 2010).

Theme 2: Learning from this world and place

O'Brien (2018) calls for climate transformation that addresses technological, political and personal spheres of our society. As a continuation of O'Brien's (2018) model, the second stage covers critical technological, material and spatial theories enabling students to address the first sphere (technology) of climate transformation. During this second stage, the educational programme introduces and unpacks the knowledge and skills related to sustainable design principles and climate change theory, to ultimately inform the students' design decision-making framework. Within the ESD Roadmap (UNESCO, 2020) the learning outcomes and content in this stage are fundamental to climate awareness and action. The programme is structured along personal experiences, prompting learning through empirical observations, as well as introducing knowledge and strategies based on externalised theory and principles.

The first phase of the programme takes a macro-view of the inhabitant's wellbeing in the built environment, addressing alternative collective narratives of place and the importance of social-ecological systems, and principles to improve the built environment performance and the larger system within which it functions, as defined by De Kay et al. (2018). Concomitant to developing individual and collective transformation by addressing the climate crisis, adjusting the behavioural (performance characteristics) with a clear systematic understanding of the context is critical within the programme discourse.

Consequently, the second phase addresses several practical and theoretical aspects related to the individual wellbeing within the built environment, promoting integrated holistic design processes and engendering positive user experiences. During this phase, the knowledge conveyed is purposefully contextualised in the student's own environment (Figure 9.3). Using their immediate environment as context for inquiry presents opportunities for students to reflect on the conditions and understand the application of the various response measures from empirical and experiential perspectives.

Learning from one's immediate environment, starting with the individuals' own residence, is a vital teaching modality we use. This approach shifts the pedagogical environment (UNESCO, 2020), laying the foundation for learning and adaptation as a life-long process.

Theme 3: A collective vision for this world

As students transition into the third stage of the transformative learning process, they are required to work collaboratively with their peers towards addressing urban and

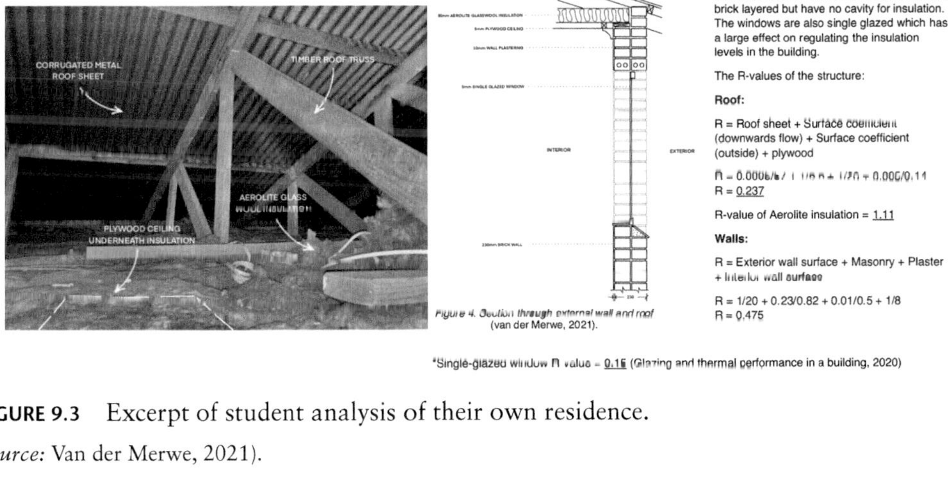

FIGURE 9.3 Excerpt of student analysis of their own residence.

Source: Van der Merwe, 2021).

climate-related crises within their local context as part of a nested social-ecological system. This third stage of the process is both reflective and an exercise in projecting their ethical values as spatial designers (reflective on action and reflective for action) (Schön, 1983; Thompson & Thompson, 2008). Students are encouraged to move beyond a human-centred approach to a life-centred approach in which people are in service of nature and the community, and design interventions strive to build resilience, heal fractured systems, and regenerate webs of connection and reciprocity.

In keeping with the ESD Roadmap, students are engaged through "interactive, project-based, learner centred pedagogy" (UNESCO, 2020, 8). Through an embedded approach, students consider the lessons from the two previous stages, as well as specialist knowledge delivered in seminar format by experienced sustainable built environment practitioners, which they collaboratively test in studio-based charrettes (Figure 9.4). As experiential learning to promote sustainability, this phase engenders collaborative learning (Kricsfalusy et al., 2018, 594). The project teams typically consist of students in the different architectural disciplines (i.e., architecture, interior architecture and landscape architecture), working collaboratively and across their disciplines to contribute to a multi-scalar approach according to their particular expertise and focus.

While the scenarios are hypothetical, the project contexts are real and accessible to the students. The project problems are therefore representative of the South African urban

FIGURE 9.4 Collaborative exploration with practitioners, academics and students.
Source: Van Aswegen, 2019.

condition. They require students to develop project informants generated from an understanding of 'place', and to solve the problems identified within the given context. Students are expected to identify and delineate problems contributing to the climate crisis, envision solutions to promote resilient and sustainable urban environments, and develop criteria and strategies to achieve their vision. They are guided, through studio-based workshops, on how to apply ecosystemic thinking and design principles to built environment projects. As part of the process, students are required to reflect on and measure their projects against their own criteria. In keeping with stages one and two, the focus is less on checklist-design processes, and more on systems-orientated, ecosystemic design principles which promote urban (and climate) resilience and cohesive urban environments supporting social-ecological well-being.

Working as teams, students are required to develop design visions and collective philosophical approaches. This challenges them to consider alternative positionalities, or frameworks, and to develop and collaboratively iterate a unique position to guide their own design praxis and inform the architectural product.

Students work collectively and collaboratively towards articulating alternative urban environments and scenarios (Figure 9.5). In addition, students are required to apply ecosystemic principles and test their own products against a collective vision. These processes of collaborative design equip students for their entry into the built environment industry.

FIGURE 9.5 Collective vision towards a sustainable urban environment.

Source: AAL 320 group project, 2019.

Theme 4: Finding my role as practitioner in the real world

Students are accepted into the fourth stage of the curriculum after a year of practical work experience, travel and reflection. The honours year is equivalent to a first year in a two-year Master's programme. A selection process provides candidates the opportunity to articulate their ethical values and worldviews around design issues and to indicate a personal motivation for re-entering the academic environment. The ability to exercise choice is encouraged throughout this stage, where a selection of studios is offered every quarter under the guidance of academic staff and research partners. Studio choices support interdisciplinary engagement and expose students to other processes and methods across various spatial design projects.

Studios are closely related to ongoing research in the department, with students participating in complex and current questions such as those related to climate change. Students are able to integrate and apply learning from the research into their design projects, which bring deeper engagement and connected learning to inform their ethical value transformation. Opportunities are created for students to engage with local communities directly affected by the climate emergency, rapid urbanisation, the challenges of inadequate housing, service delivery failure and food insecurity (Figure 9.6). Through experiencing these contexts first-hand, heightened levels of empathy result in more meaningful engagement with the complex layers of understanding needed to arrive at deep-seated solutions (Leal Filho et al., 2021; Stevenson & Kwok, 2020).

During this stage the projects aim to develop collaboration skills, and group work and process is favoured over a competitive product-oriented approach. This enables the student to be confronted with complex emergent problems at a higher level of complexity without becoming overwhelmed. Alternative approaches to design and practice are explored, with responsive conclusions being encouraged over predetermined or prescriptive outcomes.

The search for appropriate responses is further encouraged by shifting perspectives to that of the 'other' by way of normative dissociation (Butler, 2006; Panero et al., 2019). As a voluntary process of dissociation, students deliberately detach from their own preconceptions to focus on another's reality and experiences to gain understanding and insight of unfamiliar situations. Relational understanding through the development of empathy encourages students to assimilate diverse perspectives rather than becoming entrenched in their own biases. The focus at this point is on the interface between the user and the proposed climate change response measures, with an emphasis on appropriate response strategies and technologies.

In the last quarter of this stage, critical reflection workshops and iterative design explorations are used to facilitate triple-loop learning to focus on the ethical values of assumptions in design projects (Roux et al., 2008:15). During this final phase students are required to work independently rather than in groups, so that there is an opportunity to develop their unique research and design commitment. This prepares the students for the final, more independent, stage of the transformative education journey.

FIGURE 9.6 Engaging with participants to understand and learn from them.

Source: Carin Combrinck, 2023.

Theme 5: My professional responsibility towards others in this world: change agent

During the final stage, year five of their studies (Professional Masters), students are required to develop a project brief and inquiry based on their personal ethical values as developed during their studies. These inquiries often reflect contextual problems, consider the technological application of certain solutions, or use specific theoretical lenses. Although these projects are rarely explicitly articulated as climate change adaptation projects, they deal with the sensitivities, exposures and often adaptive responses of specific communities or sites to adverse climate change impacts (Simpson et al., 2021). This sensitivity to contextual, user-centred approaches to the 'place' and its users often results in students being aware of the latent and far-reaching impacts of climate change on local communities (IPCC, 2021). As the department's ethos prompts students to undertake an ecosystemic approach, questioning impact drivers, technological applications, and the long-term success of their solutions, we often find students moving beyond the mandate of the architectural profession. Their solutions are often inter-disciplinary, using appropriate technologies that straddle infrastructural solutions, spatial resolutions and user-participation processes, which are integrated within the spatial design disciplines of architecture, interior architecture and landscape architecture. Consequently, the student work often calls into question the limits of practice: highlighting the need for deep-seated systemic change in our societies, economies and spatial design professions, as well as the limited influence of current architectural practice. This awareness is most acutely highlighted when the students work in informal, emergent communities in South African cities.

The emphasis on contextually appropriate solutions forces students to grapple directly with the local heterogeneous urban context. In the Southern African region, formal urban conditions that conform to conventional regulated practice coexist with diverse informal self-built environments. These urban conditions are typically unregulated and adhere to few standards, yet accurately manifest and conform to the user needs (Okyere & Kita, 2015). Often these spaces are not clearly formal or informal but oscillate along these conditions in complex ways (Okyere & Kita, 2015). While these informal conditions are typically considered marginal, they often present the primary urbanisation processes in the Southern African region (Levy, 2021; Pieterse 2011) and comprise at least 14% of South African cities (SERI, 2018). Students will now, and in the future, be confronted with highly regulated and formal conditions where practice and the responses are predictable (typically an exclusive portion of our society), while simultaneously working in contexts that are informal and unregulated, exhibiting significant developmental and wellbeing needs (Figure 9.7).

This contextual complexity confronts designers with climate action within the formal built environment, focusing predominantly on the material and technological climate change adaptation strategies, whilst concurrently practising in informal contexts where the appropriate climate change resilience responses need to address climate change adaptation, mitigation and developmental needs (Roberts, 2010). These complex informal contexts force us to shift our responses, questioning the nature of appropriate design solutions and prompting designers to consider actions which often extend beyond the mandate of the architectural profession. Whilst the solutions still

FIGURE 9.7 Students participating in projects undertaken in informal contexts.

Source: Eyescape Photography, 2021.

require design thinking, the application thereof is often more process-oriented, with the architect performing a facilitation role (Combrinck, 2021; Combrinck & Venter, 2020). However, although informal or less affluent segments of our society are the most vulnerable, conventional architectural practices typically have limited opportunities to function in such contexts.

It is therefore during this stage that students are confronted by the tensions between the product-oriented outcomes currently valued by the built environment industry, and the need for alternative process-oriented solutions (Combrinck, 2021). For our students to truly become agents of change, their transformational journey needs to transfer successfully into the world of work. This is the biggest challenge, for the industry itself is slow to transform, and tends to embrace popular sustainability concepts, without attempting to understand the transformation in thinking that is truly required (Du Plessis, 2022). However, several alumni are forging new paths for the industry in the cracks created by larger systemic disruptions.

Reflecting on the departmental transformation and evolution over time

The Department itself is a complex, adaptive system – a responsive, perpetually learning, living entity engaging with a pedagogic ecosystem that is continually shifting and adjusting. Its evolving pedagogical and curriculum structures have deep roots in a long tradition of contextually responsive and responsible design practices that over time established the ethos of the department.

Since the inception of the department, several individuals nurtured a contextually responsive design approach in the built environment (Figure 9.8). The founding lecturers set out to create a regional pragmatic language for the local Modernist architecture (Barker, 2019), an Afro-pean architecture (Joubert, 1994) that is relevant to the cultural and climatic context of South Africa. This was augmented by very early sustainable and

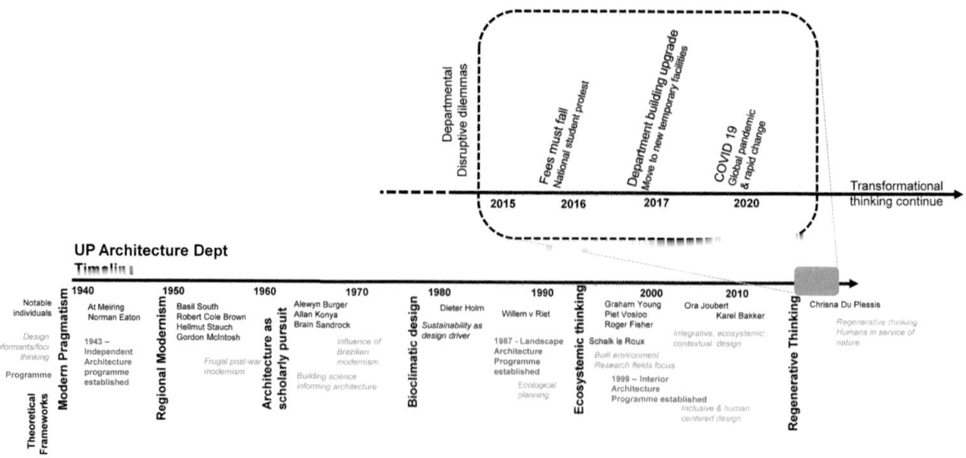

FIGURE 9.8 Timeline of the transformation and disruptions experienced at the UP Architecture department.

bioclimatically responsive design investigations by, amongst others, Dieter Holm. The incorporation of the landscape architecture programme, and later the interior architecture programme, paved the way for an ecosystemic approach to design that integrated the three spatial design disciplines and encouraged staff and students to work across multiple scales.

However, a number of disruptive dilemmas experienced within the department, and in the larger pedagogical ecosystem (Figure 9.8), accelerated the transformation and pedagogic evolution taking place in the department by forcing us to reconsider the pedagogic context, our role in the immediate community, and a socio-economic shift in our student cohort.

During this period of turmoil, a new Head of Department was appointed. Chrisna du Plessis came with a solid theoretical foundation in resilience thinking and regenerative design and development, and she saw the disruptions as an opportunity to break down old habits of thought and outdated pedagogical practices to transform the department and its curriculum. Through a process of collaborative co-creation she encouraged staff to formulate a new vision, mission and ethos for the department. This common ethos is grounded in a belief that the role of the designer in the twenty-first century is to co-create a better future for the environments and societies within which we work, while honouring the diverse legacies that shaped the present, and it informs the regenerative design and development approach that has become the hallmark of the department.

To embed this ethos and achieve the necessary curricular transformation, a two-pronged approach was followed. Firstly, a concerted effort was made to adjust the theoretical focus of the Earth Studies stream, which consisted of a number of undergraduate modules introducing concepts of sustainability, bioclimatic and inclusive design, and ecosystem services. This stream was broadened to introduce climate change theory, resilience thinking and regenerative design. Secondly, the newly created Unit for Urban Citizenship established a cumulative transformation agenda within the department that instilled an awareness of and/or collaboration in participatory co-design practices in the design studios. These changes introduced a focus on regenerative design thinking with a clear understanding of the climate crisis, as well as an awareness of public interest design, and enabled the new transformational pedagogical practices.

The result is a transformative design practice premised on inter- and transdisciplinary thinking, including both a scientific understanding of climate change and systems dynamics, and other forms of knowledge such as local user-centred insights, processes to acknowledge interbeing, holistic design responses, learning from vernacular solutions, and developing interdisciplinary thinking. Concurrently, this education ecosystem engenders critical values such as taking individual and professional responsibility, including and harnessing empathetic design methods and participative design resolution, and critically engaging with alternative design practices to develop solutions commensurate with a seva society (a Sanskrit phrase referring to selfless service) (Black, 2012) in service of nature. While these strategies, knowledge content and alternative learning environments are often presented in theoretical modules, their application manifests in design projects that are responsive solutions to diverse contexts. These solutions address processes of engagement, alternative spatial solutions, and innovative means to promote equitable, thriving and resilient urban environments.

Reflecting on the educational ecosystem and transitioning into practice

As climate change presents an existential crisis, our response as educators cannot be limited to presenting a single module introducing climate theory and knowledge. We propose instead an ecosystem of teaching modalities exploring diverse attitudes, content and skills to establish a sound basis for climate action. This curricular and pedagogical ecosystem is structured along a transformative journey, where a range of skills and attitudes is cultivated to achieve certain design products, but also to engender processes of critical, systematic and user-sensitive inquiries. As a complex emergent issue, climate change resilience cannot be defined within a predetermined scope of response measures, or even known impacts. These cascading and compounding impacts rather manifest in unexpected ways, requiring specific contextual solutions (Simpson et al., 2021). This prompts us to foster an open-ended, inclusive range of skills, knowledge and values that prepare the individual for unexpected and novel crises. As the need for transformational learning is significant, it presents an exciting opportunity to redefine architecture in service of communities as prompted by the climate emergency.

While the formal educational input stops after five years, the transformational journey starts in earnest once students transition from academia into practice. This brings its own disruptive dilemma, both for the students who must navigate a shift from an inclusive process-orientated educational environment to participating in a product-focused industry, and for a profession that is not quite ready yet for this new breed of professional. This transition forces both students and established practitioners to question their own position in practice, as their own inquiries into their professional responsibility towards others in the world is tested in an exclusive industry with a limited mandate and forms of practice. While South African academic institutions support community engagement, the profession still requires specific core competencies. This often promotes conformist students while we recognise the need for ontological diversity to nurture responsive individuals to the crises at hand.

During their education, students are often exposed to, and participate in interdisciplinary inquiries into deep-seated climate change adaptation response measures. This form of engagement with their immediate environment crosses professional boundaries and challenges preconceived notions of stakeholder participation in solution-identification and implementation processes, but is critical in developing individuals who can drive climate action, and guide and grow climate awareness during their interaction with stakeholders and communities.

While the modes of thinking and designing required to respond effectively to climate change challenges the current modes and norms of practice in the architectural profession, this certainly is not a hopeless situation. We have noted a transition in critical voices and leaders of climate awareness groups at the recent COP meetings, and new voices and actors are entering the built environment industry. Shifts have already been noted locally where the codes of conduct of the architectural profession have been adjusted to include a responsibility towards our environment (SACAP, 2021). As our current students are confronted with the limitations of singular modes of practice, we hope that their transition to practice as an emergent disruptive dilemma becomes the final stage towards engendering alternative modes of practice and climate action. As educators, we hope that this becomes the catalyst that changes the values and role of our profession.

KEY TAKEAWAYS FROM CHAPTER 9

- Education for climate change action should follow a transformative educational journey, and it continues after graduation when students transition into working environments.
- Pedagogic strategies should be intentional in their transformation agendas and responsive to the growth trajectory of students.
- A transformative educational journey requires an integrated education ecosystem that enables non-linear, iterative learning processes.

Bibliography

Alrøe, H.F. & Kristensen, E.S. 'Toward a system ethic: In search of an ethical basis for sustainability and precaution.' *Environmental Ethics*, 21(1), 59–78, 2003. Available at: https://philpapers.org/rec/ALRTAS-3 (Accessed: 13 January 2022).

Barker, A. History of the department. In *Boukunde. 75th Anniversary of the Department of Architecture, University of Pretoria* (ed. A. Barker). University of Pretoria, 2019.

Belenky, M. F, Clinchy, B. M, Goldberger, N. R. & Tarule, J. M. *Women's way of knowing: the development of self, voice and mind*. Basic Books, 1986.

Black, I. Progression and Conceptual Adjustment. Accessed 21 February2021 from https://glancesideways.com/2012/10/progression-and-conceptual-adjustment/

Bushe, G. R. 'The appreciative inquiry model'. In E.H. Kessler, (ed.) *Encyclopedia of Management Theory*, Sage Publications, (Vol 1), 41–44, 2013.

Butler, L. D. 'Normative dissociation'. *Psychiatric Clinics of North America*, 26, 45–62, 2006. DOI: 10.1016/j.psc.2005.10.004

Combrinck, C. & Venter, M.A. 'Embedded participation in the architectural curriculum towards engendering urban citizenship in graduates'. *Image and Text* (34), 01–29, 2020. DOI: http://dx.doi.org/10.17159/2617-3255/2020/n34a2

Combrinck, C. 'Is SACAP ready for Public Interest Design?'. *Architecture South Africa* 100, 40–41, 2021. Available at: https://saia.org.za/assets/docs/archsa/ASA_issue_100.pdf

De Kay, M., Roetzel, A., Nakai Kidd, A., Zinkiewicz, L. & Klas, A. 'An Integral Sustainable Design Approach to Human Inhabitation of Architectural Spaces: Theory and Project Design'. In Integral European Conference in Hungary. 1–25, 2018.

Didau, D. n.d. Didau's Taxonomy. Available at: https://learningspy.co.uk/featured/didaus-taxonomy (Accessed: 16 March 2020).

Du Plessis, C. 'The city sustainable, resilient, regenerative – a rose by any other name?' In Roggema, R. (ed.) *Design for Regenerative Cities and Landscapes: Rebalancing Human Impact and Natural Environment*, 23–48. Springer. 2022..

Folke, C., Carpenter, S.R., Walker, B., Scheffer, M., Chapin, T. & Rockström, J. 'Resilience thinking: Integrating resilience, adaptability and transformability'. *Ecology and Society*, 15(4), 2010. DOI: 10.1038/nnano.2011.191.

Grant, E.J. 'Mainstreaming environmental education for architects: the need for basic literacies'. *Buildings and Cities*, 1(1), 538, 2020. DOI: 10.5334/bc.41.

Hornsey, M., Harris, E., Bain, P. & Fielding, K.S. 'Meta-analyses of the determinants and outcomes of belief in climate change'. *Nature Clim Change*, 6, 622–626, 2016. https://doi.org/10.1038/nclimate2943

IPCC. 'Summary for Policymakers. In: Climate Change 2021: The Physical Science Basis. Contribution of Working Group I to the Sixth Assessment Report of the Intergovernmental Panel on Climate Change'. Cambridge: Cambridge University Press, 2021. DOI: 10.1260/095830507781076194.

Joubert, O. 'Afro-pean symbiosis'. *Africa Insight*, 24(4), 250–255, 1994.

Killip, G. 'A reform agenda for UK construction education and practice'. *Buildings and Cities*, 1(1), 525–537, 2020. DOI: 10.5334/bc.43.

Kricsfalusy, V., George, C. & Reed, M. 'Integrating problem- and project-based learning opportunities: assessing outcomes of a field course in environment and sustainability'. *Environmental Education Research*, 24:4, 593–610, 2018. DOI: 10.1080/13504622.2016.1269874

Leal Filho, W., Sima, M., Sharifi, A., Luetz, J.M., Salvia, A.L., Mifsud, M., Olooto, F.M., Djekic, I., Anoholin, A., Rampasso, I., Donkor, F.K., Dinis, M.A.P., Klavins, M., Finnvenden, G., Chari, M.M., Molthan-Hill, P., Mifsud, A., Sen, S.K. & Lokupitiya, E. 'Handling climate change education at universities: an overview'. *Environmental Sciences Europe*, 33(1): 1–19, 2021. DOI: 10.1186/s12302-021-00552-5.

Levy, M. 'The Art of Urban Generativity: an analytical case study documenting the process and impact of a-formality in Mamelodi East, Gauteng'. University of Pretoria: Master of Architecture dissertation, 2021. https://repository.up.ac.za/handle/2263/31741/browse?value=Levy%2C+Maxine&type=author

Mataya, D.C., Vincent, K. & Dougill, A.J. 'How can we effectively build capacity to adapt to climate change? Insights from Malawi'. *Climate and Developmen*t, 12(9), 781–790, 2020. DOI: 10.1080/17565529.2019.1694480.

Me dema, W., Wals, A. & Adamowski, J. 'Multi-loop social learning for sustainable land and water governance" towards a research agenda on the potential of virtual learning platforms'. In *NJAS Wageningen Journal of Life Sciences*, 6, 23–38, 2014. Available at: https://dx.doilorg/10.1016/j.jnas.2014.03.003

Meyer, J. H. F. & Land, R. 'Threshold concepts and troublesome knowledge: Linkages to ways of thinking and practising within the disciplines'. In Rust, C (ed.). *Improving student learning: improving student learning theory and practice – ten years on, 412–424*. Oxford: Oxford Centre for Staff and Learning Development, 2003.

Meyer, J. H, F & Land, R. 'Threshold concepts and troublesome knowledge (2): epistemological considerations and a conceptual framework for teaching and learning'. In *Higher Education*, 46, 373–388, 2005. DOI: 10.1007/s10734-004-6779-5.

Mezirow, J. 'Perspective transformation'. *Adult Education Quarterly*, 28(2), 100–110, 1978 DOI: 10.1177/074171367802800202.

Mezirow, J. 'Transformative learning: theory to practice'. In *New directions for Adult and Continuing Education*, 74, 5–12, 1997. DOI: 10.1002/ace.7401.

Nyka, L. 'Bridging the gap between architectural and environmental engineering education in the context of climate change'. *World Transactions on Engineering and Technology Education*, 17(2), 204–209, 2019.

O'Brien, K. 'Is the 1.5°C target possible? Exploring the three spheres of transformation'. *Current Opinion in Environmental Sustainability*, 31, 153–160, 2018. DOI: 10.1016/j.cosust.2018.04.010.

Okyere, S.A. & Kita, M. 'Rethinking urban informality and informal settlements growth in urban Africa: A literature discussion'. *Journal of Sustainable Development in Africa*, 16(6), 102–124, 2015. DOI: 10.1377/hlthaff.2013.0625.

O'Sullivan, E. V. Morrell, A. & O'Connor, M. A. *Expanding the boundaries of transformative learning: Essays on theory and practice*. Palgrave, 2002.

Panero, M. E., Michaels, L. & Winner, E. 2019. 'Becoming a character: dissociation in conservatory acting students'. In *Journal of Trauma & Dissociation*, (21)1, 1–16, 2019. DOI: 10.1080/15299732.2019.1675220.

Perry, W. G. *Forms of intellectual and ethical development in the college years: a scheme*. Hold, Rinehart and Winston Inc., 1970.

Pieterse, E. 'Rethinking African urbanism from the slum'. *Cities, Health and Well-Being*. (November), 1–4, 2011.

Roberts, D. 'Prioritizing climate change adaptation and local level resilience in Durban, South Africa'. *Environment and Urbanization*, 22(2), 397–413, 2010. DOI: 10.1177/0956247810379948.

Roux, D.J, Murray, K & van Wyk, E.. 'Learning to learn for social-ecological resilience: balancing strategy options in public sector organisations'. In Burns, M. and Weaver, A. (eds.). *Exploring Sustainability Science: A South African perspective.* 599–621 Stellenbosch: Sun Press / African Sun Media, 2008.

Salama, A. M. *Spatial design education: New directions for pedagogy in architecture and beyond.* Ashgate, 2015.

SACAP. Code of Conduct For Registered Person. No. 44190, Board notice 7 of 2021. South African Council for the Architectural Profession, 2021.

Schön, D. A. *The Reflective practitioner: How professionals think in action.* New Basic Books, Inc. Publishers, 1983.

SERI. *Informal Settlements and Human Rights in South Africa.* Johannesburg: Socio-Economic Rights Institute of South Africa, 2018.

Simpson, N.P., Mach, K.J., Constable, A., Hess, J., Hogarth, R., Howden, M., Lawrence, J., Lempert, R.J., Lempert, R.J., Muccione, V., Mackey, B., New, M.G., O'Niell, B., Otto, F., Portner, H., Reisigner, A., Roberts, D., Schmidt, D.N., Seneviratne, S., Stongin, S., v Aaslt, M., Totin, E., & Trisos, C.H. 'A framework for complex climate change risk assessment'. *One Earth*, 4(4), 489–501, 2021. DOI: 10.1016/j.oneear.2021.03.005.

Srivastava, M. 'Cooperative learning in design studios: a pedagogy for net-positive performance'. *Buildings and Cities*, 1(1), 594–609, 2020. DOI: 10.5334/bc.45.

Stevenson, F. & Kwok, A. 'Mainstreaming zero carbon: lessons for built-environment education and training'. *Buildings and Cities*, 1(1), 687–696, 2020. DOI: 10.5334/bc.84.

Stupar, A., Mihajlov, V. & Simic, I. Towards the conceptual changes in architectural education: Adjusting to climate change. *Sustainability* (Switzerland), 9(8), 1–19, 2017. DOI: 10.3390/su9081355.

The World Bank. 'Climate Change Adaptation and Economic Transformation in Sub-Saharan Africa'. Washington DC: The World Bank, 2021.

Thompson, S. & Thompson, N. *The Critically Reflective Practitioner.* Palgrave Macmillan, 2008.

UNESCO. Education for sustainable development. A Roadmap. Paris: UNESCO, 2020. DOI: 10.4324/9781003022763-5

UN-habitat. 'The State of African Cities. Re-imagining sustainable urban transitions'. UN-Habitat: Nairobi, 2014.

10

CHANGING PARADIGMS

Educating for the climate emergency

Elizabeth Donovan

Introduction

Uncertainty, unpredictability and shifting paradigms are all words which frame the climate emergency. Or, as Naomi Klein (2015, 25) writes: '[…] nothing is inevitable. Nothing except that climate change changes everything.' Raising the question, how do we design for the future in a world where nothing is certain, and everything is unpredictable? Yet, architecture is frequently static despite the dynamic nature of our environments and contexts – designed for a stagnant time and place. Often this is also the case for architectural education, where students are taught to design objects or buildings frozen in their intended time. Some educators already acknowledge that traditional pedagogies may not sufficiently prepare architecture students for the future challenges of the changing climate (Grover et al., 2019; Hemsath and Bandhosseini, 2017). However, this is not a simple challenge. Sustainability alone is an essentially contested concept (Guy and Farmer, 2015), there are plural and competing discourses (Guy and Farmer, 2015), and it is holistic and complex. Additionally, it is multi-scalar over different periods, and the climate crisis has no borders.

Thus, it requires radical, dynamic, inclusive and transdisciplinary modes of practice and education, as well as a critical approach to find a way through the complexity, analysing and assessing not only the competing interpretations of the discourse but also multifaceted collaborations to navigate through multi-disciplinary knowledge (Grover et al., 2019; Gürel, 2010; Guy and Moore, 2007) blurring the boundaries that define the field.

Faced with a bleak future, students and younger generations have inherited the problems of the past myopic global decision-makers. The context of their future is often framed by increased extreme weather, heatwaves, droughts, heavy rain, flooding, sea-level rise, food, water and human insecurity, not to mention the loss of biodiversity and ecosystems. There is uncertainty, not if, but when this will occur, its frequency and intensity (Tsioumanis, 2020). Some predict that it will take irreversible damage to the climate beyond the tipping point for a genuine change of course to occur. In this unfair waiting game, where past generations have not taken ownership of their mistreatment

DOI: 10.4324/9781003293903-12
This chapter has been made available under a CC-BY-NC-ND 4.0 license.

of the planet, the youth and current students are called to action and praised for being the change we need. As Greta Thunberg said in her 2019 speech at the United Nations climate summit: 'I should be back in school on the other side of the ocean. Yet you all come to us young people for hope. How dare you. You have stolen my dreams and my childhood with your empty words' (Thunberg, 2019).

While this is unfair and unmotivating, educating in the context of climate anxiety is essential. When students are aware the most sustainable thing to do is not to build at all, how do we motivate engagement in design studios? How do we support students to find opportunities in what is often only framed as negative challenges? While putting our hope in future generations, as educators, how can we best equip them to deal with uncertain futures?

This reflective chapter is framed by these questions and divided into two main sections. Firstly, exploring how several significant paradigm shifts have shaped the field of sustainable architecture, outlining the fluidity and complexity of the field. Secondly, questioning and speculating on how shifts in pedagogies can help educate critical thinkers and global citizens with solidarity towards humans and nonhumans, finding opportunities in the insecurity of constantly shifting paradigms.

Historical shifts in environmental, social and resource paradigms

A paradigm shift can be understood as a fundamental change or revolution in the usual way of thinking, assuming or doing, replacing it with a new or different approach (Merriam-Webster, 2022). Thomas Kuhn (1962) explains that paradigm shifts occur when the main paradigm framing normal science is deemed unsuitable with a new phenomenon, enabling the uptake of a new paradigm or theory (Kuhn, 1962). An example of this could be CO_2 emissions. From the industrial revolution until the middle of the twentieth century, CO_2 emissions were not considered a significant concern until scientists documented the negative change in the Earth's environmental and climatic conditions. It became apparent that human industrial behaviour was the direct cause of greenhouse gases, global warming and climate change. Thus, a paradigm shift took place. This specific paradigm shift has impacted many fields of science and is a foundational shift which partially developed sustainable architecture. Within the field of architecture, small shifts concerning environmental thinking, society and resources have occurred. They are presented here as a limited Western background to understand the continuously changing nature of the profession. And that while students are faced with pressing concerns, there are moments and examples from the past which can be reused and reintroduced into current thinking – illustrating there is *already* sufficient existing knowledge to make change.

Shifts in environmental paradigms

Despite the modern notions of environmental design being associated with the post-war period, shifts in environmental thinking (in a Western understanding) can be traced back to the eighteenth century. In a broader perspective, it can be argued that Western science has been relatively late to adopt environmental thinking, while the relationship between natural systems is at the core of many indigenous schools of thought and cultures that have often been ignored or discredited until recently (Jessen et al., 2021).

Early Western biologists such as Ernst Haeckel and Henry David Thoreau can be considered as the start of this thinking – connecting living organisms and their habitats – defining the field of ecology in 1866 (Kallipoliti, 2018). During this period, nature was still considered the sublime, and primarily wilderness was considered separate from the artificial or human-made (ibid.). Since then, various reinterpretations of ecology have emerged; despite this, they all recognise the interrelationships between organisms and the environment (Andrewartha & Birch, 1986).

Despite these considerations in the field of Western science, the acceptance of ecological or systems thinking was not as widely understood, especially by the general public, until the publishing of images of the Earth. The *Earthrise* image series by Apollo 8 in 1968 and the 'Blue Marble' image in 1972 created significant environmental and societal shifts (Rice and Littlefield, 2014). These two iconic images changed the perspective of our planet, spawning Earth Day in 1970 and facilitating the holistic thinking of the globe as a single living organism with finite space and resources (Whitehead, 2007).

Rachel Carson's *Silent Spring* (1962) indicated the drastic collision course we were – and still are on – with the natural world. It sowed the seeds for the development of modern Western environmentalism and the many progressive changes which have occurred in the last half-century (Griswold, 2012). This publication alerted (especially American) audiences to the impact and harm human activity was causing the natural world. Carson advocated against human dominance over nature's design through technology and instead sought to challenge authorities who imposed these ideas. The cultural paradigm shifts of *Silent Spring* stirred many environmental movements and further shifts in the field, especially policy changes and the development of environmental agencies (Griswold, 2012).

Arne Næss, the Norwegian philosopher, was influenced by Carson's writing, going on to coin the term 'deep ecology' in 1973. He proposed a philosophy of ecology as a holistic worldview in which all living beings are equal with substantial value (Naess, 1973). Wherein Thoreau placed greater meaning on human life in relation to the values of wilderness, Næss's philosophy of deep ecology, similar to many indigenous thoughts, encouraged a holistic approach decentralising humans, recognising we are intrinsically connected to the Earth's natural systems with equal significance for all other non-human systems (ibid.). This shift in thinking has given way to many design responses and more recent posthumanism theories and writings from Rosi Braidotti (2013a), Gregory Bateson (2000), Bruno Latour (2007) and Donna Haraway (1991).

Several architects, such as Buckminster Fuller and Ian McHarg, were inspired by this shift and paradigm, considering the design of nature as important as the design of the built environment. In the late 1990s, Sim van der Ryn and Stewart Cowan coined the term 'Ecological Design'. This approach is built on previous thinkers and designers to advocate for minimising adverse environmental impact through integrating human activities with natural processes. The authors define ecological design as 'any form of design that minimises environmentally destructive impacts by integrating itself with living processes' (Van der Ryn and Cowan, 1996, 18).

Later, in the 2000s, a significant paradigm shift for the scientific community occurred with biologist Eugene Stoermer coining the term Anthropocene, which atmospheric chemist Paul Crutzen later popularised (Masin and Lewis, 2015; Zalasiewicz et al., 2019). This signified a new geological epoch in which humans have dominated the Earth, including the climate and environment. With this comes what Earth system scientists call a 'state shift' in which the Earth is transformed rapidly and irreversibly (Moore, 2016). This includes many planetary boundaries being crossed. The acknowledgement of the vast human impact on the globe has influenced many understandings of human's role in planetary systems. However, some debate and criticism have also occurred, with scholars being unsatisfied with the 'human-centred' underpinning of the term, as well as the created dualism of humans and nature (Büscher and Fletcher, 2020). Thus, some scholars now advocate using the term 'Capitalocene' over the Anthropocene to describe the pervasive influence of human dominance over non-human systems (Büscher and Fletcher, 2020). Further, criticising the human-centred association of the term, authors such as Donna Haraway instead advocate for the Chthulucene to be used instead. Haraway explains this new term is to 'name a kind of timeplace for learning to stay with the trouble of living and dying in responsibility on a damaged earth' (Donna Haraway, 2016, 2).

This renewed paradigm that decentralises humans as the protagonist of the planetary systems within the Chthulucene epoch builds on previous thinking of Deep Ecology. Arne Næss proclaimed that all designs should support nonhumans as well as humans because they are all inherently of equal value (Naess, 1973; Kallipoliti, 2018), which resonates with current post-human theory and consequent design approaches. Ideas such as post human sustainability abandon human exceptionalism and instead advocate for the thriving of all life and recentres ecosystems (Cielemęcka and Daigle, 2019). Further, Rosi Braidotti, in her publication *Posthuman* (2013b), argues that we are entering a post-human age that blurs the nature–culture dichotomy, which has previously been considered polarised. This new shift in thinking drastically impacts sustainable architecture and all design thinking. It shifts previous human-centred and (human) user-driven design approaches to a new paradigm of thinking where multi-species need to be considered.

Shifts in social paradigms in the West

Parallel and interconnected with environmental paradigm shifts have also been societal and political changes. While the Blue Marble photo went on to become the poster child for the modern environmental movement (McDonald, 2020), with Stewart Brand using it as the cover of the first *Whole Earth Catalogue* (Cadwalladr, 2013), it also resulted in social shifts. Brand was part of a group of environmental activists that believed the image of Earth would encourage societal activism and care for the planet and its precious resources – understanding that micro-actions have macro planetary consequences. Later in the 1970s, English biochemists such as James Lovelock shaped the theory that the Earth was a single, self-regulating superorganism which he later named 'Gaia' after the Greek goddess of the Earth (Aitkenhead, 2016).

The 1960s and 1970s were also some of the most turbulent and divisive periods in the twentieth century in the Western world, with prevalent social and political discourse marked by the Vietnam War and anti-war protests, civil rights, and counterculture movements (Deviren and Tabb, 2014). This also extended to social and political concerns for an ill-managed planet running out of resources with escalating alerts of global pollution levels, the downfall of economic growth, excessive waste flows, 'dropping out' of Western social structures (Matthews, 2010), and the redistribution of global resources.

In a Western context, urban life was abandoned by many young intellectuals (despite accelerated urbanisation at a global level) rejecting the establishment for alternative communities in rural areas – one of the most famous being Drop City, built in Colorado in 1965 (Sadler, 2006). The series of these communes were self-sustaining living laboratories that tried to restore a balance with nature through waste recycling and producing their own energy. Similar to the alternative 'drop' communities, architects in the USA and Europe strived to test thoughts of autonomy. With events such as the oil crisis in the 1970s, new modes of living away from the reliance of particularly energy grids became popular. Additionally, these projects became an anti-consumerism and anti-capitalist political statement. Furthermore, particularly in the USA, the urban retreat was a substantial driver for self-sufficiency, especially in alternative energy experiments. Design approaches such as 'Earthships', bioshelters, autonomous houses and off-grid architecture were all developed in different ways to address the concerns of how we live. During these early years, architects joined with biologists, ecologists, farmers and engineers striving to create self-sufficient habitats which were seen as living laboratories. Together they strived to create self-reliant homes that integrated life-supporting systems (i.e., food production, farming, repossessing systems and recycling etc.) to redevelop energy and resources (Farallones Institute, Van der Ryn 1982). While extremely alternative and somewhat utopian or ideological, these social and resource experiments shifted ecological thinking and paved the way for many other political and resource-efficient pilot projects still under-represented in the field today (Scott, 2016).

Shifts in resource paradigms

Despite some environmental progress, such as the reduction of acid rain and the hole in the ozone layer, there has still been irreparable damage to the Earth's climate and resource banks. At the same time, the human population has doubled since the first Earth Day, and the sixth mass biodiversity extinction is occurring. Despite warnings becoming louder and scientific evidence demonstrating planetary-scale destress, our global footprint still exceeds the Earth's resources three times over (Wackernagel, 2019). Over the last decades, there have been numerous attempts (e.g., environmental footprinting, cradle to cradle and Life Cycle Analysis) towards a shift in resources; however, this transformation requires considerable behaviour change away from consumerist culture. Consequently, profound paradigm shifts concerning resources are needed at a faster rate than current progress.

The present energy and carbon paradigm stems from the reaction to the oil crisis in the 1970s and the development of environmental footprinting in the 1990s (Tabb and Deviren, 2014). No longer was it possible for buildings to ignore the energy consumed to heat and cool them. This sparked alternative technologies and many new terms, as previously mentioned, including self-sufficiency, self-reliance, and living autonomy, which later developed into current approaches of passive house, net zero energy and carbon, as well as energy plus. These early designs strived for buildings to be self-sufficient ecosystems detached from infrastructural contexts. The development of the off-grid design approach (described earlier) provided some technological foundation for these approaches, which aimed to address energy concerns. Thus, technology and architecture became intrinsically linked, especially when discussing green, biomorphic or eco-tech architecture.

In turn, many so-called sustainable buildings have been considered performative machines to address the lack of resources (e.g., energy and CO_2). Different from ecological approaches, techno-heroism or a technocentric understanding that technology will solve all issues underpins many of these designs. Technological innovation has impacted energy production and material performance use and helped regulate embodied carbon and emissions (Amale Andraos, 2016). While these buildings have drastically influenced the notions of energy and carbon, which was very important, many were not successful in architectural terms – often continuing as usual with the addition of technology to offset problems such as increased building footprints and excessive lifestyles. Consequently, a paradigm shift occurred in the 1990s, where sustainable certifications were developed to overcome greenwashing. While successful in some regards, they have created a negative paradigm shift towards shallow tick-box approaches to sustainable architecture, which do not encompass critical whole-systems thinking and can even exacerbate greenwashing.

Later in the 1990s, these ideas shifted towards footprinting not only individual consumption but also materials, their extractions, production, transportation and use. Tools like LCA (Life Cycle Assessment), which was previously mainly used for energy, developed a greater material focus – assessing the environmental impact of materials from 'cradle to grave'. Following the increase in awareness, William McDonough and Michael Braungart published their fundamental manifesto book *Cradle to Cradle* in 2002. It popularised and shifted the linear logic of 'cradle to grave' to a circular political economy. Recognising that resources are limited, environmental systems can no longer cope with a linear resource model of waste (Braungart and McDonough, 2002). This publication has impacted the architectural profession, organisations, companies and governments globally. Similar design approaches based on the circular economy have emerged, responding to limited resources, such as adaptive reuse and design for disassembly.

It is no surprise that architectural and environmental thinking responds to the associated social and political context. It is clear that paradigms and concepts that have framed the profession have undergone significant changes in our worldview over the last 70 years. It is only fitting that many architectural responses follow suit in attempting to address the 'wicked' and explicit problems of the time. Despite the ongoing challenges faced, there are always ways in which designers can attempt to make changes or address issues. However, this requires that architectural education responds dynamically to these paradigm shifts and that pedagogical methods are resilient enough to allow for the rapid responses needed to design for the climate emergency.

Educating for paradigm shifts towards the climate emergency

As educators, we are responsible for ensuring students are equipped with the skills for the future. Which, as illustrated in the previous section, is an ever-changing field of growing complexity. Thus, students must be educated as lifelong learners that are instilled with innovation and creativity to address global challenges (Elrayies, 2017). As technology advances and required knowledge increases, it is even more critical that students have the skills for cross-disciplinary collaborations (Nicol and Pilling, 2005), as architectural education is already too full to include extensive knowledge from different disciplines (Altomonte et al., 2014). Instead, architects should be trained in facilitation skills, as active listeners who respond, collaborate and use their specific proficiencies to create sustainable solutions (Elrayies, 2017).

While it is not possible to cover the magnitude of ways in which education can help equip students for the future, this section will present different strategies.

Uncertainty, critical thinking and learning to learn

One of the certainties of the future is that it will be filled with uncertainty, which is an immense consideration for architects. How do we live, plan and design for this uncertainty? As Amale Andraos (2016, 301) poetically phrases the need for a shift in architecture's temporality and performance: 'architecture becomes an expanded, perforated, and porous object, whose edges in space and time are always in flux, we can plan for redundancies and design for resiliency, or at times, engage in mad alchemy as we rewrite architecture as the art and science of the unknown.' Yet, this raises the question, how do we educate students for the unknown? This is a challenge which is faced by all disciplines but is particularly important for architecture. Our scarce resources are used to build spaces (often exacerbating the problem) we hope will last for generations, but as the climate emergency continues, the future context becomes more and more unpredictable. In turn, if we are to ensure the use of these resources is not frivolous; we need to explore what is architecture of the unknown. This requires knowing enough to tackle this ambiguity, which would require students to know everything – an unfeasible possibility.

In the past, design knowledge and design have been framed by problem-based learning with well- or ill-structured problems. Traditionally, the design studio has often followed a master–apprentice or Beaux Arts model, with knowledge being transferred from instructors to students (Salama, 1995). Recently, and in correspondence with societal and environmental uncertainty, twenty-first-century design has been about embracing change and eliminating boundaries – even between teaching and learning (Aydınlı and Kürtüncü, 2018). Dialogue between teacher and student changes the learning environment to one of 'discovery of knowledge' or learning dialogues (constructivist approach) (Armstrong, 2011) rather than the transmission of it. Goldschmidt (1994) elucidates that design problems are indeterminant in nature, and indeterminate knowledge cannot be transmitted. Furthermore, design knowledge is contextual and changes correspondingly. Thus, the discovery of knowledge becomes essential – fostering a culture of 'learning by doing' in which students learn-to-learn and develop knowledge specific to their requirements – emphasising the importance of learning methods rather than teaching methods (Aydınlı and Kürtüncü, 2018), unfolding the need for a paradigm shift in architectural education to address the complexity of the future.

Learning to learn

The ability to be a lifelong learner is essential to tackle the broadness of the climate emergency, the contextual nature of architecture and the unpredictability of the future. Learning to learn is a holistic learning strategy and a necessary skill for student-centred learning and teaching, as well as critical thinking. It requires students to be curious, constantly adjust themselves, and learn to take positions and develop contextually relevant tactics (Aydınlı and Kürtüncü, 2018). In addition, it puts emphasis on the research phases of the design process, as students with increased learning-to-learn skills can better inform the brief and early design decisions.

Learning often begins with the whole and expands to include the parts, prioritising the student's interest and position. However, for students, this compels a personal responsibility and increased engagement in the educational environment (e.g., participating in studio discussions as active learners rather than passive recipients). Subsequently, this change in educational paradigms towards a student-centred one requires a rethinking of studio culture towards students 'learning how to learn' rather than transferring specific knowledge or competencies. While also important for other learning environments (e.g., lecture-based courses), it is especially relevant for studio-based learning as it already fosters individual and problem-based learning (Clune, 2014). Additionally, studio teachers often 'teach as they were taught' (Ballantyne et al., 1999, 237; Quinlan et al., 2007), continuing with traditional master–apprentice models where knowledge (tacit and explicit) is often transferred from tutor (expert) to student (learner) through individual tutoring and 'crits' (Salama, 1995) without reflection on whether these methods actually result in an increase in student learning (Quinlan et al., 2007). This knowledge is often transmitted with high authority and based on personal opinions concerning individual design development (Webster, 2004). Aptly, in many institutions there is still a reluctance to change teaching methods and move away from the perception that a teacher's expertise is central to the design studio education (Dutton, 1991; Webster, 2004, Quinlan et al., 2007).

However, Hemsath (2017), Olweny (2018) and Grover et al. (2019) imply that this traditional way of educating students – especially when deep reflection is needed – is no longer productive, especially concerning sustainable architecture education. And while it may be difficult for some teachers to adapt to a more democratic process, it is essential. Quinlan et al. (2007, 1), summarising others' research, (e.g., Biggs, 2003; Prosser and Trigwell, 1999; Barnett 2000; Kreber 2002; Trigwell and Shale 2004), states that students are more motivated to learn when 'the educational environment is student-centred, they are empowered to construct their own understanding of disciplinary knowledge and teachers demonstrate their own learning in their teaching practice'. Thus, with a student-centred approach the design studio learning environment becomes one of an intellectual atmosphere in which the communicative relationship between teacher and learner is continuous, and the teacher's role turns to one of facilitation of a student's construction of knowledge rather than that of master–apprentice (Kolb, 2014; Monroe, 2017; Mezirow, 2018). Quinlan et al. (2007, 5) also acknowledge that this may be difficult and propose that until teachers are ready to reflect and accept change, we can empower students to 'act on their beliefs' and 'initiate tutors into the preferred student-centred world of learning'.

Thus, through this approach, students question their own knowledge level, engage in critical thinking towards their learning (expanded on in the following section); and generate knowledge by experiencing, reflecting and doing (Mezirow, 2018; Grover et al., 2019). Nonetheless, learning by doing or making should not be seen as a representation tool to validate the thinking process but in itself as a way of thinking. This requires a further shift in education models, to focus more on the process of education and less on the object at the end. An analogy could be made with an education in mathematics: grades are given for showing the working of a problem (i.e., demonstrating understanding, critical thinking and applied knowledge) not just for the correct answer at the end. This same approach is needed in architecture education to align with a student-centred approach, celebrating the learning process rather than primarily the conclusion.

Critical thinking

Critical thinking is essential for architectural education, especially in enhancing creative thinking, new ideas and relations within the new climate emergency paradigm. Each new context requires new knowledge; thus, critical thinking is essential to understand and learn what is unknown (Grover et al., 2019). Criticality and reflexivity are the foundation for all phases of the design studio and process (Stead, 2003). It is in no way a new concept and is already an aspiration for many educators; however, it is not always fostered with confidence, especially within master–apprentice teaching frameworks (Stead, 2003). Stead (2003, 05), explains that universities are sites where 'responsible, critical, engaged citizens' are educated rather than simply imparting knowledge or training skills and competencies. As a result, students are more aware of the complexity and hidden assumptions of the world, questions, motives and ideologies (Stead, 2003).

Critical thinking is often linked with intuition, curiosity, scepticism and discovery. It requires more experiential and experimental educational approaches to enable students to develop their positions and ideas. To address the climate emergency, the rapidly changing profession and to find unpredictable or innovative outcomes, non-linear and creative thinking is essential, both of which are learned through critical thinking (Grover et al., 2019). Additionally, critical thinking cultivates an ability to embrace failure through experimentation, building a creative ability to generate solutions that did not exist before (Charalambous & Christou, 2016).

To achieve critical thinking for architectural education, deep learning (i.e., learning with understanding) or critical pedagogy are reflective approaches that are required (Grover et al., 2019). Deep learning requires a meta-reflective process that critically challenges underlying values and assumptions to provide a deeper understanding (Grover et al., 2019). It requires iterations of action and reflection where assumptions are questioned. Additionally, deep learning is particularly relevant to sustainable architecture education as it is holistic and interdisciplinary in nature (Buckingham-Hatfield & Evans, 1996). Furthermore, deep learning is closely related to critical pedagogy (Pettit, 2010) and experiential learning (Kolb, 2014). They embrace competing interpretations, and experiential learning specifically includes cycles of reflection (perceived knowledge) and action (processed knowledge), constantly referring back to personal experiences (Grover et al., 2019; Kolb, 2014). However, for all of these approaches to be successful, they require a level of internal motivation from both the learner (i.e., they must desire to understand) (Warburton, 2003) and the educators who need to be reflective and engage with student-centred pedagogies (Quinlan et al., 2007). While it requires both the student and teacher to take responsibility, a student-centred approach can allow students to find an educational pathway which is of interest to them, and thus find internal motivation.

Scenario-based learning and design

Within the built environment, actors and users, both human and nonhuman, are dynamic and constantly changing, especially when broadening the discussion to the context and changing climate. Nonetheless, as mentioned, architecture and its education are often considered a static built object finished and then moved on to the next. However, the challenges faced by the profession are increasingly complex, entailing wicked problems (Burry, 2012). For climate emergency design, the uncertain future and flux are crucial to consider both in the education and design of architecture. Thus, teaching and learning methods such as scenario-based learning (SBL) and scenario-based design (SBD) offer an opportunity to include these dynamic aspects. In turn, giving chances for students to identify real-life opportunities which can foster creative, alternative and, most importantly, optimistic responses (Charalambous and Christou, 2016).

SBL supports active learning strategies through interactive scenarios often centred around complex or wicked problems that students are required to solve. In this, students are required to engage in critical thinking and problem-solving skills while applying subject-based knowledge in a real-world context. Further, SBL is grounded in theories of situated learning (Lave & Wenger, 1991) that argue for learning to occur in the context in which it will be used. It is useful when simulating or dealing with real-world problems where students can find their own way or make their own choices while engaging with a somewhat real situation (Seren Smith et al., 2018). While SBL is not often explicitly used in architecture education but in other fields (e.g., medicine), it is a valuable tool when applied to decision-making, critical thinking and complex situations, all of which are crucial within the context of sustainability.

In contrast, SBD is more familiar to the field of architecture education, in which scenarios are used as tools for exploring future circumstances within the potential 'what-if' setting (Eilouti, 2018). This is especially relevant in sustainable architecture, which requires even more thought into the dynamic future performance and lifecycle concerning flexibility, transformability, adaptability and functionality (Eilouti, 2018). This requires describing the setting, functions, users (actions and interactions), and climatic environments (Lüley et al., 2019).

However, various key scenarios must be considered early in the design process to effectively imagine future temporal and spatial contexts throughout the predicted lifecycle (Eilouti, 2018). Thus, SBD allows for the possibility to explore and predict different possibilities and potentials, visualising performance and post-occupancy of the project in use beyond the initial static design context. However, this requires knowing various future settings and paradigm shifts, both social and environmental (including the climate). Ultimately, it is about imagining stories and narratives which provide inspiration for uncertain futures.

As scenarios are not static, the use of stories based on the users and their activities defines them (Carrol, 1999). Further scenarios can be defined by three types: the *Normative, Predictive* and *Explorative* (Börjeson et al., 2006). Normative scenarios start with well-defined targets and seek to sustain them. The predictive scenario is data-driven and envisions depicting future situations or performance. Lastly, explorative scenarios intend to discover situations that may detect what could happen in a given context. Explorative scenarios are often used in sustainable architecture education as a constructive approach to learning, allowing students to unfold possible potentials that arise due to changing contexts and climates. Storytelling and narratives are especially beneficial for building scenarios and communicating ideas or expressing spatial experiences (Thompson, 2016). This further enhances the need to encourage students to establish critical and reflective thinking concerning the indeterminacy and fluid nature of architectural concepts and perspectives to find various solutions (Lüley et al., 2019). SBD methods aim to stimulate creativity and analytical tools while encouraging exploration and escaping pragmatic constraints in favour of visionary pursuits. It reintroduces students to exploration, speculation, discovery and position themselves in political, societal and environmental upheaval (Kallitsis & Andrews, 2020).

Engaging in a change in studio culture and architectural education requires both students and educators who are curious, flexible, innovative and willing to fail while challenging and changing the values of the architectural profession in the face of the climate emergency and its consequent paradigm shifts. Deep learning and SBL pedagogies grounded in creating self-driven life-long learners that are creative, critical thinkers are some of the most essential skills for the twenty-first century (Armstrong, 2011).

Towards a new paradigm

The accelerating climate emergency has upended the somewhat stable paradigms which framed sustainable architecture. Amale Andraos (2016, 300), declares that this 'presents an incredible chance to reimagine what architecture is, as discourse and as practice'. However, as discussed in the preceding sections of this chapter, architectural education will demand considerable change. Accordingly, the educational environment of the design studio requires new methods of teaching and learning which educate students with the values and skills needed to address the complexity and uncertainty of the climate emergency. The past paradigm shifts concerning the environment, society and resources show that complexity is ever-increasing and the knowledge needed to address this complexity is inconceivable within the existing architecture education framework. However, other skills, such as the ability to work interdependently, critical thinking and 'learning-to-learn', can foster the interdisciplinary complexity which is needed within the education. Resultingly, this change towards student-centred learning creates curiosity, awareness as well as critical and creative thinking as the foundation of learning (Aydınlı & Kürtüncü, 2018). These abilities form life-long learning and develop 'learning as inquiry, learning how to embrace the change, learning by experience, learning by doing' (ibid., 93).

Some of the major areas where there have been significant paradigm shifts are in the nature–culture relationship – shifting from one of dominance to one of interconnectedness and systems thinking. This has many challenges for education as it requires a certain level of knowledge of many different fields, while at the same time, there is not adequate space in the education to become an 'expert' in other fields (Altomonte, et al., 2014). Instead, there should be a focus on consideration and collaboration (learning from and with) not only other disciplines and experts but also non-humans who are without a 'voice'. It changes the design process, away from a stagnant object to one which is part of a context, considering users beyond human who are equally entitled to inhabit natural and human-made environments (Amale Andraos, 2016).

Similarly, societal shifts are constant – and architecture is constantly responding to social and political concerns. This often results in exploring how we live together, especially in recent years with non-humans. This paradigm is more difficult to predict, unlike resource paradigms, where there is a more clearly understood quantitative measure for the number of planetary resources. Thus, in response, how we live, whether, for example, in communities, communes, or in tiny homes, needs to be flexible and adaptable and become more locally resource-reliant for unknown future scenarios. Therefore, scenario-based learning is especially relevant to foster creative thinking, exploring and imagining new solutions to different dwelling patterns and ways of life (e.g., integrating places to work, live and produce resources within the built environment). However, this requires interdisciplinary knowledge similar to the natural-culture paradigm, which entails particular knowledge and competencies. Early examples described in the first section indicate that it is possible to collaborate with other fields (rather than needing to be the expert) to test how we can live a more holistic and sustainable lifestyle. Nevertheless, this necessitates an education in which students can make political statements or amplify the role politics has on the field. Scenario thinking can be used to imagine futures and alternative ways to respond to challenges while maintaining ethical and healthy living environments.

Resource and material paradigm shifts are often the consequence of more significant environmental or societal revolutions. It is a problematic paradigm as it requires a change in behaviour and lifestyle, moving away from a consumer culture by architects and citizens as a whole. It directly influences the architecture profession, as material choices and construction processes openly contribute to the (un)sustainability of a design. Despite shifting toward more circular and renewable resources, implementing this with a change in how we live (e.g., smaller housing or alternative ways of living etc.) is often challenging. Critical thinking and deep learning are especially needed to address this as it requires considerable questioning and reconsidering of existing designs to find innovative, or not yet utilised, solutions.

When architecture is reconsidered in a new paradigm as an integrated (negatively or positively) part of the Earth's natural systems, human and non-human users, resources and planetary politics – it requires a new type of architect with a reprioritising of competencies (i.e., critical and creative thinking) and additional new skills (learning-to-learn, embracing the unknown, etc.) to make change. This underscores the necessary shift of values within education, advocating for a transition from the traditional master–apprentice mode, emphasising a student-centred learning environment. Embracing the unforeseen future and recognising architecture as a dynamic form of knowledge is essential for addressing the ever-evolving challenges brought about by the climate emergency.

KEY TAKEAWAYS FROM CHAPTER 10

- By framing learning environments and activities as opportunities (within new value frameworks), a scenario or design can be viewed differently or more innovatively rather than a restrictive endeavour framed by outdated values. Thus, it can create unexpected perspectives and emergent and innovative architecture that is responsive to complex and dynamic scenarios.
- Creating value changes the role of the architect and reduces the likelihood of egotistical design endeavour, such as the master–apprentice studio culture. And instead towards a student-centred learning environment where students are interdependent team-workers, who value the wider society, the users of their designs, and the non-human context.
- Encourage and foster critical thinking as an essential foundation of the design education. Giving students the skills to question and challenge existing norms and go beyond to create the emergent solutions required for the climate emergency.
- Educating students to learn-how-to-learn creates the knowledge to become lifelong learners who can tackle new and unforeseen future challenges.
- Embracing the unforeseen future and seeing it as part of any given context. Consider SBL and SBD to work with narrative and storytelling, unfolding the potential of the future or the unknown. Use real (local when possible) environments in design briefs which can encourage real-life learning rather than abstract to the learner (Armstrong, 2011).
- Understanding that architecture is not a single fixed object but a form of knowledge (Amale Andraos, 2016) which can address continuous paradigm shifts and the challenges presented by the climate emergency.

Bibliography

Aitkenhead, D., 2016. Interview: James Lovelock: 'Before the end of this century, robots will have taken over'. The Guardian. Available at: www.theguardian.com/environment/2016/sep/30/james-lovelock-interview-by-end-of-century-robots-will-have-taken-over (accessed 21 December 2022).

Altomonte, S., Rutherford, P. & Wilson, R., 2014. Mapping the way forward: Education for sustainability in architecture and urban design. *Corporate Social Responsibility and Environmental Management* 21, 143–154. https://doi.org/10.1002/csr.1311

Amale Andraos, 2016. What Does Climate Change (For Architecture)?, in: *Climates: Architecture and the Planetary Imaginary*, Columbia Books on Architecture and the City. Lars Müller Publishers, pp. 297–301.

Andrewartha, H.G. & Birch, L.C., 1986. *The Ecological Web: More on the Distribution and Abundance of Animals*. University of Chicago Press.

Armstrong, C.M., 2011. Implementing education for sustainable development: The potential use of time-honored pedagogical practice from the progressive era of education. *The Journal of Sustainable Education* 2.

Aydınlı, S. & Kürtüncü, B., 2018. Paradigm shift in studio culture. *AZ ITU journal of Faculty of Architecture* 15, 3, 91–108. https://doi.org/10.5505/itujfa.2018.20053

Ballantyne, R., Bain, J. & Packer, J., 1999. Researching University Teaching in Australia: themes and issues in academic reflections. *Studies in Higher Education* 24, 2, 237–257.

Barnett, R., 2000. *Realizing the University in the age of Complexity*, The Society for Research into Higher Education and the Open University Press, Buckingham, UK.

Bateson, G., 2000. *Steps to an Ecology of Mind: Collected Essays in Anthropology, Psychiatry, Evolution, and Epistemology*. University of Chicago Press.

Biggs, J. (2003) *Teaching for Quality Learning at University*, The Society for Research into Higher Education and the Open University Press, Buckingham, UK.

Börjeson, L., Höjer, M., Dreborg, K.-H., Ekvall, T. & Finnveden, G., 2006. Scenario types and techniques: Towards a user's guide. *Futures* 38, 723–739. https://doi.org/10.1016/j.futures.2005.12.002

Braidotti, R., 2013a. Posthuman Humanities. *European Educational Research Journal* 12, 1–19. https://doi.org/10.2304/eerj.2013.12.1.1

Braidotti, R., 2013b. *The Posthuman*, 1st edition.

Braungart, M. & McDonough, W., 2002. *Cradle to Cradle: Remaking the Way We Make Things*, 1st edition. North Point Press.

Buckingham-Hatfield, S. & Evans, B. (Eds.), 1996. *Environmental Planning and Sustainability*, 1st edition. Wiley.

Burry, M., 2012. Towards meeting the challenges of facilitating transdisciplinarity in design education, research and practice, in: *Design Innovation for the Built Environment: Research by Design and the Renovation of Practice*. Routledge, pp. 53–66. https://doi.org/10.4324/9780203129746-5

Büscher, B. & Fletcher, R., 2020. *The Conservation Revolution: Radical Ideas for Saving Nature Beyond the Anthropocene*. Verso Books.

Cadwalladr, C., 2013. Stewart Brand's Whole Earth Catalog, the book that changed the world. The Guardian.

Carrol, J.M., 1999. Five reasons for scenario-based design, in: Proceedings of the 32nd Annual Hawaii International Conference on Systems Sciences. Presented at the P32nd Annual Hawaii International Conference on Systems Sciences., Maui, HI, USA. https://doi.org/10.1109/HICSS.1999.772890

Carson, R., 1962. *Silent Spring*. Penguin Classics.

Charalambous, N. & Christou, N., 2016. Re-adjusting the Objectives of Architectural Education. Procedia – Social and Behavioral Sciences. Presented at the 2nd International Conference on

Higher Education Advances, HEAd'16, June 2016, Valencia, Spain 228, 375–382. https://doi.org/10.1016/j.sbspro.2016.07.056
Cielemęcka, O. & Daigle, C., 2019. Posthuman sustainability: An ethos for our anthropocenic future. *Theory, Culture & Society* 36, 67–87. https://doi.org/10.1177/0263276419873710
Clune, S. (2014). Design for sustainability and the design studio. *Fusion Journal* 3, 1–16.
Deviren. S. & Tabb, D., 2014. *The Greening of Architecture: A Critical History and Survey Of Contemporary Sustainable Architecture and Urban Design*. Ashgate Publishing.
Dutton, T., 1991. *Voices in Architectural Education*. Bergin and Garvey.
Eilouti, B., 2018. Scenario-based design: New applications in metamorphic architecture. *Frontiers of Architectural Research* 7, 530–543. https://doi.org/10.1016/j.foar.2018.07.003
Elrayies, G.M., 2017. Flipped Learning as a Paradigm Shift in Architectural Education. *International Education Studies* 10, 93–108.
Farallones Institute & Van der Ryn, S., 1982. *The Integral Urban House: Self-reliant Living in the City*, 1st edition. Random House, Inc.
Goldschmidt, G., 1994. On visual design thinking: The vis kids of architecture. *Design Studies* 15, 158–174. https://doi.org/10.1016/0142-694X(94)90022-1
Greta Thunberg UN speech, 2019. United Nations Climate Action Summit.
Griswold, E., 2012. How 'silent spring' ignited the environmental movement. *The New York Times*.
Grover, R., Emmitt, S. & Copping, A., 2019. Reflecting on sustainability: coproducing a critical framework for sustainable design in the architectural studio. *Higher Education Pedagogies* 4, 41–63. https://doi.org/10.1080/23752696.2019.1624589
Gürel, M.Ö., 2010. Explorations in teaching sustainable design: a studio experience in interior design/architecture. *International Journal of Art & Design Education* 29, 184–199. https://doi.org/10.1111/j.1476-8070.2010.01649.x
Guy, S. & Farmer, G., 2015. *Reinterpreting Sustainable Architecture: Theories, Discourses, Practices*, Routledge.
Guy, S. & Moore, S.A., 2007. Sustainable Architecture and the Pluralist Imagination. *Journal of Architectural Education* 60, 15–23. https://doi.org/10.1111/j.1531-314X.2007.00104.x
Haraway, D., 1991. A cyborg manifesto: Science, technology, and socialist feminism in the late twentieth century, in: *Cyborgs and Women: The Reinvention of Nature*. Routledge, pp. 149–181.
Haraway, D. 2016. *Staying with the Trouble: Making Kin in the Chthulucene*. Duke University Press.
Hemsath, T. & Bandhosseini, K.A., 2017. *Energy Modeling in Architectural Design*. Routledge. https://doi.org/10.4324/9781315712901
Jessen, T.D., Ban, N.C., Claxton, N.X. & Darimont, C.T., 2022. Contributions of Indigenous Knowledge to ecological and evolutionary understanding. *Frontiers in Ecology and the Environment* 20, 93–101. https://doi.org/10.1002/fee.2435
Kallipoliti, L., 2018. History of Ecological Design. Oxford Research Encyclopedia of Environmental Science. https://doi.org/10.1093/acrefore/9780199389414.013.144
Kallitsis, P. & Andrews, M., 2020. Architectural Responses in Alternative Realities: The Politics of Space through Fiction in Architectural Education. *SFRA Review* 50, 8.
Klein, N., 2015. *This Changes Everything: Capitalism vs. The Climate*. Simon & Schuster.
Kolb, D., 2014. *Experiential Learning: Experience as the Source of Learning and Development*, 2nd edition. Pearson.
Kreber, C. (2002) 'Controversy and Consensus on the Scholarship of Teaching', *Studies in Higher Education* 27, 2, 151–167.
Kuhn, T.S., 1962. *The Structure of Scientific Revolutions*. University of Chicago Press.
Latour, B., 2007. *Reassembling the Social: An Introduction to Actor-Network-Theory*, 1st edition. Oxford University Press.
Lave, J. & Wenger, E., 1991. *Situated Learning: Legitimate Peripheral Participation*. Cambridge University Press. https://doi.org/10.1017/CBO9780511815355

Lüley, M., Pifko, H. & Špa, R., 2019. Adaptability and a scenario-based design methodology for architectural education. *Global Journal of Engineering Education* 21, 97–102.

Maslin, M.A. & Lewis, S.L., 2015. Anthropocene: Earth System, geological, philosophical and political paradigm shifts. *The Anthropocene Review* 2, 108–116. https://doi.org/10.1177/2053019615588791

Matthews, M., 2010. *Droppers: America's First Hippie Commune*, Drop City, 1st edition. University of Oklahoma Press, Norman Okla.

McDonald, B., 2020. Two early images of Earth that bolstered the environmental movement. CBC radio. Available at: www.cbc.ca/radio/quirks/two-early-images-of-earth-that-bolstered-the-environmental-movement-1.5534843 (accessed 21 December 2022).

Merriam-Webster, 2022. Definition of paradigm shift. Merriam-Webster Dictionary. Available at: www.merriam-webster.com/dictionary/paradigm+shift (accessed 19 December 2022).

Mezirow, 2018. Transformative Learning Theory. In K. Illeris (Ed.), *Contemporary Theories of Learning*, 2nd edition, 29–45. Routledge. https://doi.org/10.4324/9781315147277-8

Monroe, M.C., Plate, R.R., Oxarart, A., Bowers, A. & Chaves, W.A., 2017. Identifying effective climate change education strategies: a systematic review of the research. *Environmental Education Research* 25, 791–812. https://doi.org/10.1080/13504622.2017.1360842

Moore, J.W., 2016. *Anthropocene or Capitalocene?: Nature, History, and the Crisis of Capitalism*. PM Press.

Naess, A., 1973. The shallow and the deep, long-range ecology movement: a summary. *Inquiry: An Interdisciplinary Journal of Philosophy* 16, 95–100. https://doi.org/10.1080/00201747308601682

Nicol, D. & Pilling, S., 2005. *Changing Architectural Education: Towards a New Professionalism*. Taylor & Francis.

Olweny, M., 2018. Introducing sustainability into an architectural curriculum in East Africa. *International Journal of Sustainability in Higher Education*, 19, 6, 1131–1152.

Pettit, J., 2010. Multiple Faces of Power and Learning. IDS Bulletin 41, 25–35. https://doi.org/10.1111/j.1759-5436.2010.00134.x

Prosser, M. & Trigwell, K., 1999. *Understanding Learning and Teaching: the experience in higher education*, Society for Research into Higher Education. Open University Press, Buckingham, UK.

Quinlan, A., Corkery, L. & Marshall, N., 2007. Positioning the Design Tutor's presence in the Design Studio for successful student design learning. CONNECTED 2007 International Conference on Design Education, 9–12 JULY 2007, University of New South Wales, Sydney, Australia.

Rice, L. & Littlefield, D., 2014. *Transgression: Towards an Expanded Field of Architecture*. Routledge.

Ryn, S.V. & der, Cowan, S., 1996. *Ecological Design*. Island Press.

Sadler, S., 2006. Drop city revisited. *Journal of Architectural Education* 59, 5–14.

Salama, A.M., 1995. *New Trends in Architectural Education: Designing the Design Studio*. Tailored Text and Unlimited Potential Publishers.

Scott, F.D., 2016. *Outlaw Territories: Environments of Insecurity/Architecture of Counterinsurgency*. Zone Books.

Seren Smith, M., Warnes, S. & Vanhoestenberghe, A., 2018. Scenario-based learning. In: Davies, J.P. and Pachler, N, (Eds.) *Teaching and Learning in Higher Education: Perspectives from UCL*. 144–156. UCL IOE Press.

Stead, N., 2003. Producing critical thinkers, designing critical objects: re-examining the role of critique in architectural education. The Second International Conference of the Association of Architecture Schools of Australasia.

Thompson, P.B., 2016. The many meanings of sustainability: a competing paradigms approach, in: Moore, S.A. (Ed.), *Pragmatic Sustainability*. Routledge, 16–29. https://doi.org/10.4324/9781315648545-13

Trigwell, K. and Shale, S., 2004. Student learning and the scholarship of university teaching, *Studies in Higher Education* 29, 4, 523–536.

Tsioumanis, A., 2020. Climate change policy and the need for paradigm shift – Thessaloniki Office, Heinrich-Böll-Stiftung. Available online: https://gr.boell.org/en/2020/02/05/climate-change-policy-and-need-paradigm-shift (accessed 22 March 2022).
Wackernagel, M. & Beyers, B., 2019. *Ecological Footprint: Managing our biocapacity budget*. New Society Publishers.
Warburton, K., 2003. Deep learning and education for sustainability. *International Journal of Sustainability in Higher Education* 4, 44–56. https://doi.org/10.1108/14676370310455332
Webster, H., 2004. Facilitating critically reflective learning: excavating the role of the design tutor in architectural education. *Art, Design & Communication in Higher Education*, 2, 3, 101 –111.
Whitehead, M., 2007. *Spaces of Sustainability: Geographical Perspectives on The Sustainable Society*. Routledge.
Zalasiewicz, J., Waters, C. N., Williams, M. & Summerhayes, C. P. (eds), 2019. History and Development of the Anthropocene as a Stratigraphic Concept, in *The Anthropocene as a Geological Time Unit: A Guide to the Scientific Evidence and Current Debate*. Cambridge University Press, 1–4.

PART 3

Architectural practices to address the climate emergency

Justice, interdisciplinarity, co-design and decolonisation

Part 3 of the book focuses on architectural thinking in practice. As in the previous two parts of the book, value changes, rethinking the status quo (of how things are versus what they could be) and different ways of working are highlighted.

Interdisciplinary working has been elevated in Chapter 11 by Cara Mulholland, Jenny Hughes and Flora Samuel. They set out the need for a framework for mapping of social value to be co-created with communities, also referred to as 'active' mapping datasets.

The chapter opens by highlighting – as in other chapters – that climate change cannot be addressed 'without revealing and tackling the inequalities within society and where they are happening' (p. 194) and why therefore any research, actions and discussions around climate change must take place at the local level. The chapter focuses on social value mapping and how social value can feed into formal planning processes. To do so, the authors argue that any 'active datasets' must be combined with 'passive datasets' (i.e., the combination of the 'active' citizen-co-created data with passive national socio-economic, environmental data etc.).

Key to this happening is new ways of working and collaborating: not only to collect local active data with citizens and communities, overlaying with other data and interpreting together with other expert stakeholders. The authors highlight that an obstacle to effective climate adaptation planning is the compartmentalisation at an even larger scale than that described in Chapter 4, where in England, for example, agriculture and urban planning are separate policy jurisdictions, yet agricultural land management affects flood events in urban areas (e.g., where flood waters run own the fields and overwhelm urban infrastructures lower down the hill). Data is often also collected according to local administrative boundaries, but climate risks do not change or stop at these boundaries, and this creates a lack of coordinated action; this will be returned to in Chapter 14 where Maria Vidali discusses a community of local farmers working together to make best use of the resources in the given climate and with the land's constraints.

Key questions of inclusivity and democratic process are also raised in this chapter, similarly to other chapters elsewhere: how can more voices be heard and underrepresented groups be included in the planning process, which needs to be democratised to

DOI: 10.4324/9781003293903-13
This chapter has been made available under a CC-BY-NC-ND 4.0 license.

capture and target local and small-scale climate interventions. This also includes issues such as how data is collected, its affordability and ease of access and understanding the data. Without this, action for climate change will disproportionately affect left-behind communities if it is not known where to best allocate scarce resources for climate adaptation planning and other actions.

This requires different professionals working together, and becoming guardians of both hyperlocal 'active' data and 'passive' data but also built environment professionals getting involved in active data collection with communities and with other large-scale passive data sets to ensure they can be mapped back to spatial aspects and that they can be used for evidence-based decision making in climate mitigation action.

The use of social value maps, which build on community knowledge and experience can include information about deprivation, health, security, location of services, general wellbeing, etc., and together with scenario mapping can highlight and illustrate the impact of different transformation and adaptation actions. In Chapter 15, Renata Tyszczuk illustrates the value of scenarios beyond forecasting in more detail.

In Chapter 12, McAleer-Harding, Abbott, Kiddle and Pedersen Zari extend the argument for hyper-locality to Aotearoa (New Zealand), emphasising the uneven distribution of disasters and the vulnerability of social and physical systems that support societal functioning. As highlighted in previous chapters in this book, marginalised communities are particularly vulnerable to the impacts of climate change and disasters, including pandemics, earthquakes, floods, cyclones, storms etc.

Due to the unpredictability of disasters, some disaster response management strategies have shifted from focusing on controlling disasters to managing their impacts by developing community resilience. This approach prioritises improving pre-disaster resilience to reduce recovery time after disasters. It aligns with the concept of adaptive capacity, emphasising a society's or community's ability to adapt to change rather than relying solely on post-disaster responses. However, the authors highlight that definitions of resilience and corresponding responses are often top-down and centred on individual persons or buildings. This neglects the social and cultural values that form the social fabric of communities, which are crucial for post-disaster resilience. As a result, resilience strategies often miss opportunities to truly benefit the communities they aim to serve. To address this, the authors call for greater inclusion of communities to understand their knowledge, identities, needs and ideas.

Specifically, this chapter tackles the issue of colonisation and how this has affected the resilience of some communities, especially Māori (Indigenous peoples of Aotearoa). This poses challenges for architecture to ensure that physical resilience structures are designed and created *with* and *for* the benefit of the communities they are intended to serve, rather than perpetuating colonial paradigms. As in Chapter 11, the co-design of resiliency strategies with local communities is emphasised, alongside embedding appropriate cultural practices. This approach ensures Māori communities and their knowledges are authentically represented rather than reinforcing dominant *Pākehā* (New Zealand people of European descent) worldviews. When working with such an approach it is crucial that 'causes, consequences, and ways to address disasters must be informed by an analysis of power and intersectional identities, and understandings of who benefits from, pays for, and must live with impacts' (p. 214).

The authors emphasise that technology-driven solutions may make it more difficult to create community-focused approaches. This is a particularly important issue for Indigenous communities as they are typically more vulnerable to the impact of disasters and are often exposed to (or forced into) situations of increased risk, further perpetuating and exacerbating inequalities and impacts of colonisation. Colonisation is an ongoing process where societal and political structures deny Indigenous peoples' rights and erase their perspectives, identities and knowledges in favour of the majority's worldview and structures and systems that uphold this. The authors highlight that this challenge is further exacerbated by the architecture profession, which predominantly represents the majority worldview and is largely non-Māori.

The authors acknowledge their own positionality, recognising that the work we do as researchers in the built environment affects and is driven by our and others' worldview, and can perpetuate or reduce climate and social injustices. The chapter highlights our interconnectedness, asserting that diminishing the rights of any group of people in society, undermines the well-being and resiliency of the entire community. The authors argue that resilient architecture is 'more than just the design of a strong physical structure' (p. 223), and go on to powerfully argue that decolonised pathways can foster greater social resilience and inclusivity for all Indigenous peoples and vulnerable communities, ultimately enhancing the resilience of the built environment.

The authors propose a framework, *The Ethics of Restoration*, to be adopted within built environment discourse. This framework aims to create space for Māori people to '"Indigenise" the built environment on their own terms and from own worldview' (p. 219). It calls for non-Māori architects to step back, actively remove barriers for Māori to enter the profession, and work to eradicate power imbalances and colonial practices. The coming together of two knowledge bases that are valued equally, involves 'decolonisation co-design' processes where the design team and community build relationships, from the very beginning before any design propositions are made. To facilitate this, the authors propose a *Decolonisation Toolbox* for architects. The first is the creation of a 'third space', a shared platform where Indigenous and non-Indigenous stakeholders can engage in respectful conversations and reimagine architectural practices in ways that align more closely with Indigenous worldviews.

Faith Ng'eno Chelang'at's chapter 'Situating the concept of sustainable design in Africa' (Chapter 13) further delves into issues of colonial and ongoing international influences in the imposition of knowledge and standardisation imposed of sustainable design in the Global south. Particularly the author highlights the enduring legacy of colonial practices in Africa (Kenya in this case), where regulations, standards and theories are applied (and imposed) from Western countries or are influenced by them, instead of concepts of sustainability coming from local cultural practices and being constructed locally. The author asks, 'sustainable for whom?' and explores why and how this international influence continues to repress local knowledge and concepts of sustainability in Africa. Chelang'a also raises that the continent of Africa has hardly contributed to global CO_2 emissions compared to the global north, yet Africa can significantly shape the climate crises given that it is engaged in new building activity and how sustainably it does so will affect the future course of global emissions and ecological impacts. As the author puts it: 'In the case of sustainability, knowledge is largely constructed in terms of CO_2 emissions, global warming, and climate change, which is grounded in Western science and thus proposed solutions, prioritise adapting wealthy, culturally rich Western lifestyles, without endangering that wealth. In doing so, other contexts, especially African contexts, are deemed irrelevant and often excluded.' (see p. 229).

But, as the author argues, these international approaches and solutions do not respond to the local challenges and priorities, and a tension exists between local understanding and approaches versus pursuing legitimacy from the west which has set the standards and makes claim as global gatekeeper of knowledge and solutions and educational systems.

As Chelang'a writes, this reliance and monopolisation of the west's foreign knowledge and solutions that are considered aspirational cripples local innovation, colonises Indigenous imagination, creating barriers to meaningful local contributions when entrapped in foreign, universalist knowledge presented as the only form of valid knowledge.

This also links to capitalism and the exploitation of people and planet for profit as discussed in other chapters; but here the author raises the issue of the commodification of sustainability knowledge, embedded in Western technological solutions and rating systems and certifications that come at a premium price point.

Instead, the author argues for situated knowledge that is context specific rather than being a universalist construction of (Western) knowledge. Self-determination is crucial in this, as also argued in the previous and subsequent chapters, where local people hold the power to define solutions, what knowledge is, how to assess and certify it, and also to define what problems need solving in the first place. This will require deconstructing and reconstructing knowledge, recognition of colonial history and power structures and interests, and how knowledge was shaped (imposed) through the dialogue between local stakeholders. This should lead to a reconstruction of concepts of sustainable design, highlighting the need for a plurality of contextual responses based on physical and socio-political differences. Without this local, situated framing and solving of problems, a sustainable world is simply out of reach.

Maria Vidali also highlights that this holds true in the West itself and in Chapter 14 discusses 'Boundaries and bonds in a boundless world: the relationship of place and community to landscape' and illustrates the power of local narratives to understand local issues and highlights the social and ethical role architecture plays within this.

Maria Vidali describes the case study of Kampos, a village on Tinos, a Cycladic island in the Aegean Sea that was occupied by the Venetians (1390–1715 AC) to defend against the Ottoman Empire. The island's rocky terrain was terraced for farmland and it also defined boundaries set out in contracts. The way farmland was measured had to change when Greece joined the European Union (from the time it took two oxen to plough and the crops and trees that had been planted to measuring land in meters).

The author uses fictional narratives based on villager's metaphors of their stories, spaces, life with each other and animals, and illustrates how this way of working 'reveals the complexity of life with nature and its changes' (p. 244) and questions how we as architects often do not allow room for people's stories and local cultures before we shape space. Instead, indigenous people usually have a close understanding and awareness of their landscape, its environment, the seasonal and climatic changes happening to it, the needs for it, the sharing and exchanging of resources with others (water, machinery, land, knowledge, farming products), and giving it meaning. The use of new technologies and imposed solutions by others may take this away, disconnecting from environmental and human reality. Instead, the author argues that the use of local narratives can aid architects to connect to the everyday reality, rather than architects – as is typically the case – creating abstract and stylistic interventions that are removed from their context. As Vidali concludes: 'This can lead to interpretations in architecture and design that are more intimate and connected to human living, closer to the actual complexity of life and the environment, to an awareness of seasonal changes, and to an understanding of the climate emergency as something that we all face and care to respond to' (see p. 257).

This leads to the final chapter of this book, in which Renata Tyszczuk uses scenarios as part of 'An architecture of paying attention: scenarios for rehearsing climate futures otherwise'. Tyszczuk uses scenarios as a mode of rehearsal – 'for improvising climate futures that are uncertain and unknowable' (p. 263) in the here and now. The chapter starts by reiterating the urgent issues at stake framed through the Anthropocene and the injustices associated with the uneven distribution of impacts, including that for future generations. The chapter also cautions that any international agreements to contain climate change, environmental pollution and ecosystem degradation have been based on best-case scenarios. Instead an 'architecture of paying attention' is about 'rehearsing the future by attending to the present' and asks six questions, along which the chapter is also structured: what if, what now, what works, what matters, what for, what next?

In architecture, the author argues that scenarios, instead of being tools to forecast and predict, can be used by architects as a potential way to enable collective improvisation based on paying attention by all those taking part at times of uncertainty to improve the future (i.e. rehearsing climate futures) rather than aiming to 'capture the climatic future in advance, or "stop the future happening"' (p. 263). As other chapters also highlighted, architecture is a social activity, i.e., working (rehearsing, improvising) together in democratic conversations to work out what to do next, collectively and with curiosity, allowing space for unexpected and contradictory perspectives and configurations. The rehearsal of future climate scenarios also is a rehearsal of the skills needed and working out connections and roles of and between people, systems, species infrastructures, etc. As the author writes, it is 'about testing how a more improvised and speculative undertaking might challenge preconceptions, accommodate differences, and at the same time respond to widely diverging experiences of climate change' (see p. 264).

An architecture of paying attention ultimately also means an architecture of care, i.e., all that is undertaken to maintain, continue, and repair where all are concerned with caring for others and the world. This is a messy process – as also iterate in other chapters – as it is a moving away from 'the way things are' to 'how they could be otherwise'. To conclude with the author's own words: it 'attends to climate futures, as a collective endeavour, as a matter of urgency, in trial and error, by taking the time it takes, and not giving up. It is a provisional practice agile enough to cope with the unruly temporalities of climatic disruptions, and does so with care and humility' and 'It is a matter of paying attention to the future in the present' (p. 268).

11

PLANNING FOR CLIMATE JUSTICE

Digital mapping as an interdisciplinary tool for alleviating inequalities

Cara Mulholland, Jenny Hughes, and Flora Samuel

Introduction

Climate change is a social justice issue (Dasgupta, 2021). It cannot be addressed without revealing and tackling the inequalities within society and where they are happening. This, we argue, is why geographic data is so critical to any discussion of climate change. Only when we know what is happening, and where, can we make strategic, holistic decisions about the best use of resources for people and planet. This is going to be particularly urgent and crucial given that 'leaving no one' behind in the green transition is part of policy in the EU, the UK and other countries around the world. This chapter examines the role of digital map making in spatial planning to address climate change. Although the focus of the discussion is on the UK, in particular the distinct neoliberal planning system of England, the issues raised are equally relevant to other parts of the globe, particularly the Wellbeing Economy Governments (WEGOs) that prioritise wellbeing over economic measures.[1] The chapter also describes the advent of social value mapping, making maps with communities about the things that they value in their area, to feed into formal planning processes. This is a growing field led by the Participatory Mapping Institute at Aalto University in Finland.

The Intergovernmental Panel on Climate Change (IPCC) notes that 'collective action & strengthened networked collaboration, more inclusive governance, spatial planning and risk-sensitive infrastructure delivery' are necessary for reducing climate change risk (2022, p. 110). Digital technology and big data advancements will increasingly allow us to spatialise, and understand in real time, changes in resource use and value, making the use of fixed plans redundant. We are moving to a situation of 'adaptive planning' for planning on the move in response to critical situations like the pandemic (Harrison, Galland and Tewdwr Jones, 2021). Unfortunately, lack of investment in the planning sector means that spatial planning is lagging behind other fields in its use of data (Wilson and Tewdwr Jones, 2021), in particular spatialised data. Whilst the UK has remarkable Ordnance Survey maps built on the work of our Victorian ancestors, investment in co-ordinating our

DOI: 10.4324/9781003293903-14
This chapter has been made available under a CC-BY-NC-ND 4.0 license.

twenty-first-century digital mapping infrastructure is lacking.[2] The focus of this chapter is on what needs to be done to improve the digital mapping infrastructure of England and is built on a series of interviews with industry leaders in this area.[3]

The chapter begins with a brief introduction to the planning system in England before moving to a discussion of the usage of existing 'passive' data sets, for example national meteorological, environmental and socio-economic data that is gathered through the census and other government data gathering exercises; and 'active' data sets, which are community context-specific 'subjective' data that is generated directly with citizens. We highlight that the passive and the active must be brought together as a whole if we have any chance of capturing 'social value' and including it in decision making in planning for a resilient, climate-just future. This will require new kinds of professional clusters bringing together data analytics, GIS and programming, and skills for working in a respectful way with communities. The findings were developed through desktop research, embedded action research through stakeholder advisory and feedback groups through Stantec UK[4] and a series of interviews with experts working at the fore of the intersecting fields of planning, social justice, stakeholder engagement and digitisation (Mullholland et al., 2022).

The planning system in England as a tool for addressing climate justice

The planning system in England focuses on built development and specifically excludes agriculture and forestry activity (Shoard, 1997), meaning that the landmass of England is rarely looked at holistically as a system of spatial inputs and outputs (Perrotti and Stremke, 2018). Furthermore, agriculture and urban planning fall under different policy jurisdictions. Examples of the consequences of this compartmentalised way of thinking include the flooding that happens in settlements on river valleys due to the denuding of land in the hills above it. An example of this is in Hebden Bridge in Yorkshire, where the community have been fighting for decades to have some say in the way the land above them is managed (Monbiot, 2015). It is also well understood that there is overall inequality in terms of social deprivation and flood risk exposure from all sources of flooding in the UK, which would be further affected by future climate change. The UK is notably poorly prepared for climate change, with a widening gap between the risks faced and the investment and the implementation of adaptation measures to increase resilience to climate impacts (CCC, 2021).

The social implications of how the environmental data management is carved up is symbolic of the status of planning in the UK. That the planning system in England is on its knees due to public sector cuts is widely recognised (RTPI, 2019). Public sector planners, whether strategic or in development control, are barely able to achieve their caseload and have almost no time for professional development or resources to update antiquated digital systems (Boland et al., 2020). Nor do they have access to the necessary range of in-house expertise necessary to make truly informed decisions. Much of what happens in English local authorities is focused on the Local Plans that are developed on a five-yearly basis, but which are arguably out of date almost as soon as they are made (Harrison, Galland and Tewdwr-Jones, 2021). They are often associated with static maps that fail to allow for interaction or further analysis. These Local Plans are based on blunt measures of housing need assessment (Clapham, 2019), economic assessment and 'sustainability assessment' with less weighting and consideration of explicit social value when compared with economic and financial valuations. Hence the need for a Geographic Information System (GIS), a data-led system that makes it very clear how the long-term triple bottom line of sustainability – environmental, social and economic value – is being achieved.[5] This triple bottom line approach is embedded in the *UK Treasury Green Book* (2018) that sets out the rules for government spending but there appears to be no correlation between this type of thinking and government statements on the future of planning released at similar times (MHCLG, 2020). Even in the most recent National Planning Policy Framework (MHCLG, 2024) there are stipulations of the use of digital tools to engage with the public, but still only mention of static policy maps.

Some key questions related to inclusivity and understanding and ensuring social justice include: How do we better democratise the planning process? How do we make it easier for more voices to be heard? How do we reach more underrepresented groups? How do we reach more young people, make it easier for them to engage? How do we reach more ethnic minorities that aren't normally engaged and other hard-to-reach groups? (adapted from Government PropTech representative, 2021).

The UK Government repeatedly acknowledges the need to address inequality within the planning system, though quite how it proposes to do this measurably when the existing data management and mapping system of England at the local level is so fragmented and incomplete is unclear. The renaming of the Ministry of Housing Community and Local Government to the Department of Levelling Up, Housing and Communities, then back again only three years later demonstrates the lack of strategic ambition which is exasperated by changing leading political parties. Even the register of land ownership has gaps (Shrubsole, 2019).

Although the UK government is very alive to the need to develop a geospatial strategy (UK Gov, 2023), there is considerable work to be done if the UK is to take a consistent approach to data and planning both at the macro and micro scale. It is for this reason that the Digital Task Force for Planning is setting itself up to create an open source 'national trust' for maps. Scotland has proposed a joined-up digital planning strategy in line with its enlightened approach to land and land access (Scottish Government, 2020). The government in England is aware that investment will be needed if its 'PropTech', the digital transformation of the property industry aspirations are to be achieved. The emphasis on real estate terminology, rather than the terminology of holistic systems thinking, should be noted. The government has released four rounds of Proptech Engagement funds from 2021–2024 to support pilot projects in this area as part of an effort to make planning more systematic and digital, including a new emphasis on the use of Design Codes for planning. These National Model Design Codes (MHCLG, 2021a) focus on issues of form and aesthetics rather than value, but have the potential to be used to encode social, environmental and economic value into a map-based planning system, in other words sustainability.

Collating and sharing existing (passive) datasets on digital maps

There are some very useful existing datasets already which begin to unpack the stories of social value of places. We call them 'passive' as they already exist and do not actively have to be collected by a project team through new active engagement. These are ways that communities can passively contribute data and knowledge.

In terms of the data that is regularly collected about the UK, the UK Statistics Authority 'promotes and safeguards the production and publication of statistics that serve the public good' (Statistics Authority, 2021). Much of this is collected by the Office for National Statistics,[6] the recognised national statistical institute of the UK. It is responsible for collecting and publishing statistics related to the economy, population and society at national, regional and local levels, which it gathers using its own stated 'Census Geography' (ONS, 2021), its methodology for dealing with things like boundaries.

Another common data source used by built environmental professionals is the Department for Environment, Food and Rural Affairs' MAGIC[7] online platform, detailing the natural environment of Great Britain, which is being expanded by Natural England with their newly established Green Infrastructure database.[8] Building on these, we can include the MHCLG open data platform[9] sharing their significant data used, or the more general data.gov.uk[10] portal, which stores all available data from the UK Government.

These robust, standard datasets set a precedent for strong analytical practice in planning for the built environment, with increasingly mature data science telling a rich story for acting on the climate emergency. They provide data on social value and social inequality such as deprivation, health, security, location of services, general wellbeing, and others. Which as standalone data may not convey a full picture of social value, but when interrogated together begin to detail the richer story.

National datasets are relatively straightforward. It becomes more challenging when working on a local level as these datasets often depend on how data is structured depending on who owns or manages it, and on specific projects this needs to be supplemented by the professionals through ground surveys or further collation and analysis. This is where there is space for variation in quality with prescribed or standardised approaches being asked for more widely. A further issue is lack of the hyperlocal detail necessary for capturing and targeting small-scale climate interventions. Another challenge of the localised data is that it is often clustered according to political boundaries and is spatialised using several levels of output areas (such as a local authority, ward), which may not reflect the fluid boundaries in real life, even though these are regularly reviewed by the Boundary Commission. Further, there is very little cross-correlating of data sets (Browne-Gott, 2019). This is because approaches have emerged organically according to need. "A lot of the data that is out there or could be out there better is published by local authorities. And I don't think they're trying to be cagey about it. They all just are given the freedom to publish it in different ways in different places" (Government PropTech representative, 2021). So complex is this world of data that there a range of data subscription services which charge considerable sums for access to advice and maps. Anecdotally, local authorities rarely use these services as they have to be paid for as an expensive subscription service. Considerable public investment is now being used through the Geospatial Commission to begin breaking down these barriers (Geospatial Commission, 2023). To enable better decision-making, this data should be available to local authorities and other stakeholders for free. But the challenge will remain to encourage the uptake of these new data-mapping methodologies across industry. None of this is helped by the fact that digital strategy in local authorities has grown organically, meaning that they use different, and sometimes incompatible, systems. All of these are common barriers to a consistent approach, and overcoming these would help the development of big data streams, which is now slowly being addressed in the development of new social value measurement methods.

Data access is a highly political issue and while 'digital exclusion' has become a buzz word during the pandemic, it lacks a solid definition (Greater Manchester Council, 2021). 'Digital exclusion is everyone's problem and nobody's responsibility' (Greater Manchester Researcher, 2021). Although Britain espouses to being an advocate of 'open data' enabling ordinary people to understand what is happening in their environment, the reality is that most data is extremely inaccessible (Wang and Shepherd, 2020). This is exacerbated by current lack of trust in how location data is gathered from people, with a need for greater transparency, accountability and clarity on how people can be involved not only as data sources but as 'data citizens' (Maxwell et al., 2021). Giving access to well-presented, transparent and reliable data on places is key to democracy and is crucial if communities are to make informed decisions about how to alleviate climate change.

Despite all this, there are some local authorities that are facing the data challenge head on, a notable example being the Greater Manchester Combined Authority. Back in the 1960s Ian McHarg, author of *Design with Nature* (1969), advocated a system of planning that involved the use of overlaid mapping techniques which would enable us to see ecosystems at work in a holistic way. Manchester city are using a similar technique in creating a framework of overlaid maps as a basis for feeding into decision making across the city, The Greater Manchester Infrastructure Framework 2040 (GMCA, 2019). Their starting point was to create digital maps that contained the often needed and used existing datasets across departments, but this was expanded to be viewed openly via the internet when the wider value was realised, to public and to stakeholders across the industry. This further developed into several communication projects that directly used the maps to gather information on particular sites and potential infrastructure projects, leading to more informed engagement strategies at later stages. 'We've got something there that we can just keep building on. And I think I'd like to see something where we start to bring together more information and make it more consistent. There's an opportunity to expand that nationally' (Greater Manchester Mapping researcher, 2021).

Further work is needed for local spatial analysis and mapping of climate change risk and adaptation for holistic decision-making at the local level. Each map and spatial presentation of data needs objective assessment. Algorithms can be used to give weightings to different map layers based on the quality of the data sources used and the challenge of spatialising data that is collected by political wards. Work is needed to cluster maps together into a single schema to allow a cluster of maps to collectively represent, for example, environmental value. It is only a matter of time before algorithms are built into the maps to factor in things like accessibility – for example, the way in which a major road prevents children from accessing play facilities. The passive data maps can fairly readily be clustered to give maps of environmental and economic value. What is, however, more difficult is to consistently and coherently map social value. This is in essence the missing area of spatial analysis.

Developing new (active) social value data with community maps

Environmental, social and economic value are intimately linked, as this diagram by the UK Green Building Council makes clear (Figure 11.1). They all have to be taken into account if we are to have any chance of tackling climate change. Unlike environmental value, which has some reasonably well-established methods of calculation (e.g. CO_2 and

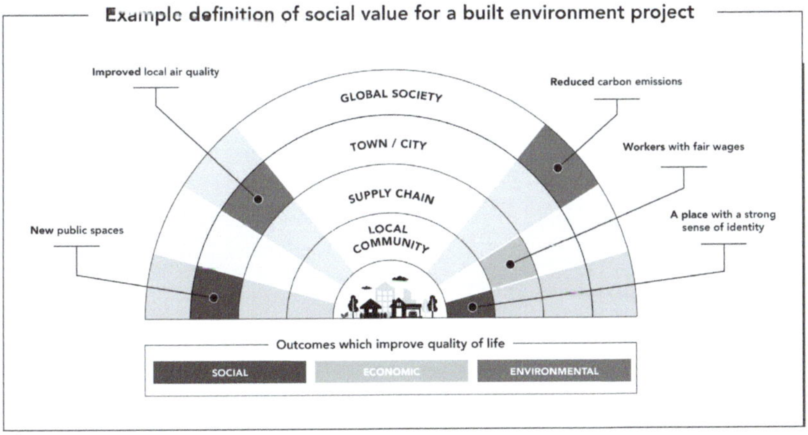

FIGURE 11.1 Social Value Framework, UKGBC.

biodiversity) or economics, social value is an unexplored area, partly because it relates to intangible qualitative outcomes that are difficult to measure but also partly because it is strongly aligned to the unpaid activities of women, which are traditionally left out of calculations of Gross Domestic Product (Criado Perez, 2019). In an increasingly data-driven environment, we have to find ways to capture intangible impacts or they will not figure in digital twin city models, Building Information Modelling or value-based procurement (CIH, 2021).

Social value[11] is key to achieving and operationalising the triple bottom line of sustainability, and ultimately needs to be better understood to address climate inequality and ensure a just transition to a net-zero carbon economy. Creating social value through the planning of new communities relies on responding closely to community need, building what fulfils the current needs while anticipating future need (Social Value Portal, 2021; Fitton and Moncaster, 2021); engagement with communities is key to this process.

Social value has been on the ascendant in the UK since the inception of the Social Value Act 2012 and the Future Generations Act (Wales) 2015 (Samuel and Hatleskog, 2020), but it is a complex picture of growth. The momentum and support of social value has continued with the Labour Government in 2024 supporting more policy and legislation to push the work. This seems at odds with other UK Government policy, which does not necessarily offer dignity and opportunity to every individual (Nussbaum, 2011). And 'social value' is often positioned at odds with other sustainability-focused work, such as the UN Sustainable Development Goals as the *RIBA Sustainable Outcomes Guide* makes very clear (Clark, 2019). This can be seen in organisations having separate strategies or teams to deal with frameworks, but they have the same fundamental goals. At times, the 'social' is separated from the environmental and technical considerations, or the social can be fragmented and minimised in itself.

We are all by now familiar with the potential of GIS technology to spatialise and geo-locate all manner of experiences and how this is increasingly being utilised in a variety of creative ways. Here holds the potential to bring complex and interdependent social value considerations into decision making through the medium of digital maps. The 'Know your Place' maps of Bristol are a case in point. They allow visitors to explore their neighbourhoods through historic maps, images and linked information (Bristol, 2021).

However, the spatial variations of social value across a site, project or asset are still under-explored and under-utilised in the maturing practice of social value. The spatial data are sorely needed as more large projects are attempting to communicate their value and provide enhanced outcomes (Mulholland, Ejohwomu and Chan, 2019). Innovation, research and development is needed to explore new ways of working to meet the potential of meeting capabilities through digital social value processes.

> And that's why we've almost got failed places because from a technical perspective, we think we know what works. But from a social perspective, it doesn't... start talking to those communities. Get them on board, get those little bits of insight and those little bits of knowledge that we can't get from a desk study.... that we can't get from just looking at big data. And using that to enhance the project.
>
> *(Social value and planning expert, 2021)*

Trialling spatialised social value data creation with communities

In the UK built environment industry, social value is typically ascribed to dimensions of the construction process, for example job creation. The *Social Value Toolkit (SVT) for Architecture* was developed as a prompt for architects to promote the social value of buildings and places in use (Samuel, 2020). One of the key aims of the SVT was in helping architects to navigate the social value terrain, making the best of existing systems. It was while developing the SVT that the team came to realise that social value has to be spatialised; we have to know where it happens, if it is going to be of real use in decision making.

The spatialising of social value was explored through the Mapping Eco Social Assets (MESA) project,[12] an in-depth study of a housing estate in Reading, UK (Hatleskog and Samuel, 2021). Hand-drawn social value maps were developed in consultation with community members to create multi-layered mappings of social value, each layer referring to one of the social value toolkit outcomes – for example, places of 'connection' (Figure 11.2). Data was collected on aspects including active lifestyles, positive emotions and safety that community members attributed to the place they lived. This was achieved through a series of mapping workshops with a variety of different stakeholders who drew, wrote and placed stickers on the large map before them (Figures 11.3 and 11.4). The findings were translated into digital heat maps, which were presented back to the community for checking. The result was a snapshot of social value on the estate at the time of the consultation. It was around this time that the local authority was closing down a local swimming pool. It became apparent that the social value heat maps could be used to capture the social value impact of closing the pool and indeed that scenario mapping could help local authorities project the implications of their actions on social value in the community.

The MESA maps were laboriously hand made with the researcher, Eli Hatleskog, interpreting the information given by the community on the hand-drawn maps in a potentially subjective way. Clearly digital intervention is needed to spatialise the opinions of community members in a more robust and economical way, while taking into account

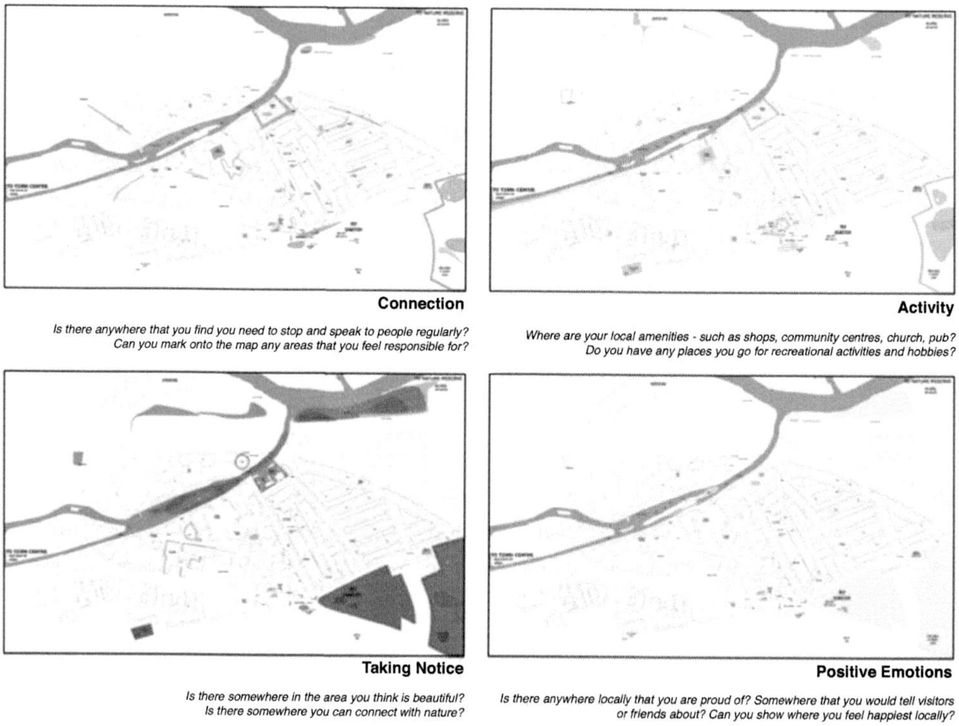

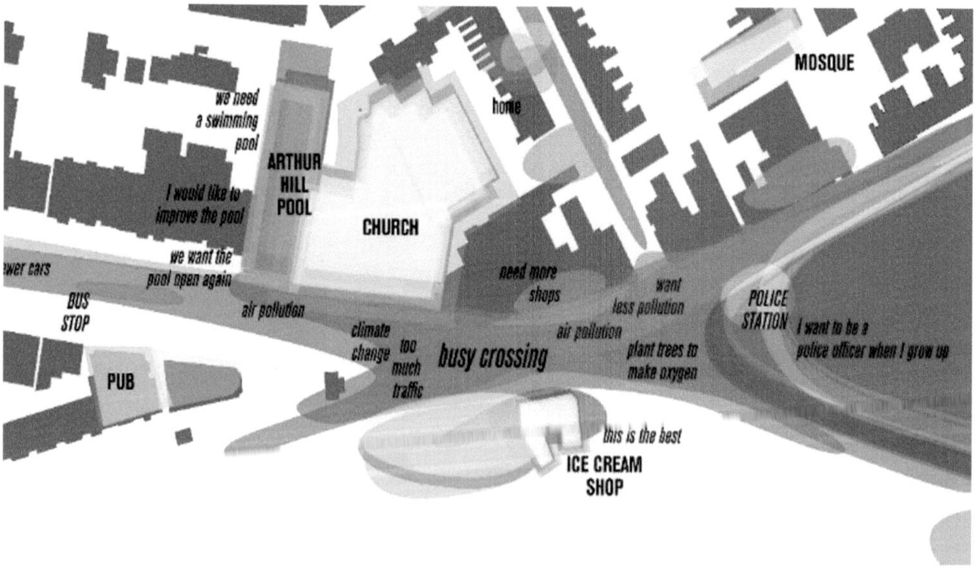

FIGURES 11.2, 11.3 AND 11.4 Mapping Eco Social Assets workshop (credit: Flora Samuel); Mapping Eco Social Assets, social value layers (credit Eli Hatleskog and Flora Samuel); Mapping Eco Social Assets heat maps of social value on a Reading housing estate (credit: Eli Hatleskog and Flora Samuel).

issues of digital exclusion. This is the focus of the Community Consultation for the Quality of Life Project, which is being developed with the digital consultation firm Commonplace (CCQOL, 2021). The project, with pilots in all four nations of the UK, is exploring best practice in terms of inclusive community map making, both face to face and digital to improve the stop-start, wasteful and adversarial process of community consultation in planning. Building on Commonplace's digital format for community consultation, the team anticipate the possibility of developing social value maps that can be continuously updated by the community in real time, allowing continuous consultation for communities to get involved in participatory budgeting (Cabannes, 2018) and other democratic processes. In developing social value maps, the maps that are currently 'missing', the MESA maps underpin for the Better Places research, the aim of which is to make it easy for local authorities and other stakeholders to make informed spatially based decisions about land use in their jurisdictions. This requires the collection of hyperlocal data made by and with communities on where best to target scarce resources, building on community knowledge and science to address problems caused by climate change.

Bringing together active and passive data for the creation of digital social value maps

The Better Places Knowledge Transfer Partnership was developed in response to the obvious need to include social value in land valuation and planning set out above. The intention of the team was to unpack the thorny area of spatial analysis and mapping for social value, to describe the state of the art in this area and to build an enabling framework to enable social value to be standardised in spatial analytics, digital twins, maps and other digital models. This is emphasised further as methods of analysing and mapping economic and environmental data are becoming increasingly crucial in discussions on climate change, both in relation to reducing climate impacts and adapting to climate change. The work builds on the activities of the Stantec GIS portal (Figure 11.6), which

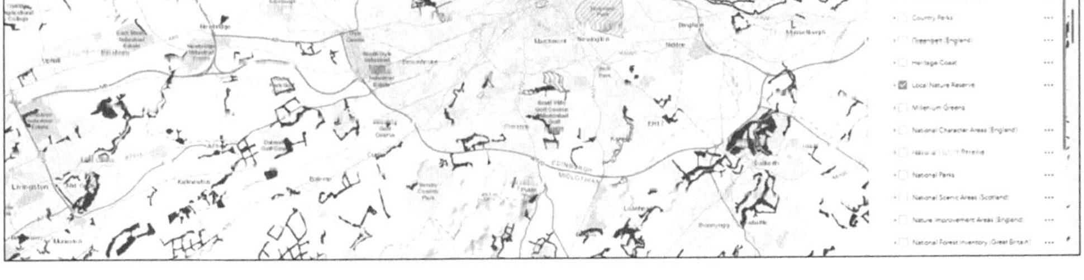

FIGURE 11.5 Stantec GIS Portal.

is collating business and project knowledge in a spatial environment to enable better data access and analytics to underpin work with clients.

The first step to establish the state of the art and to potentially identify the need for a new social value mapping methodology for strategic decision making was to review existing tools which have been used on projects. We did this by looking widely at a longlist of 'tools' to determine what exists already, what is commonly used and where the gaps are. Most notably, very few existing tools or frameworks aim to spatially map social value, and the spatial consideration of where social value is created was not yet included as a standard across nearly all tools reviews. There are no widely used or standard approaches to metrics, methods or processes other than pushing to move from outputs to outcomes-thinking. This aligns with the immaturity of the field, which is setting higher level foundational thinking. The outcomes of this review concluded that there is the need for more spatialisation of, when possible, quantified social value; the need to increase the use of large and existing datasets; the importance of creating something that can be openly shared to have most impact; and the significance of continued engagement to give communities a voice which underpins the whole ethos of the work.

What quickly became apparent was to distinguish between the data that exists already, often in large datasets, and data that will be collected through new engagement processes. These are two sources of data which are often entangled as they bring different things to a project. Collecting existing, or 'passive', datasets is valuable in gaining wider information often using fewer resources, and is more common amongst certain professionals in digital planning but not done in detail for social value. Collecting new, or 'active', data gives deeper understanding but can often be resource intensive through engagement processes with stakeholders and communities, which is a more common process within social value practices. Both sources of data together will produce richer data and can sense check how social value has been interpreted spatially. How both data sources will be utilised through digital interfaces, processes and platforms will have substantial differences but ultimately the spatialisation and mapping of data is intended to merge these two areas in a systemised way, to provide fresh insights. Advances in natural language processing mean that, increasingly, quantitative data can be generated from qualitative statements during consultation, as utilised by the Community Consultation for Quality of Life[13] research team.

Advances in data science are leading to the development of new kinds of data analysis, for example smart data and deep data, which are capitalising on the potential of machine learning. Spatial planning has a great deal of catch-up work to do in this area. However, it is pertinent to acknowledge the social construction behind every digital model (Nochta and Oti-Sarpong, 2024). In the ever changing landscape of AI development in 2024, it has become increasingly clear that the ethics of data management need to be at the fore of any discussion of social value in a just climate transition. Any data collection process has the risk of bias that can have significant negative consequence. So transparency and clarity on ethical considerations of all digital data mapping needs to be available to relevant stakeholders.

There are many barriers to developing cohesive social, environmental and economic value mapping. This is where the potential of collaboration, cooperation and partnerships across disciplines and organisations come in. Further, new kinds of partnerships will be needed that bring together large well-known brands and 'the smaller company with the better capabilities' (Government PropTech Representative, 2021).

Facing the data challenge: the need for interdisciplinary professionals

> How do you make it easy as possible for those professions to use it? And to speak their language... These are the planners. These are the real, you know important people, if you like.
>
> *(Social Impact PropTech Lead, 2021)*

A move towards a digital future of planning will mean the need to move away from the traditional skillsets used in planning authorities, with those in private organisations needing to upskill to keep up. As the world, and the built environment industry, becomes more digitised, there is more data being collected through projects and organisations that need to be utilised: data needs to be collected more systematically; data needs to be analysed more strategically; data needs to be checked for quality, used for advantage, and revisited to check changes and updates (Construction Innovation Hub, 2021b).

The kinds of data sets that do exist have, very often, been generated without the input of built environment professionals and are therefore of limited use for a discussion of the built environment. In England, the National Panel Household Data Survey is an illustration. Lack of data on the impact of the built environment makes it difficult to write proxies and algorithms to predict the likely performance of places. The issue that underpins the situation is a lack of leadership, an over-reliance on the private sector to produce systems that were traditionally the remit of the public sector, and – until recently – political will. These issues can be exacerbated further through the frameworks used to structure data. The way data is collected is often skewed in favour of certain demographics (Criado Perez, 2019). Inclusive data is an important consideration, as the effects of climate change disproportionately affect those in poverty, and can exacerbate economic, gender and other social inequalities and impacts. Professionals are needed who can help collect the stories of how people are adapting to climate change and what they need to continue to do so.

In terms of the practicalities of digital tools and data being created, there are challenges identified through existing tools application and iterations of successful digital uses. The outcome of working with data and digital tools can have multiple benefits, including knowledge exchange across skillsets and specialities (Construction Innovation Hub, 2021a). A well-managed data analysis process requires interdisciplinary engagement, which in turn has the potential to influence organisational strategy, meaning up-front investment is made in digital tools, saving money and creating value in the long term (Fitton and Moncaster, 2021).

> Especially in the research team, we don't want just specialists. It's great having someone who knows everything about environment policy. But if they then can't manage a project to use environment data then that's not as helpful. What we want is a mix and it's very difficult to find that at times.
>
> *(Greater Manchester Mapping Researcher, 2021)*

There is an opportunity here for new kinds of built environment professionals who have a foundation in sustainability, data, GIS and interdisciplinary research.

Conclusion: digital spatial planning for climate justice

This chapter has argued the importance of digital spatial planning for the green transition. Every country needs data base of maps that can be brought together to chart progress on social, environmental and economic value. These maps need to be made with communities who can provide vital detail at the hyperlocal level as a basis for policy action. An agreed format is needed for a map-based digital infrastructure, which offers a transparent and robust foundation for evidence-based decision making as well as a very public information source on progress towards climate justice, ideally in real time. This can be done through the careful and critical collation of passive and active data sets. Agreeing and developing the foundations and principles in place for a climate justice driven planning infrastructure would facilitate, in turn, strategy around a range of other issues such as net zero carbon targets, land reform and housing. This has ramifications for planning professionals who need to become guardians of data (Wilson and Tewdwr Jones, 2021) working together to provide the spatial evidence base for climate change.

KEY TAKEAWAYS FROM CHAPTER 11

- Digital spatial plans, including social, environmental and economic spatialised datasets are vital for decision making on where to use scarce resources in a climate change emergency.
- Digital spatial planning is a nascent field which requires leadership and intergovernmental action on data standards.
- Intangible and qualitative wellbeing data can be spatialised through maps.
- Hyperlocal maps are needed to capture hyperlocal issues and opportunities for climate change mitigation and adaptation.
- Communities need to be involved in the collection of hyperlocal data about their places, uploaded onto digital maps.

Notes

1 https://wellbeingeconomy.org/wego
2 For a comparison between spatial planning in England, Europe and Asian countries, see 'An Overview of Spatial Policy in Asian and European Countries, www.mlit.go.jp/kokudokeikaku/international/spw/general/uk/index_e.html
3 The research was undertaken for the Better Places Research Project, an Innovate UK Knowledge Transfer Partnership between Stantec and the University of Reading. We are particularly grateful to Helen Cole for her assistance.
4 Stantec is a global sustainable engineering, architecture and environmental consulting company.
5 In some cases, as with the Future Generations Act (2015) Wales, culture is added as a key indicator.
6 https://geoportal.statistics.gov.uk/
7 https://magic.defra.gov.uk/

8 https://designatedsites.naturalengland.org.uk/GreenInfrastructure/MappingAnalysis.aspx
9 https://data-communities.opendata.arcgis.com/
10 https://data.gov.uk/
11 Terminology increasingly being used as an umbrella term for the social considerations of a project, primarily due to embedded UK legislation of the Public Services, Social Value Act 2012.
12 This work was supported by an institutional links grant 332241573 under the Newton-Philippines partnership, a collaboration with the School of Architecture in the University of Reading (principal investigator Flora Samuel) and the University of Santo Tomas, Manila (Co-investigator Leah dela Rosa). The grant was funded by the UK Department of Business, Energy and Industrial Strategy (BEIS) and CHED and delivered by the British Council. For further information, please visit www.newtonfund.ac.uk.
13 www.ccqol.org

Bibliography

Boland, P., Durrant, A., McHenry, J, McKay, S. and Wilson, A., A "planning revolution" or an "attack on planning" in England: digitization, digitalization, and democratization, *International Planning Studies*, 27(2), pp. 1–18, 2021. doi: 10.1080/13563475.2021.1979942.

Bristol, *Know your Place*, 2021. Available at: https://maps.bristol.gov.uk/kyp/?edition=

Browne Gott, H. (2019) "Using administrative data to understand the service interactions of people experiencing homelessness", *International Journal of Population Data Science*, 4(3). doi: 10.23889/ijpds.v4i3.1333.

Cabannes, Y., The contribution of participatory budgeting to the achievement of the Sustainable Development Goals: lessons for policy in Commonwealth countries, *Commonwealth Journal of Local Governance* [Preprint], 2018.

CCQOL, Community Consultation for Quality of Life, 2021. Available at: ccqol.org.

Clapham, D. (2019). *Remaking housing policy: An international study* (1st edition). Routledge.

Clark, G., *RIBA Sustainable Outcomes Guide*. London: RIBA, 2019. Available at: www.architecture.com/-/media/GatherContent/Test-resources-page/Additional-Documents/RIBASustainableOutcomesGuide2019pdf.pdf.

CIH, *Value Toolkit*, Construction Innovation Hub, 2021 https://constructioninnovationhub.org.uk/value-toolkit/

Climate Change Committee. (2021). Progress in adapting to climate change: 2021 Report to Parliament. www.theccc.org.uk/wp-content/uploads/2021/06/Progress-in-adapting-to-climate-change-2021-Report-to-Parliament.pdf

Construction Innovation Hub, Digital Capabilities: A Framework for early career professionals across built environment disciplines, Construction Innovation Hub, 2021a. Available at: www.cdbb.cam.ac.uk/files/cdbb_routes_v12.pdf

Construction Innovation Hub, The Digital Innovation Imperative: Advancing progress towards a digital built Britain, Construction Innovation Hub, 2021b. Available at: www.cdbb.cam.ac.uk/files/digital_innovation_imperative_140921.pdf

Criado Perez, C. (2019). *Invisible women: Exposing data bias in a world designed for men*. Chatto & Windus.

Dasgupta, P., *The Economics of Biodiversity: The Dasgupta review, Biological Wealth and Other Essays*. London: HM Treasury, 2021. Available at: www.gov.uk/government/publications/final-report-the-economics-of-biodiversity-the-dasgupta-review.

Fitton, S. and Moncaster, A., Social value, infrastructure and stakeholder engagement: a complex triangle, *Proceedings of the Institution of Civil Engineers – Engineering Sustainability*, pp. 1–8, 2021. doi: 10.1680/jensu.21.00030.

Geospatial Commission, *Unlocking the power of location. The UK's Geospatial Strategy, 2020 to 2025*. UK: Geospatial Commission, 2023. Available at: https://assets.publishing.service.gov.uk/media/6489b1fb103ca6000c039ea2/2023-06-15_UK_Geospatial_Strategy_2023_.pdf

GMCA, *The Greater Manchester Infrastructure Framework 2040*, Greater Manchester Combined Authority 2019, greater-manchester-infrastructure-framework-2040.pdf

Hatleskog, E. and Samuel, F., Mapping as a strategic tool for evidencing social values and supporting joined-up decision making in Reading, England, *Journal of Urban Design* [Preprint], 2021. doi: 10.1080/13574809.2021.1890555.

Harrison, J., Galland, D. and Tewdwr-Jones, M., Regional planning is dead: long live planning regional futures, *Regional Studies*, 55(1), pp. 6–18, 2021. doi: 10.1080/00343404.2020.1750580.

HM Treasury, Central Government Guidance on Appraisal and Evaluation, The Green Book. UK: HM Treasury, 2018.

Intergovernmental Panel on Climate Change (IPCC). (2022). Climate Change 2022: Impacts, Adaptation and Vulnerability. Contribution of Working Group II to the Sixth Assessment Report of the Intergovernmental Panel on Climate Change. Cambridge University Press. Available at: www.ipcc.ch/report/ar6/wg2/downloads/report/IPCC_AR6_WGII_FullReport.pdf

Maxwell, M., McCool, S., Peppin, A., Spittle, K. and Stocks, T., *Public dialogue on location data ethics: engagement report*. UK: Geospatial Commission, 2021. Available at: https://assets.publishing.service.gov.uk/government/uploads/system/uploads/attachment_data/file/1040807/Accessible_Public_dialogue_on_location_data_ethics_Engagement_report.pdf

McHarg, I., *Design with Nature*, Wiley, 1995 (1969).

MHCLG, *National Planning Policy Framework*. UK: Ministry of Housing, Communities and Local Government, 2024. https://assets.publishing.service.gov.uk/media/67aafe8f3b41f783cca46251/NPPF_December_2024.pdf

MHCLG, *Planning for the Future: White Paper*. UK: Ministry of Housing, Communities and Local Government, 2020.

Ministry of Housing, Communities and Local Government. (2021). National Model Design Code. https://assets.publishing.service.gov.uk/media/60140c1d8fa8f53fc52c5c31/National_Model_Design_Code.pdf

Monbiot, G. (2015, December 29). Deluge: Why do we still let farmers drain the land for grouse shooting? The Guardian. www.theguardian.com/commentisfree/2015/dec/29/deluge-farmers-flood-grouse-moor-drain-land

Mullholland, C. et al, *Delivering Better Place Outcomes Through the Use of Data, Stantec.* 2022, www.stantec.com/content/dam/stantec/files/PDFAssets/2022/better-places-report.pdf

Mulholland, C., Ejohwomu, O. A. and Chan, P. W., Spatial-temporal dynamics of social value: Lessons learnt from two UK nuclear decommissioning case studies, *Journal of Cleaner Production*, 237, p. 117677, 2019. doi: 10.1016/j.jclepro.2019.117677.

Nochta, T. and Oti-Sarpong, K. Participation matters: The social construction of digital twins for cities. *Environment and Planning B: Urban Analytics and City Science*, 0(0). 2024 https://doi.org/10.1177/23998083241305695

Nussbaum, M., *Creating Capabilities*. Harvard University Press, 2011.

ONS, *Census Geography*, Available at: www.ons.gov.uk/methodology/geography/ukgeographies/censusgeography, [Accessed 7 December 2021].

Perrotti, D. and Stremke, S. (2018). Can urban metabolism models advance green infrastructure planning? Insights from ecosystem services research. *Environment and Planning B: Urban Analytics and City Science*, 47(4), 678–694. https://doi.org/10.1177/2399808318797131 (Original work published 2020)

Public Services (Social Value) Act explanatory notes, Parliament of the United Kingdom, 2012.

Royal Town Planning Institute. (2019). Resourcing public planning. www.rtpi.org.uk/policy-rtpi/2019/november/resourcing-public-planning/

Samuel, F., *Social Value Toolkit for Architecture*. RIBA, 2020. Available at: www.architecture.com/knowledge-and-resources/resources-landing-page/social-value-toolkit-for-architecture.

Samuel, F. and Hatleskog, E., *Architectural Design Special on Social Value*. 4th edn. Wiley, 2020. Available at: https://onlinelibrary.wiley.com/toc/15542769/2020/90/4.

Scottish Government. (2020). Transforming Places Together: Scotland's digital strategy for planning. www.gov.scot/publications/transforming-places-together-scotlands-digital-strategy-planning/

Shoard, M., *This Land is our Land: The struggle for Britain's countryside*. Gaia Books Limited, 1997.

Shrubsole, G. (2019). *Who owns England? How we lost our green and pleasant land, and how to take it back*. William Collins.

Social Value Portal, *Embedding Social Value into Planning*. Social Value Portal, 2021.

Stantec, *Creating Communities Fit for the Future: Repurposing the Development Plan Process by Putting Places First*. UK: Stantec, 2019. Available at: www.stantec.com/content/dam/stantec/files/PDFAssets/UK/places-first-volume-2-publication.pdf.

UK Statistics Authority. (2021). Centre for Applied Data Ethics Strategy: Enabling ethically appropriate research and statistics for the public good. https://uksa.statisticsauthority.gov.uk/wp-content/uploads/2021/04/Centre-for-Applied-Data-Ethics-Strategy.pdf

Wang, V. and Shepherd, D., Exploring the extent of open government data – A critique of open government datasets in the UK, *Government Information Quarterly*, 37(1), 2020.

Wilson, A. and Tewdwr Jones, M., *Digital Participatory Planning*. Routledge, 2021.

12

DECOLONISING RESILIENT ARCHITECTURE

Pieta McAleer-Harding, Mercia Abbott, Rebecca Kiddle, and Maibritt Pedersen Zari

Introduction

For the architectural profession, which often assumes physical structural resiliency as the primary concern of disaster-resilient architecture, there is a struggle to design architecture through 'bottom-up' approaches which support the social resiliency of communities and benefit those in vulnerable positions. We examine how architects can better reflect the values agreed upon in *Te Tiriti o Waitangi*[1] and create disaster-resilient solutions suitable for all peoples in Aotearoa that do not perpetuate colonisation paradigms.

This chapter summarises a larger design-led research project that explores decolonisation concepts and proposes a set of strategies to improve built environment disaster resiliency by considering how to increase the social resilience of communities pre-disaster, in order to support them more effectively in post-disaster contexts.

We suggest that decolonising architectural resiliency strategies in Aotearoa must focus on embedding *tikanga Māori* practices (correct/appropriate cultural or customary practices) to place community needs at the forefront. We also stress that without co-design between *iwi* (tribe), *hapū* (sub-tribe), architects and other built environment professionals, designs will likely continue to represent the dominant *Pākehā* (New Zealand people of European descent)[2] worldview and may not effectively support vulnerable communities which Māori are overly represented in.

Positionality. Ko Wai Mātou?

An important part of decolonisation practices is to make the positionality of researchers clear (Latai-Niusulu et al., 2020; Smith, 2016). This section briefly explains who the authors are.

McAleer-Harding

The lived experience of the major Ōtautahi (Christchurch) earthquakes in 2010 and 2011 shaped much of my youth and subsequent worldview, motivating me to explore how architecture for disasters could be better approached. As someone born in *Aotearoa*

DOI: 10.4324/9781003293903-15
This chapter has been made available under a CC-BY-NC-ND 4.0 license.

with English, Scottish and Irish heritage, I am *Pākehā*. I acknowledge colonisation and its impacts while supporting *tino rangatiratanga* (Māori Sovereignty) by recognising *Te Tiriti*.

Abbott

I was born in Ōtautahi but grew up in *Te Whanganui-a-Tara* (Wellington). I am of Māori descent with *iwi* and *hapū* affiliations to *Ngāti Kahungunu* and *Ngāi Tahu*. I acknowledge the inherent disconnect and mourning experienced at the loss of not truly knowing my own culture and that this is the by-product of colonisation. Challenging the locus of power and control of an imported imperial view that favours Eurocentric-led ideals as the norm, requires a process of decolonisation. My research looks at weaving climate change science, ecologies and *mātauranga* Māori (Māori knowledge) through spatial constructs. This *mahi* (work) is guided by relationships to the environment and incorporates a strong emotional and embodied Māori way of knowing that works with *kaitiaki* (guardians) and is grounded in *mātauranga* Māori.

Kiddle

I am Māori from the tribal groups of *Ngāti Porou* and *Ngā Puhi*. I have expertise in urban design. Specifically: decolonising Aotearoa New Zealand cities and Indigenous place identities; Māori housing and urban design; spatial justice; *rangatahi* (youth) involvement in built environment decision-making; urban and suburban spaces for community building and third places; and the politics of the production of place.

Pedersen Zari

I was born in and grew up in *Kirikiriroa* (Hamilton) and am now based in *Tāmaki Makaurau* (Auckland). My ancestors are predominantly Danish and Scottish. I am *Pākehā*. I acknowledge the relationship between colonisation and ecological wellbeing and understand that ecological justice cannot be separated from social justice. I am motivated by wanting to ensure that built environment climate change adaptation does not perpetuate or become a new form of colonisation.

Understanding disasters in a climate emergency

The geologically young and geographically remote islands of Aotearoa continue to be at threat from disasters (Saunders et al., 2020). The frequency of these disasters is increasing, and their unpredictability is becoming difficult to manage and mitigate (Kenney et al., 2015; King et al., 2007; Saunders et al., 2020). In 2021–2023, for example, there was a global pandemic, numerous devastating floods, storms, earthquakes and cyclones which have displaced communities. Today, society exists in a precarious place, experiencing the direct and indirect impacts of climate change, and coming to terms with the increasing instability of the environment. These disasters are created by both non-human and human forces which collectively shape Aotearoa's unique landscape.

Traditionally, disaster response techniques were considered through the lens of control, where, if a disaster could be controlled, its challenges could be alleviated (Berke & Campanella, 2006; Ginige, 2011; Hayward, 2013). However, these control-based approaches are becoming redundant due to the lack of predictability caused by current driving forces of change. Consequently, approaches to disaster management have shifted to developing community resilience (Allen & Roberts, 2009). In 1999, the United Nations General Assembly adopted the International Strategy for Disaster Reduction, which reflected the major shift in disaster recovery practices that previously centred disaster response, and instead now places resiliency into discourse (Burnside-Lawry & Akama, 2013). Taking steps to improve resilience pre-disaster has been identified as a key indicator for reducing community recovery time post-disaster (Daly et al., 2009).

Building resilience

Resiliency

Resiliency refers to the capacity of human and physical systems to respond to disruptions or extreme events (Bosher & Dainty, 2011, p. 2). Many definitions also highlight the critical nature in which resiliency can anticipate change and maintain the system's core purpose even under radically changed circumstances (Platt, 2018, p. 370). As a concept, resiliency draws heavily on influences from ecology and biology (Blakeley, 2016). The concept of 'adaptive capacity' can help to distinguish an aspect of resiliency in which a society will draw on its cumulative competence and utilise individual, collective, and/or external institutional resources in disaster contexts (Daly et al., 2009). Considering 'adaptive capacity' within the context of resiliency helps to move the discourse away from focusing on the threat (a post-disaster response) towards developing the capacity of that society to adapt to the change (pre-disaster resiliency) (Allen & Roberts, 2009; Blakeley, 2016; Daly et al., 2009). Definitions, language, and jargon associated with resiliency often neglect non-material values of a community. Namely, relationships, taonga (tangible and conceptual treasures) and *wahi tapu* (sacred spaces) (Hayward, 2013). These definitions drive architecture's understanding of resiliency practices, which are often 'top-down', and technology driven. Without the engagement of vulnerable communities and their values in resiliency practices, architectural interventions may be of diminished benefit to the communities they seek to serve (Burnside-Lawry & Akama, 2013).

The principles of equity, fairness and equal access to resources are all essential in generating a resilient community (Cutter, 2016). Therefore, recognising the inherent social component of resilience (and resilient architecture) highlights the ways in which communities can withstand a variety of external shocks to their social infrastructure[3] (Hayward, 2013). A strong social fabric is built by the strength, trust and equity found within the interrelationships that form between individuals and their community, natural and built environments, and social institutions. When emphasis is placed upon an individual's resilience, whether a person or a singular building, it undermines those community bonds ignited by shared experiences within the social fabric, which are vital in post-disaster situations (Hayward, 2013).

Visualising the traditional, tacit, immaterial or informal knowledge associated with communities, including *mātauranga Māori* in an Aotearoa context, will prove valuable in indicating the strength of social resilience. Burnside-Lawry & Akama (2013) state that this process helps to understand the community's interactions and assists in 'bridging relationships between neighbours that can lead to better preparation for all hazards' (Burnside-Lawry & Akama, 2013, p. 32). England (2016, p. 29) points out that '[architectural] visions must rest on their [a community's] beliefs, shared identity and ideals if they are to have any meaning for people.' Definitions of resiliency that favour technology-driven solutions make it more difficult to transition to design approaches that are community-focused as well as technologically effective.

Vulnerability

Disaster vulnerability is understood as the set of conditions that 'affect the ability of countries, communities and individuals to prevent, mitigate, prepare for and respond to triggering agents' (Ginige, 2011, p. 288). The concept is multi-layered, determined by the political, economic and institutional capabilities of individuals in a specific place at a specific time. Simply put, vulnerability is the predisposition to be adversely affected (Cutter, 2016). Vulnerability is not evenly distributed throughout society; suffering in post-disaster contexts occurs at disproportionate rates (Bosher & Dainty, 2011). As Rae and Thompson-Fawcett (2018, p. 86) state: 'marginalisation is a concept that is arguably inseparable from the notion of disaster vulnerability. The more marginalised a group is, the more likely they are to be vulnerable to disasters.'

Vulnerabilities are largely dependent on development practices that often occur without taking pre-existing hazard susceptibility into account (Ginige, 2011; Thornley et al., 2015). Typically, these development practices are driven by economic growth and population expansion, which push vulnerable individuals into locations where disaster risk is exacerbated and its consequences exaggerated (Bosher & Dainty, 2011; Hayward, 2013; Rose et al., 2018). For people with marginalised realities, the absence of resources often means those who are already in situations that perpetuate poverty, move into inherently unsafe environments (Lizarralde et al., 2009). It is often Indigenous communities who are forced into situations of increased risk (Rae & Thompson-Fawcett, 2018, p. 87). Dynamic pressures created by societal inequalities, including the impacts of colonisation, encourage the maintenance of social structures which uphold these conditions of disaster risk (Lizarralde et al., 2009).

Once the social sphere in which vulnerabilities exist is understood, the occurrence of a disaster can be identified as the combination of hazard and exposure to that social vulnerability. Social vulnerability[4] is inherently linked to health vulnerabilities, which are linked to spatial or built environment vulnerabilities in a kind of 'triple jeopardy' situation (see, for example, Gray, 2017). Understanding these connections implies that vulnerability may emerge as the controllable component of a disaster, and its regulation can help to determine risk reduction practices and build disaster resilience (Ginige, 2011). However, current disaster management strategies tend to, 'extend and perpetuate political exclusion through new decision-making structures that fail to accommodate and account for marginalised realities' (Rae & Thompson-Fawcett, 2018, p. 87). The common lack of understanding of Indigenous groups' needs in current disaster response methods often exacerbates the inequality and political exclusion they faced prior to the disaster (Rae & Thompson-Fawcett, 2018, p. 87). Disasters are recognised as another major struggle for Indigenous communities that builds on the already poor relationship between 'Indigenous groups, the colonisers and the national apparatus' (Rae & Thompson-Fawcett, 2018, p. 87). Without addressing this, vulnerability will threaten the potential of sustainable and regenerative development, and climate change adaptation, and as a consequence, positive progress towards decolonisation will be limited (Rose et al., 2018).

Relationships between disasters, vulnerability, resilience and colonisation

While resilience and vulnerability are conceptually related, one is not the opposite of the other (Cutter, 2016). Highly vulnerable social groups do not inherently lack resiliency (Cutter, 2016; Thornley et al., 2015). The term resiliency neglects to recognise the political power, injustices and inequalities that create vulnerabilities for those communities and may reduce their resiliency. Urban settings are inherently political spaces (Ward, 1996), and some colonial cities may be more vulnerable to disasters, exactly because of the rapid process of colonisation (Vandertop, 2019). Causes, consequences, and ways to address disasters must be informed by an analysis of power and intersectional identities, and understandings of who benefits from, pays for, and must live with impacts (Abbott et al., 2022).

A disaster must be understood as the combination of a 'hazard or triggering agent and vulnerabilities' (Ginige, 2011, p. 287). Without the existence of humans, an earthquake is simply land movement, and a cyclone is merely a weather event (McKinnon & Cook, 2020). While the origins of a hazard may be natural, their consequences are brought on by several factors to create 'disasters'. A disaster exists as a result of the precarious relationship between hazards, social vulnerability and the built environment (Bosher & Dainty, 2011).

Social vulnerability is not evenly distributed throughout society and those who are living in states of poverty or who are socially marginalised are disproportionately impacted (Bosher & Dainty, 2011). The relationship of poverty to vulnerability is often disregarded in Aotearoa, and is intimately linked to the ongoing impacts of colonisation. Cutter (2016) notes that a key component of vulnerability continues to be the long progression from its initial roots to its contemporary manifestation in those existing societal constructs, systems and contexts. It must be acknowledged that colonisation and associated structural racism maintains those power structures which are responsible for disaster vulnerability inequitably throughout contemporary Aotearoa. For example, according to Houkamau and Sibley (2015), the extent to which people think they are visibly identifiable as Māori, significantly predicted decreased rates of home ownership. Less than half of Māori live in their own homes, compared to two-thirds of the general population (Goodyear, 2017). A lack of tenure correlates to increased social vulnerability in relation to coping with disasters.

The arrival of the British Crown to Aotearoa is widely recognised as the advent of colonisation, but colonisation is not a past historical event. Colonisation is an ongoing continuing process upheld by societal structures which deny Indigenous peoples the right to govern themselves (Huygens, 2011; Jackson, 2020). In Aotearoa, colonisation exists as an oppressive structure where the coloniser remains in the colony and establishes a society, centring their perspectives and dominating society (Tuck & Yang, 2012). This is known as 'settler colonisation' and its processes continuously erase marginalised perspectives and dominate Indigenous identity (Tuck & Yang, 2012). Colonisation continues to harm communities, specifically Māori (Matunga, 2018) and upholds oppressive structures that privilege the worldview of *Pākehā* (and to a lesser extent *Tauiwi*[5]) (Amundsen, 2018; Kiddle, 2020; Thomas, 2020). It actively maintains those regimes, structures and concepts that disproportionately harm Māori, perpetuating inadequate living conditions and contributing to or preserving poverty and poor social and political outcomes (Amundsen, 2018; Matunga, 2018; Tuhiwai Smith, 2012).

Architecture for resiliency: What are architects missing?

While resiliency has become more prevalent among architectural discourse in recent years, its praxis typically emphasises the resiliency of physical structures rather than the people within them and the communities they are part of. The success of architecture remains measured by its ability to recognise those impacts of disasters which can be directly measured. This includes solutions such as raising or moving buildings to address rising sea levels, or improving seismic performance of buildings to prevent catastrophic collapse during earthquakes. While we cannot discredit the value of mitigating these direct impacts of disasters, built environment professionals can often forget to discuss or design for the resiliency of the people who use the buildings.

For architects, effectively contributing to community resiliency practices means recognising that colonisation, and the social structures it perpetuates, continues to play a role in reducing the resiliency of communities, both pre- and post-disaster. Aiming just for 'resilience' without an analysis of social justice is not enough. Compounding matters in *Aotearoa*, the architectural profession is heavily representative of a *Pākehā* worldview, which has and continues to benefit from colonisation. The demographic break-down of architects in *Aotearoa* consists primarily of *Pākehā* and *Tauiwi* practitioners (Matunga, 2018). While an important conversation about diversifying the architectural workforce exists, non-Māori architects do have a crucial role to play in understanding how their practice can support *Tangata Whenua* (Māori) in decolonisation efforts (Ward, 1996). Architects must understand how to be astute *Te Tiriti o Waitangi* partners to Māori (Huygens, 2011). It is important to understand that failing to honour *Te Tiriti* maintains the status quo which perpetuates racial disparities throughout all facets of *Aotearoa's* society, including in the built environment (Newcombe, 2019).

This recognises that equity between diverse groups has wider implications for the wellbeing of an entire community (Kiddle, 2020). Boag, in Kiddle (2020, p. 100), states that 'anything that enhances, or diminishes, the wellbeing of Māori enhances or diminishes the wellbeing of the community'. Taking active steps to understand community wellbeing will allow for an improved understanding of the process required to design fair and equitable spaces. It will ensure disaster vulnerability can be effectively supported by resiliency practices that work towards decolonial outcomes and recognises that decolonisation is a pathway towards greater social resiliency and inclusive practices for all Indigenous and diverse peoples (Bosher & Dainty, 2011; Burnside-Lawry & Akama, 2013; Daly et al., 2009; Kiddle, 2020). Greater social resiliency, which must take into account ongoing impacts of colonisation, is likely to result in greater built environment resiliency (Castaño-Rosa et al., 2022).

To support vulnerable communities, built environment professionals must reframe their discussions of resiliency. Conversations must transition from colonial constructs which drive individual technological agendas and should instead examine the potential benefits decolonisation practices may hold for an appropriate disaster resilience practice for the built environment.

Decolonisation, disasters and architecture

If disaster vulnerability is determined in part by the impacts of colonisation, then introducing decolonisation as a framework for resiliency practices in *Aotearoa* may provide a more appropriate response to disasters. Decolonisation favours accessibility to those resources which are necessary for disaster response, especially for Māori in Aotearoa. This theoretical framework begins to examine how decolonisation architecture, and its practice, can build more resilient communities pre-disaster, to support vulnerable people and improve post-disaster recovery.

Decolonisation is complicated and deeply political but is hugely important. It is a process which must occur at every level of society and continue through to large hegemonic structures built by colonial institutions. Decolonising Aotearoa requires a number of complex steps (Tuck & Yang, 2012). This begins with processes such as the return of land to *iwi* and *hāpu*, an improved form of government which is underpinned by *tikanga Māori*, a transfer of power (Kake, 2021, p. 129), and overall, a more equitable society (Jackson, 2020). A process of decolonisation is slow but any transition towards the values agreed upon in *Te Tiriti o Waitangi* will benefit all of *Aotearoa* (Kiddle, 2020).

To honour *Te Tiriti o Waitangi*, *Pākehā* must actively work to amend the wrongs Māori encounter/ed through colonisation by undoing those power structures which privilege *Pākehā* worldviews (Huygens, 2011; Jackson, 2020; Mercier, 2020). *Pākehā* and *Tauiwi* must take responsibility for their own active part in decolonisation in order to better consider the values expressed in *Te Tiriti* through all facets of society, including the design and occupation of the built environment (Huygens, 2011; Mercier, 2020; Thomas, 2020). However, *Pākehā* and *Tauiwi* must be guided by *Tangata Whenua*-led agendas for decolonisation (Elkington & Smeaton, 2020; Newcombe, 2019; Thomas, 2020).

Actively engaging in decolonisation can introduce ways for Indigenous people to gain (or begin to gain) sovereignty from colonial power structures (Tuck & Yang, 2012). In *Aotearoa*, the term decolonisation does not necessarily suggest the removal of settler colonisers (non-Māori) (Mercier, 2020). *Te Tiriti o Waitangi* implies that both Māori and non-Māori can co-exist within a shared place (Mercier, 2020). Therefore, an 'Aotearoa-specific' decolonisation will restore Indigenous land and life to shift the 'ideas, knowledges and value sets that underpin the systems which shape our country' rather than advocate for the removal of non-Māori (Mercier, 2020, p. 52).

As the dominant culture, non-Māori worldviews are understood by the general population as 'normal' and are rarely questioned, re-evaluated, or reflected upon (Kiddle, 2020). Equally, challenging worldviews can be confronting, especially when those perspectives have actively maintained colonial hegemony and racially oppressive structures which alienate Māori in their own lands (Amundsen, 2018; Hutchings et al., 2016; Huygens, 2011). Often, when confronting current discourse, complex and deeply antagonising emotions will arise and can be why personal reflection for non-Māori is often difficult (Kiddle, 2020). However, non-Māori must recognise that exclusionary reflection processes will drive white defensiveness and create 'settler moves to innocence' (Harris in Thomas, 2020; Tuck & Yang, 2012). For *Pākehā*, positive roles within a decolonisation methodology are to constantly become the 'allied other', and to clearly position their work as non-Māori in reference to work by Māori who are seminal in directing the work of decolonisation and Indigenisation (Denzin & Lincoln, 2008).

In Aotearoa, decolonisation in architecture is not widely discussed in discourse. There are few who recognise its significance, so examples of decolonisation architecture are not abundant (Brown, 2018; Smith, 2016). There are many projects which are led by *iwi* and *hapū* that navigate the complexities of decolonisation within *Te Ao Māori*, such as Te Kura Whare by Ngāi Tūhoe and Jasmax and Ngā Purapura by Te Wānanga o Raukawa and Tennent Brown (Jasmax, n.d.; Partington & Pedersen Zari, 2020; Tennent Brown Architects, n.d.).

The term decolonisation may not be the most appropriate word for the remedy sought by Māori to ease the hurt of colonisation in Aotearoa, because ultimately this term came from elsewhere (Jackson, 2020).

Jackson (2020) refers to a remedy that draws on the hopes and stories of the land of Aotearoa and the people who wish to live there, *Pākeha* and *Tauiwi* alike, which he calls the 'ethic of restoration':

1. *The value of place* – the need to promote good relationships with and ensure the protection of *Papatūānuku* (the living earth entity).
2. *The value of tikanga* – the core ideals that describe the 'ought to be' of living in Aotearoa and the particular place of Māori within that *tikanga*.
3. *The value of community* – the need to facilitate good relationships between all peoples.
4. *The value of belonging* – the need for everyone to have a sense of belonging.
5. *The value of balance* – the need to maintain harmony in all relationships, including in the exercise of constitutional authority.
6. *The value of conciliation* – the need to guarantee a conciliatory and consensual democracy.

'Restoration' is not any less difficult than 'decolonisation' but it recognises that 'decolonisation is not a metaphor' (Jackson, 2020). The Ethics of Restoration still requires the upheaval of colonial structures, alongside the promotion of Māori perspectives but it provides actionable steps through a defined framework in which a more just society can be achieved (Jackson, 2020) Figure 12.1 suggests how the Ethics of Restoration could be implemented within built environment discourse. The Ethics of Restoration provides non-Māori architects an opportunity to explore, utilise and manifest these values throughout their design processes and in the creation of the built environment. For example, the value of place in architectural design will strengthen relationships between land and building, symbiotically benefitting both elements and establishing more successful architectural interventions.

A deliberate generation of counter-hegemony by non-Māori through applying a framework of the Ethics of Restoration could combat the normative supremacy which exists throughout Aotearoa. This may result in simultaneously defining a cultural

FIGURE 12.1 The Ethics of Restoration and its potential implementation in the built environment discourse.

identity which exists symbiotically with Māori, and not in opposition to them (Huygens, 2011; Colquhoun in Kiddle, 2020, p. 86). The responsibility therein lies with *Pākehā* to become 'the *Pākehā* that Māori had in mind when they signed *Te Tiriti*' (Nairn in Huygens, 2011, p. 77).

Decolonisation and its role in disaster resilient architecture

The architecture of post-colonial indigeneity became that of extreme vulnerability and had to facilitate survivability and uphold an inherent resiliency for these communities (Matunga, 2018). For example, the acts of generosity provided by *marae* (Māori communal meeting spaces) for entire communities, particularly in times of crisis has redefined the role of the *marae* as a place of refuge and resilience for pan-ethnic communities under stress (Brown, 2018). Hence, Indigenous architecture is innately resilient due to the transferrable survival skills and community relationships built and adapted by Indigenous communities (Thornley et al., 2015).

However, decolonising resilience practices in architecture cannot be done by (re) creating Indigenous architecture. While Indigenous architecture is largely successful for disaster resiliency, creating Indigenous architecture is not the role of *Pākehā*. Indigenous architecture is created by Indigenous architects for their people (Matunga, 2018). But, for this to happen there must be a transformation of architectural practice to remove barriers to Māori entering the profession. Therefore, a crucial role of *Pākehā* architects is to step back and provide spaces within design processes where power imbalances are eliminated and colonial hegemony is eradicated. This means that Māori artists, creators, builders, architects and designers have fair opportunity to 'Indigenise' the built environment on their own terms and create a better space for all, but more importantly, have a chance to create a place reflective of *Te Ao Māori* (the Māori world) (Kake, 2021; Matunga, 2018). It is here that deliberately creating space in design for 'Indigenisation' can become a step towards decolonising disaster-resilient practices in Aotearoa (Kake, 2021).

Decolonisation and co-design

This research employed a design-led methodology (McAleer-Harding, 2022). Due to the inherent complexities which arise when examining social issues, a process of reflection was undertaken, in which the current paradigms that harm parts of the community were examined. Often in academic discourse, Māori are regarded as 'producers of culture rather than of knowledge' (Cooper in Smith et al., 2016, p. 133) and consequently *mātauranga* Māori is often disregarded as an academic methodology (Tuhiwai Smith, 2012). Implementing *He Awa Whiria* (a braided river model) shifts thinking away from that 'one stream' paradigm which prioritises a dominant knowledge basis such as a Western worldview, to a 'two stream' approach where both knowledge systems are valued equally (Arago-Kemp, 2018). The model dictates that when the streams converge, it is a place of learning and not assimilation, in order to support the integrity of both streams and the well-being of the peoples who created them (Arago-Kemp, 2018). This approach formed the basis of a series of design explorations of the concepts discussed in this chapter (Figure 12.2).

It became apparent through the design process that there is true value in engaging designers of the built environment in preparing cities and citizens for natural disasters (Mazereeuw & Yarina, 2017, p. 65). However, the process of incorporating *mātauranga*

FIGURE 12.2 A series of conceptual images from the design process and the range of projects explored throughout the design-led research process.

Māori into disaster resilience begins with the generation of a collective responsibility for community resiliency, one in which Māori can express and utilise knowledge they possess for hazard preparedness, mitigation and recovery (King et al., 2007).

A second key finding was that at many stages of the design process, a point was reached when input from *iwi* or *hapū* was necessary to create a new kind of community-resilient architecture. Typically, non-Māori architects would respond to this by holding a consultation period with *iwi* and *hapū* to find answers and seek advice. However, a co-design process between *iwi, hapū* and design professionals must be initiated at the beginning of the project to ensure the collaboration is genuine, respectful and focuses on building relationships before architectural outcomes can be suggested. Collaboration should not be limited to a consultation period, which may often only serve the architecture practice and not build valuable relationships between the design team and community.

Finally, a co-design process between *iwi, hapū* and the broader architecture team is an appropriate approach to ensure that the built environment more respectfully reflects *Te Ao* Māori (Burnside-Lawry & Akama, 2013; Daly et al., 2009), and meets expectations to become more resilient as the climate continues to change. A co-design process, based on decolonisation values that encompasses an understanding of power dynamics and emphasises action to redress injustice, is more likely to allow for diverse individuals to hold agency within the design process. Such 'decolonisation co-design' is part of a necessary transition towards design processes that uphold decolonisation agendas and acknowledge (and work to change) positions of power that *Pākehā* architects benefit from, even if unconsciously.

Recommendations: A decolonisation toolbox for architects

The findings from this research can be distilled into a series of three recommendations, which create a decolonisation co-design process 'toolbox'. This toolbox can be translated into architectural practice and used by non-Māori architects in order to begin to take steps towards better representing the values agreed upon in *Te Tiriti o Waitangi*, and to becoming the 'allied other'.

The first tool, or strategy, is the creation of what Mossman (2018) calls a 'Third Space'. This space is created between Indigenous and non-Indigenous peoples to hold respectful, meaningful and beneficial conversations which progress towards better futures. In these spaces, Mossman suggests that the role of the architect must be to 'provoke a reconsideration of architecture's past and present narratives in order to envisage more inclusive and productive directions, better suited to the reality of Indigenous world views and practices' (2018, p. 202). For *Aotearoa, Te Tiriti o Waitangi* is the basis of New Zealand as a nation state and the values and principles expressed in *Te Tiriti*[6] must take centre stage for any design team working within the 'Third Space' (Mossman, 2018; Tau et al., 2016, p. 95).

The second tool is within the creation of the 'Third Space' representative of *Te Tiriti*. In these spaces, non-Māori architects must work with *ngākau tapatahi* (integrity) and *māhaki* (humility) (Kake, 2021). The typically more individualistic approach of *Pākehā* can be challenging for Māori when working in coalition, so *Pākehā* need to be conscious of how their world views could manipulate or impact Indigenous concepts (Huygens, 2011). Additionally, Māori are at times dismissed or patronised in ways that *Pākehā* may not be, especially in conversations which discuss racism and the ongoing processes of colonisation (Thomas, 2020). Therefore, *Pākehā* architects must find opportunities to stand back and give power to Māori designers who have the knowledge to create culturally significant architecture in *Aotearoa* (Kake, 2021). In this conscious 'standing back', non-Māori architects can reflect upon the integrity of their personal and professional *Te Tiriti* partnership, and can take the opportunity to act as allies by centring Māori designers and concepts. At the same time, they are afforded space to actively refute oppressive sentiments about such a process if they arise from their non-Māori peers (Thomas, 2020).

The final tool or strategy is for non-Māori architects to spend time genuinely exploring and learning about *Te Ao* Māori in respectful ways. Māori have a complex grasp on *Pākehā* and *Tauiwi* New Zealand culture, identity, heritage, values and customs but few design teams have that same collective understanding of *Te Ao* Māori (Tau, 2016). Māori design consultation group, Matapopore Trust, state that 'it is necessary to spend some time considering the heritage of both cultures so that a synthesis of values creates a communal focal point. Drawing an empty space and calling it a "square/plaza/*marae*" is not enough. It needs to incorporate the values of a *marae* rather than the design principles' (Tau, 2016, p. 13). The process of learning cannot be rushed and it must, in part, allow for a challenging of perspectives alongside a redirection towards Indigenous agendas, and therefore greater whole of community resilience. These strategies are illustrated in Figure 12.3: the decolonial architect's table.

As *Pākehā* and *Tauiwi* begin to better understand decolonisation practices, the discourse will develop, and actionable steps towards decolonisation will become more embedded within societal norms. Therefore, to better understand the decolonisation tools and strategies as presented above, a series of thought-provoking 'decol-cards' were created that could assist in creating a decolonisation design and dialogue process for developing architectural practices that build resilience throughout the built environment of Aotearoa (Figure 12.4). These cards are the beginning of understanding how decolonisation paradigms can be implemented into architectural workplaces and practices. The intention is that they are used by architects, designers or those engaging with the built environment in Aotearoa throughout the planning, designing and developed design stages of a design-led project. This will encourage engagement with decolonial agendas at the earliest stages of the process. The cards do not solely traverse decolonisation at the surface level, instead they allow designers to collectively explore who they are, what their intentions with the project may be and how to challenge their unconscious biases. The intention is that they facilitate the beginning of open and honest conversations as part of the design process.

Conclusion

Decolonisation is a process that is necessarily painful, but it will lead to an Aotearoa, which will focus on and realise the commitments outlined in *Te Tiriti*. Approaching conversations about decolonisation will often emphasise feelings of guilt and unease

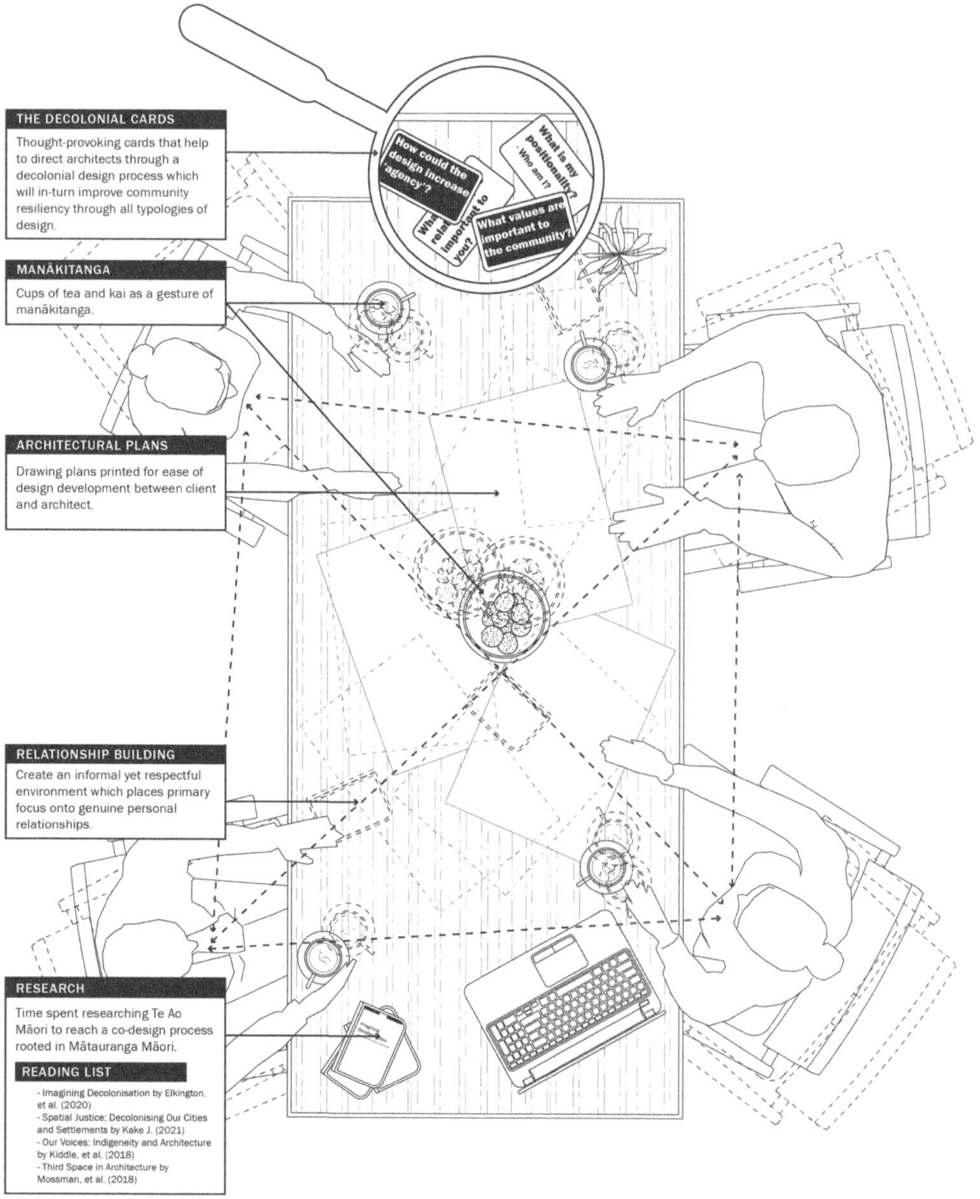

FIGURE 12.3 The decolonial architect's table. Illustrating relationships between architects, participants and a series of important elements of a decolonial practice.

related to responsibility for non-Māori, but decolonisation demands growth, redirection, reflection, and crucially, action. As Aotearoa moves towards decolonisation agendas and resulting practices, it must be stressed that the process is long and cannot be 'solved' within the practice of architecture alone. This work suggests one way that architects can better approach a singular part of the built environment framework:

FIGURE 12.4 Decol-cards to facilitate valuable conversations among designers during the design process as well as between designers and the community.

disaster resiliency. However, if non-Māori architects engage with decolonisation-based resiliency strategies in all facets of architectural design then there will be opportunities for communities within Aotearoa to build upon their existing resilience and drive better relationships which will strengthen the collective social fabric and make it more capable of wrapping around marginalised communities in both normal times and in times of crisis. This is important as the peoples of Aotearoa begin to understand how to integrate decolonisation agendas into all areas of its social structures, systems and practices, including architecture, and is of particular importance as the climate continues to change and the impacts are more frequently felt. Individuals must utilise their knowledge sets, positionality and roles within the social fabric to collectively work towards supporting the needs of all communities within Aotearoa, particularly those that will be most impacted by disasters of all kinds. Architects must understand that the design and creation of resilient architecture is more than just the design of a strong physical structure.

Our findings and recommendations relate specifically to Aotearoa, but some of the underlying concepts and values may be useful in other international contexts, particularly those relating to settler-colonial communities. It is important that as built environment professionals grapple with responding to current and future impacts of climate change that adaptation does not become a new form of colonisation, or a vehicle of further social injustice. In fact, if approached with humility, and an acknowledgement of the past while looking to the future, adaptation could become the basis for reconciliation. Decolonisation is not a clearly defined path, but that must not stop architects and other building professionals from beginning the journey.

KEY TAKEAWAYS FROM CHAPTER 12

- Climate resilience is one of the most important factors that architects must consider. However, architects are neglecting key factors which make communities resilient and instead tend to focus on technological solutions that do not build resilience, particularly for vulnerable communities.
- The way forward in developing community resilience strategies is through applying a decolonisation approach. A decolonisation approach will benefit all cultures, including the settler coloniser.
- As an architect, it is paramount that an individual makes their positionality transparent when co-designing alongside Māori.
- Several important factors should be integrated into the design process to develop a decolonial approach. For instance, open dialogue and thoughtful critical reflection can help dismantle prejudice, individual ego and unjust hierarchies, creating a space where genuinely socially conscious design can emerge. In the context of Aotearoa, this could be by beginning with a pepeha, integrating principles of tikanga or exploring Te Tiriti.
- These tasks/approaches should be undertaken by Pākehā and Tauiwi architects without expecting Māori to hand-hold throughout.

Notes

1 *Te Tiriti o Waitangi*, first signed in 1840, is the *te reo* (Māori language) version of the founding document of Aotearoa New Zealand. It is a document of central importance to New Zealand historical and contemporary governance of relationships between 'the crown' (meaning in practice the government of New Zealand) and Māori. It sets up the basis for non-Māori to be able to live in the country. A historical and political overview of *Te Tiriti* can be found in Orange (2015).

2 The term '*Pākehā*' is a te reo term that provides the opportunity to root cultural identity in Aotearoa for New Zealanders of European descent, many of whom can trace New Zealand-born ancestors back for many generations. Claiming a *Pākehā* identity is for some a political act, which comes with a moral obligation to engage with Māori 'for the purpose of decolonisation and reconciliation' (Amundsen, 2018, p. 146).

3 'Social infrastructure' is typically thought of as facilities, spaces, services and networks that support the quality of life and wellbeing of communities. In the Aotearoa context this can be expanded to include 'features of social organisation, such as trust, norms and networks that can improve the efficiency of society by facilitating coordinated actions' (Treasury New Zealand, 2018).

4 Social vulnerability refers to the capacity of groups of people with specific characteristics, or whole communities to withstand adverse impacts from multiple human made (like colonisation or climate change) or natural (like floods, or pandemics) stressors. Characteristics may be related to gender, ethnicity, socio-economic status, employment status, age, cultural values etc. Environmental Health Intelligence New Zealand (EHINZ) has devised a set of social vulnerability indicators and maps for natural hazards, pandemics and other disasters in Aotearoa and provides further detail (EHINZ, 2021).

5 *Tauiwi* is the *te reo* term used for New Zealanders who are not Māori and includes non-*Pākehā*.

6 Broadly, one interpretation of the values or principles of Te Tiriti o Waitangi are: Tino rangatiratanga (self-determination) and equity, active protection, participation (related to proper resourcing and access), and partnership. Further discussion and interpretation of the meaning, practice and direction of Te Tiriti as a living document can be found in Came and Zander (2015).

Bibliography

Abbott, M., Bakineti, T., Carter, L., Latai-Niusulu, A., Missack, W., Puketapu, I. & Kiddle, R. (2022). Indigenous Thinking about Nature-based Solutions and Climate Change. The British Academy. Available online: www.thebritishacademy.ac.uk/publications/indigenous-thinking-about-nature-based-solutions-and-climate-justice/?fbclid=IwAR2TPu5vvn47u-4-ouxnoHMo-FyBc7NxVyvhrKOJyf11n04L-kkG-As2v0m8. Date accessed: 17 August 2023.

Allen, P. & Roberts, J. (2009). Urban resilience and the open space network. *Tephra* 22, 55–61. www.civildefence.govt.nz/resources/tephra

Amundsen, D. (2018). Decolonisation through reconciliation: The role of Pākehā identity. *MAI Journal: A New Zealand Journal of Indigenous Scholarship*, 7(2). https://doi.org/10.20507/MAIJournal.2018.7.2.3

Arago-Kemp, V. (2018). *Bridging Cultural Perspectives*. Superu.

Berke, P. R. & Campanella, T. J. (2006). Planning for Postdisaster Resiliency. *The ANNALS of the American Academy of Political and Social Science*, *604*(1), 192–207. https://doi.org/10.1177/0002716205285533

Blakeley, R. (2016). *Building Community Resiliency*. New Zealand Society of Local Government Managers.

Bosher, L. & Dainty, A. (2011). Disaster risk reduction and 'built-in' resilience: Towards overarching principles for construction practice. *Disasters*, *35*(1), 1–18. https://doi.org/10.1111/j.1467-7717.2010.01189.x

Brown, D. (2018). Contemporary Māori Architecture. In E. Grant, K. Greenop, A. L. Refiti & D. J. Glenn (Eds.), *The Handbook of Contemporary Indigenous Architecture* (pp. 107–125). Springer. https://doi.org/10.1007/978-981-10-6904-8_4

Burnside-Lawry, J. & Akama, Y. (2013). Communication research needs for building societal disaster resilience. *Australian Journal of Emergency Management*, *28*(4), 29–35. https://doi.org/10.3316/agispt.20140771

Came, H. & Zander, A. (Eds). (2015). *State of the Pākehā Nation: Collected Waitangi Day Speeches and Essays*. Whangarei, New Zealand: Network Waitangi Whangarei.

Castaño-Rosa, R., Pelsmakers, S., Järventausta, H., Poutanen, J., Tähtinen, L., Rashidfarokhi, A. & Toivonen, S. (2022). Resilience in the built environment: Key characteristics for solutions to multiple crises. *Sustainable Cities and Society*, 87, 104259.

Cutter, S. L. (2016). Resilience to what? Resilience for whom? *The Geographical Journal*, *182*(2), 110–113. https://doi.org/10.1111/geoj.12174

Daly, M., Becker, J., Parkes, B., Johnston, D. & Paton, D. (2009). *Defining and Measuring Community Resilience to Natural Disasters: a case study from Auckland. 22.*

Denzin, N. K. & Lincoln, Y. S. (2008). Introduction: Critical Methodologies and Indigenous Inquiry. In N. Denzin, Y. Lincoln, & L. Smith, *Handbook of Critical and Indigenous Methodologies* (pp. 1–20). SAGE Publications, Inc. https://doi.org/10.4135/9781483385686.n1

EHINZ (2021). *Social vulnerability indicators for natural hazards*. Massey University. Available online: www.ehinz.ac.nz/indicators/population-vulnerability/social-vulnerability-to-natural-hazards/. Date accessed: 17 August 2023.

Elkington, B. & Smeaton, J. (2020), Introduction. In: Elkington B, Jackson M, Kiddle R, et al. (eds), *Imagining Decolonisation*. Bridget Williams Books, pp. 7–20.

England, J. (2016). Principles of partnership. In *Grand Narratives* (pp. 29–39). Canterbury Earthquake Authority.

Ginige, K. (2011). Disaster risk reduction and its relationship with sustainable development. In D. Amaratunga & R. Haigh (Eds.), *Post-Disaster Reconstruction of the Built Environment: Rebuilding for Resilience* (pp. 287–303). John Wiley & Sons, Incorporated. http://ebookcentral.proquest.com/lib/vuw/detail.action?docID=693809

Goodyear, R. (2017). A place to call home? Declining home-ownership rates for Māori and Pacific peoples in New Zealand. *New Zealand Population Review*, 43, 3–34.

Gray, L. (2017). Social determinants of health, disaster vulnerability, severe and morbid obesity in adults: Triple jeopardy?. *International Journal of Environmental Research and Public Health*, 14(12), 1452.

Hayward, B. M. (2013). Rethinking Resilience: Reflections on the Earthquakes in Christchurch, New Zealand, 2010 and 2011. *Ecology and Society*, *18*(4). www.jstor.org/stable/26269450

Houkamau C.A. & Sibley C.G. (2015). Looking māori predicts decreased rates of home ownership: Institutional racism in housing based on perceived appearance. *PLoS ONE*, 10(3): e0118540. https://doi.org/10.1371/journal.pone.0118540

Hutchings, J., Lee, J. & New Zealand Council for Educational Research. (2016). Introduction: Kaupapa Māori in action: Education, research and practice. In *Decolonisation in Aotearoa: Education, research and practice.* NZCER Press.

Huygens, I. (2011). Developing a Decolonisation Practice for Settler Colonisers: A Case Study from Aotearoa New Zealand. *Settler Colonial Studies*, *1*(2), 53–81. https://doi.org/10.1080/2201473X.2011.10648812

Jackson, M. (2020). Where to next? Decolonisation and the stories in the land. In R. Kiddle (Ed.), *Imagining Decolonisation* (1st edn). Bridget Williams Books. https://doi.org/10.7810/9781988545783

Jasmax. (n.d.). Te Kura Whare. Jasmax. Retrieved 2 August 2021, from www.jasmax.com/ projects/featured-projects/te-kura-whare/

Kake, J. (2021). Spatial justice – decolonising our cities and settlements. *Counterfutures*, *9*, 123–135. https://doi.org/10.26686/cf.v9.6777

Kenney, C. M., Phibbs, S. R., Paton, D., Reid, J. & Johnston, D. M. (2015). Community-led disaster risk management. *Australasian Journal of Disaster and Trauma Studies*, *19*(1), 12.

Kiddle, R. (2020). Colonisation sucks for everyone. In R. Kiddle (Ed.), *Imagining Decolonisation* (1st edn). Bridget Williams Books. https://doi.org/10.7810/9781988545783

King, D. N. T., Goff, J. & Skipper, A. (2007). Māori environmental knowledge and natural hazards in Aotearoa-New Zealand. *Journal of the Royal Society of New Zealand*, *37*(2), 59–73. https://doi.org/10.1080/03014220709510536

Latai-Niusulu, A., Nel, E. & Binns, T. (2020). Positionality and protocol in field research: Undertaking community-based investigations in Samoa. *Asia Pacific Viewpoint*, *61*(1), 71–84. https://doi.org/10.1111/apv.12252

Lizarralde, G., Johnson, C. & Davidson, C. (2009). Rebuilding after disasters; From emergency to sustainability. In *Rebuilding after Disasters: From Emergency to Sustainability* (pp. 1–24). CRC Press LLC. http://ebookcentral.proquest.com/lib/vuw/detail.action?docID=446574

Matunga, H. (2018). A Discourse on the Nature of Indigenous Architecture. In E. Grant, K. Greenop, A. L. Refiti & D. J. Glenn (Eds.), *The Handbook of Contemporary Indigenous Architecture* (pp. 303–330). Springer. https://doi.org/10.1007/978-981-10-6904-8_12

Mazereeuw, M. & Yarina, E. (2017). Emergency Preparedness Hub: Designing Decentralized Systems for Disaster Resilience. *Journal of Architectural Education*, *71*(1), 65–72. https://doi.org/10.1080/10464883.2017.1260928

McAleer-Harding, P. (2022). *The Demographics of Resiliency: Decolonising Disaster Resilient Architecture* (Thesis, Master of Architecture, Professional). Te Herenga Waka Victoria University of Wellington. Retrieved from https://doi.org/10.26686/wgtn.20455551. 31 August 2023.

McKinnon, S. & Cook, M. (2020). Introduction. In S. McKinnon & M. Cook (Eds.), *Disasters in Australia and New Zealand: Historical Approaches to Understanding Catastrophe* (pp. 1–17). Springer. https://doi.org/10.1007/978-981-15-4382-1_1

Mercier, O. (2020). What is decolonisation? In R. Kiddle (Ed.), *Imagining Decolonisation* (1st edn). Bridget Williams Books. https://doi.org/10.7810/9781988545783

Mossman, M. (2018). Third Space in Architecture. In R. Kiddle, *Our Voices: Indigeneity and architecture.* ORO Editions.

Newcombe, N. I. (2019). How Pākehā in not-for-profit organisations implement Te Tiriti o Waitangi (Thesis, Master of Māori and Pacific Development). The University of Waikato, Hamilton, New Zealand. Retrieved from https://hdl.handle.net/10289/13373. 31 August 2023.

Orange, C. (2015). *The Treaty of Waitangi*. Bridget Williams Books.

Partington, J. & Pedersen Zari, M. (2020). Ngāi Tūhoe's Te Kura Whare: Our Living Building. In Pedersen Zari, M., Connolly, P. & Southcombe, M. (Eds), *Ecologies Design* (pp. 212–221). Routledge: Oxon.

Platt, S. (2018). Factors affecting the speed and quality of post-disaster recovery and resilience. In R. Rupakhety & S. Ólafsson (Eds.), *Earthquake Engineering and Structural Dynamics in Memory of Ragnar Sigbjörnsson: Selected Topics* (pp. 369–403). Springer International Publishing. https://doi.org/10.1007/978-3-319-62099-2_19

Rae, H. & Thompson-Fawcett, M. (2018). Kia Tahuri i te Riu, Kia Tika: Indigenous Participation in Earthquake Recovery Planning – Insights from Taiwan and Canterbury. In R. Kiddle, *Our Voices: Indigeneity and architecture*. ORO Editions.

Rose, F. T., Wilkinson, S., Sandeeka, M. & Chang-Richards, A. (2018). Post-disaster reconstruction in Christchurch: A 'build back better' perspective. *International Journal of Disaster Resilience in the Built Environment*, *9*(3), 239–248. http://dx.doi.org/10.1108/IJDRBE-01-2017-0009

Saunders, W. S. A., Kelly, S., Paisley, S. & Clarke, L. B. (2020). Progress toward implementing the Sendai Framework, the Paris Agreement, and the Sustainable Development Goals: Policy from Aotearoa New Zealand. *International Journal of Disaster Risk Science*, *11*(2), 190–205. https://doi.org/10.1007/s13753-020-00269-8

Smith, L. T. (2016). Keeping a decolonising agenda to the forefront. In J. Hutchings, J. Lee, & New Zealand Council for Educational Research (Eds.), *Decolonisation in Aotearoa: Education, research and practice*. NZCER Press.

Smith, L. T., Maxwell, T. K., Puke, H. & Temara, P. (2016). Feature article: Indigenous knowledge, methodology and mayhem: What is the role of methodology in producing indigenous insights? A discussion from Matauranga Maori. *Knowledge Cultures*, *4*(3), 131–156.

Tau, T. M. (2016). Grand Narratives for Christchurch. In *Grand Narratives* (pp. 5–29). Canterbury Earthquake Authority.

Tau, T. M., McMeeking, S., Lenihan, T. M. & Rice-Edwards, A. (2016). The justice and emergency services precinct narrative. In *Grand Narratives* (pp. 71–161). Canterbury Earthquake Authority.

Tennent Brown Architects. (n.d.). Ngā Purapura | Tennent Brown Architects. Tennent Brown Architects. Retrieved 31 August 2023, from https://tennentbrown. co.nz/nga-purapura

Thomas, A. (2020). Pākehā and doing the work of decolonisation. In R. Kiddle (Ed.), *Imagining Decolonisation* (1st edn). Bridget Williams Books. https://doi.org/10.7810/9781988545783

Thornley, L., Ball, J., Signal, L., Lawson-Te Aho, K. & Rawson, E. (2015). Building community resilience: Learning from the Canterbury earthquakes. *Kōtuitui: New Zealand Journal of Social Sciences Online*, *10*(1), 23–35. https://doi.org/10.1080/1177083X.2014.934846

The Treasury New Zealand (2018). Living Standards: A Short Guide to 'Social Infrastructure'. The Treasury New Zealand, Wellington. Available online: www.treasury.govt.nz/sites/default/files/2017-12/hls-ag-socinfr-jan13.pdf. Date accessed 18 August 2023.

Tuck, E. & Yang, K. W. (2012). Decolonization is not a metaphor. *Decolonization: Indigeneity, Education & Society*, *1*(1), Article 1. https://jps.library.utoronto.ca/index.php/des/article/view/18630

Tuhiwai Smith, P. L. (2012). *Decolonizing Methodologies: Research and indigenous peoples*. Zed Books. http://ebookcentral.proquest.com/lib/vuw/detail.action?docID=1426837

Vandertop, C. (2019). Opium cities, carbon routes: World-ecological prehistory in Amitav Ghosh's Hong Kong. *Journal of Postcolonial Writing*, *55*(4), 527–540.

Ward, A. (1996). The suppression of the social in design: Architecture as war. In Dutton, T.A and Mann, L.H. (Eds). *Reconstructing Architecture: Critical discourses and social practices*, 27–70. University of Minnesota Press.

13

SITUATING THE CONCEPT OF SUSTAINABLE DESIGN IN AFRICA

Ng'eno Chelang'at Faith

Introduction

Estimated at 2.37% of the global contribution of CO_2 emission in 2020, with the exception of South Africa and Nigeria, African countries bear an insignificant contribution to the global climate and ecological crisis when compared to Western countries (Moss, 2020). Cumulatively, buildings generate approximately 38% of the annual global CO_2 emissions (UNEP, 2020). However, the UN estimates that the building stock in developing countries like Kenya will increase by 75% compared to European countries at 25–30% by 2050 (UN Habitat, 2014). This understanding necessitates a critical consideration of sustainable built environment design by developing countries, given their significant potential to influence the future global climate change scenario. Broadly, African countries such as Kenya currently struggle with two seemingly divergent forces. On the one hand, the desire to develop as modelled by the Western (developed) world and on the other, the crucial demand for environmental and social consciousness and responsibility.

The subsequent discussion was therefore inspired by questioning popular discourse on sustainability and sustainable design. What really is sustainable design? What, who and why it (they) is being sustained, and finally how would this be achieved? In this regard, the main objective of this chapter is to assess and critique the discourse and practice of sustainable design within the African context. Arguments will engage with the larger discourse of the politics of knowledge and its construction in relation to sustainability, seeking to confront Western constructions of the concept of sustainability, and by extension sustainable design imposed on developing cities such as Nairobi.

The chapter investigates the Western epistemology of knowledge construction and its influence on other epistemologies. In doing so, it continues the arguments around anti-Eurocentric thinking previously submitted by scholars such as Ngugi wa Thiongo (1981), Chinua Achebe (1958), Wiredu (1992), Tuhiwai (1999), Mignolo (2007; 2008; 2010; 2018), Escobar (2004; 2007), Quijano (2000; 2002; 2010), and Dussel (2000), who critique epistemological impositions of knowledge and knowledge construction processes, in this case with reference to sustainability. Fundamentally, the chapter aims to

DOI: 10.4324/9781003293903-16
This chapter has been made available under a CC-BY-NC-ND 4.0 license.

establish an intersection between knowledge construction, hegemony, legitimacy, and sustainability.

There is a distinct antagonistic relationship between Western epistemologies and local epistemologies. However, local knowledge construction has significantly been influenced by Western epistemology to an extent that one could argue has become Western, therefore does not benefit local society. In the case of sustainability, knowledge is largely constructed in terms of CO_2 emissions, global warming, and climate change, which is grounded in Western science and thus proposed solutions prioritise adapting wealthy, culturally rich Western lifestyles, without endangering that wealth. In doing so, other contexts, especially African contexts, are deemed irrelevant and often excluded. Discordantly, as this chapter will demonstrate, Western epistemologies seek dominance and supremacy, while local construction of knowledge struggles to deal with local challenges and priorities on one hand, while seeking legitimacy from the West on the other hand.

This pursuit for legitimacy from the West stems from what 'Western' represents for developing countries like Kenya. Overall, two fundamental issues, which are in part correlated and in part competing, emerge: Western *intention* and Southern *response*. By way of example, in the case of the green building assessment systems predominantly developed in the West, besides the obvious economic (capitalist) gain from establishing a global footprint, like any other standard, the Western *intention* is to establish authority and control, to define, dictate, and determine what counts as "green" or "sustainable". Unfortunately, the response by cities like Nairobi, on the other hand, plays into these *intentions* by adopting these systems without interrogating them. Thus, historical and present global conditions and systems create and continue to influence this position in which cities like Nairobi find themselves inextricably entangled.

The theory of knowledge is not politically neutral, nor can it exist outside of politics. Therefore, there is the imminent need to interrogate and engage with knowledge from a less pretentious universal standpoint to one that is "more aware of its limitations, its partiality, its interests and motivations, and its positionality" (Reiter, 2018, 8). Thus, it explores the notion of decolonial thinking, together with Haraway's (1988) concept of "situatedness". Both these concepts present processes of unmasking the hegemonic nature of Western modernity and knowledge construction towards self-determination. In essence, it is the process by which cities like Nairobi construct their own meaning, in their own language while cognisant of their priorities and contextual dynamics and challenges – in this case, in an effort to create a situated sustainable built environment.

International influence on construction of knowledge

The Western influence on the built environment is not limited to sustainable design and construction practice. Nairobi's built landscape and its relationship to its environment (nature) for example, has been radically transformed by Western images of spatial form and order.

Broadly, this chapter distinguishes three phenomena that contribute to international influence in the urban built landscapes of cities in Africa like Nairobi: coloniality, modernity, and globalisation (Figure 13.1). It is imperative to question how the formation of 'new' concepts such as sustainable design sit within the existing colonial, modernity, and globalisation matrix to avoid the creation of variations of Western epistemology.

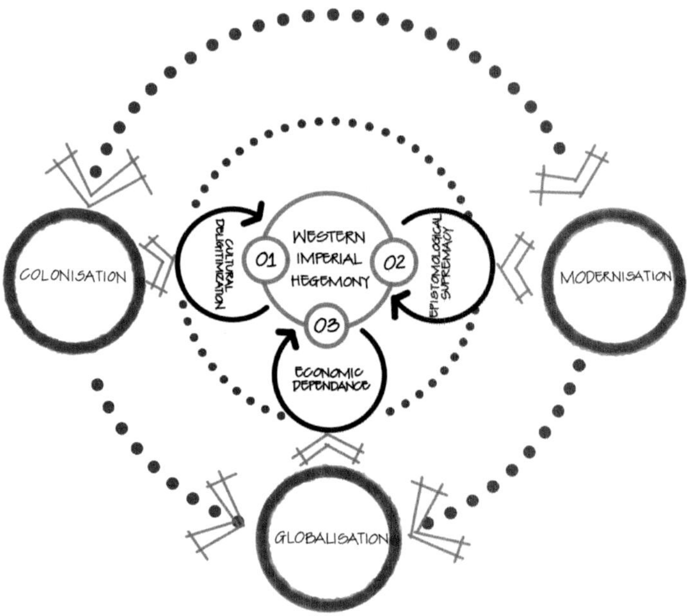

FIGURE 13.1 The relationship between colonisation, modernisation, globalisation, and Western imperial hegemony. (Author, 2019)

The looming shadows of colonial history

As stated earlier, to understand the process of knowledge construction, with reference to sustainability in Kenya, like other African countries, it is imperative to begin by situating the process of knowledge construction within the country's imperialist colonial history in an attempt to conceptualise its continued impact. Colonisation represents, as Tuhiwai (1999, 64) posits, the "imposition of Western authority over all aspects of indigenous knowledges, languages and cultures". It is evident that colonisation has had far-reaching effects not only on Nairobi's built environment but its language, culture, and identity. Beyond territorial control, colonisation meant the destruction and delegitimisation of not only indigenous culture but also established cultural and scientific knowledge ontologies and epistemologies. This involved the imposition of a world system in which, as Grosfoguel (2011, 7–8) asserts, "a European/capitalist/military/Christian/patriarchal/ white/ heterosexual/male arrived in the Americas [in this case Africa] and established simultaneously in time and space several entangled global hierarchies". This thereby created what Escobar (2007, 185) termed a "privileged epistemological and political space".

Since the inception of post-colonial "modern" African cities like Nairobi, Western influence has been a predominant driving force for development (Manuel, 2015). Incipiently, modernisation of Nairobi cannot be divorced from colonisation, which was embedded with external hegemonic influences from its former colonial powers. As in Nairobi, the presence of European influence is evident in the landscapes, regulatory systems, and education systems of several African cities. Kenya, like many African countries, was socialised to believe in the aggrandised concept of the West. This imposition of Western superiority and positionality is evident in how the Western world referred, and continues to refer, to African countries as "the ancient world, … the primitive world, the third world, the underdeveloped world, the developing world and (more recently) the global south" (Comaroff & Comaroff, 2012, 144).

The Western world actively created distorted narratives that misrepresented African realities as knowledge from indigenous societies was collected to advance the colonial agenda. In one of his speeches, *"Native policy of Africa"* delivered in 1929, Smuts describing Africans, asserts, "it (Africans) has largely remained a child type, with a child psychology and outlook… No indigenous religion had been evolved, no literature, no art… no architecture…" (Scheub, 2010, 59). Embedded within Western narratives is the contestation and disregard for indigenous forms of religion, art, and architecture. Africans were robbed of their identity, legitimacy, and knowledge forms, which paved the way for the imposition of Western forms of knowledge. While overt expressions like Smuts's are less common today, the underlying attitude of superiority persists in more insidious forms. The Western world often continues to frame Africa as a space needing rescue, intervention, or validation by external powers. This is evident in the global narrative and policies on sustainability that still disproportionately privilege Western technology, standards, materials, and expertise, over African voices and indigenous practices.

An exploration of the built environment in cities in Africa (Manuel, 2015), suggests a strong link between knowledge construction, identity, and legitimacy. The latent Eurocentrism embedded in Western knowledge epistemologies renders it difficult to dissociate knowledge from power, identity, and legitimacy. It follows that, perhaps the most significant consequence of colonialism was the deprivation of African cities' local identity, global legitimacy, and recognition. Western epistemology of knowledge sought and continues to seek to impose "universal truth", through homogenisation of knowledge using language, narratives, design standards, and assessment systems among other ways. It further seeks to regulate how this knowledge is interpreted, applied, and consequently legitimises its application based on the West's own assessment. The universalist notion as Grosfoguel (2002) posits, presents itself as a "disembodied" impartial global design disguising the epistemological position of its origin. However, despite the aim of the Western epistemology of knowledge construction that the knower assumes the position of a detached neutral observer in pursuit of objective truth, the politics engrained in knowledge construction suggests that the knower is inextricably linked to the known. With regards to sustainable design, this is manifested in the clear antagonistic relationship between the struggle to understand, prioritise, and mitigate local challenges on one hand, and the pursuit for legitimacy from the Western world, which often does not understand or even care to understand the context as they carry their own vested interests, on the other.

Furthermore, the entrenchment of Western epistemology is exacerbated by education systems in several countries in Africa that largely remain as colonial products entrenched in Western ideologies (Ngugi wa Thiong'o, 2009; Ntarangwi, 2004). Previous arguments have placed universities as "the most enduring colonial institutions" (Nhemachena et al., 2016, 12). In several developing countries, many of the Eurocentric theories remain unquestioned, and this is not only limited to architecture. Du Bois (1903) argues "education was so arranged that the young learned not necessarily the truth, but the aspects and interpretation of the truth which the rulers of the world wished them to know and follow." Maathai (2006, 60) similarly argued that the removal of culture from the school systems "contributes to the trivialization of anything African and lay the foundation for a deeper sense of self-doubt and an inferiority complex". This imposed perception of the supremacy of Western theory is influenced by the idea of what "Western" represents. Thus, it is no surprise that when the Western world introduces "new" concepts such as sustainability and by extension sustainable design or green buildings, African cities – Nairobi included – are quick to adopt this concept without critical interrogation. Yet, ironically, the climate and ecological crises the West purports to solve, can be traced back to colonisation and the continued exploitation of resources in African countries.

What "the West" represents – the chase for legitimacy

> "Modernity" was imagined as the house for epistemology.
>
> *(Mignolo, 2006, 93)*

The Western construct of modernity was, and still is, an extension of the Western hegemonic legacy that came to fruition as a result of coloniality (Quijano, 2000; Mignolo, 2007). Modernity as a Western concept gained popularity during the second half of the twentieth century, often attributed to the European Renaissance, European Enlightenment, and later the industrial revolution, as a descriptor of Western civilisation and consequently as a pathway for the rest of the world to follow. "Europe and America mutually produced themselves as the historical and the first two new geocultural identities of the modern world" (Quijano, 2002, 552). Therefore, "Modernity's Eurocentrism lies in the confusion between abstract universality and the concrete world hegemony derived from Europe's position as centre" (Dussel, 2000, 471).

Although the Eurocentric concept of modernity was presented and imposed on countries like Kenya as an emancipation from a barbaric, immature way of life, it is however difficult to ignore what Dussel (2000, 473) refers to as "the hidden 'other side' of modernity". Western modernity was (is) part of the larger colonial legacy whose epistemology was not only to deculturalise Africa but more significantly, to portray and replicate Western culture as superior, in an attempt to produce a homogenous Western culture that would ensure their continued hegemony. These "modern" values were systematically entrenched in the language, governance, religious, and education systems established by the West in their colonies.

Through disguised narratives, modernity subjectively divided the world into the modern and those seeking modernity (Mignolo, 2018), those who have developed economic and knowledge structures and those that are underdeveloped, creating a false hierarchical measure of progress and civilisation, primarily based on cultural identity. Therefore, legitimacy of African culture and knowledge forms was/is measured and determined only against or in terms of Western knowledge forms (Appiah, 1992; Eze, 2010; Etieyibo, 2016). Thus, in the words of Etieyiobo (2016, 85), Africans began to "define themselves through the eyes of the West". As Chinua Achebe submits, "until the lion learns to write, the narrative will always glorify the hunter."

Furthermore, as Western culture through the concept of modernity was and continues to be presented as an aspiration, these narratives continue to colonise indigenous imagination, crippling local innovation and increasing reliance on "foreign" knowledge and solutions. Western epistemologies relegated other (especially African) countries to being consumers of knowledge generated in the West, unable to generate their own knowledge.

The West's successful monopolisation and commercialisation of technical knowledge makes it difficult for the rest of the world to make meaningful contributions. In essence, this positionality and knowledge imposed through colonisation, and later modernity, has created the phenomenon that Ngugi wa Thiong'o referred to as "colonisation of the mind". In seeking this Western-prescribed form of modernity, the global south began to reject their own culture, religion, art, and other knowledge forms as their ambitions became entrapped in the web of Western presentation of modern knowledge and culture.

Several researchers from countries that bear the Western characterisation of "developing" (Ngugi Wa Thiong'o, 1986; Wiredu, 1992; Chatterjee, 1998; Suresh, 2002; Mignolo, 2002, 2009) have contested the hegemony embedded in the Western modernity epistemology and have questioned the nature of modernity presented and imposed by the West. It is important to note that these are not anti-modern fundamentalist arguments but rather what Grosfoguel (2002), making reference to Dussel's (2000, 26) concept of "transmodernity", argues is a "decolonial transmodern response of the subaltern to Eurocentric modernity". This argument goes beyond the singularity that Western modernity presents to a pluriverse epistemology that is inclusive of all knowledge forms. Similarly, in making a distinction between "our modernity and their modernity", Chatterjee (1998) highlights the peculiarities that evidently exist in the conceptualisation of the concept of modernity and calls for the re-contextualisation of this concept of modernity. To achieve this, it must be understood that "the modern crisis is a crisis of thought" (Escobar 2004, 209), and therefore there is a limit to using the same knowledge epistemologies that created this crisis.

Consequently, there is the need to move from a single Eurocentric "global" conception of modern to a network of alternative local constructions of modernity, based on multiple contextualised knowledge forms that are reflective of the socio-cultural and economic orders of their respective contexts. At the core of the modernism movement is the pursuit for universalism and therefore standardisation. Standardisation not only privileges those constructing "the standard" but more significantly it involves the reductive abstraction of the diverse realities that characterise different contexts, resulting in concepts being far removed from contextual realities.

Globalisation – the universalist thinking

Globalisation is the acme of the process that was initiated through the colonial/modernity Eurocentric (and more recently American) hegemonic and capitalist system. Similar to colonial/modernity, the theory of globalisation is confined to Western construction driven fundamentally by capitalistic processes of exploitation (often of the global South) and accumulation (mainly by the global North). Capitalism therefore became a model of power and control, particularly for the exploitation of labour and resources from the global South, for the production of commodities for a global market. This system continues to flourish as a result of the already established Eurocentric hierarchical cultural and epistemological system. Thus, this discussion goes back to the question of Western intent. Why globalise? Answers to this question begin to illuminate the power and agency embedded in this global system.

As Escobar (2007, 181) submits, "globalisation entails the universalisation and radicalisation of modernity". Cooper (2005, 91) argues that globalisation presents two concerns: one, the "global" suggesting universality and two, "ization" suggesting an on-going process. As a result, "the globalization of western knowledge and western culture constantly reaffirms the West's view of itself as the centre of legitimate knowledge, the arbiter of what counts as knowledge and the source of 'civilised' knowledge" (Tuhiwai, 2006, 63). This, therefore, not only disadvantages local knowledge but advances control over knowledge production, dissemination, assimilation, and application.

Perhaps a good example of this with reference to sustainable design in cities like Nairobi is the introduction and proliferation of green building assessment systems. Often, these systems are introduced to "developing" countries with the intent to increase their global footprint and further the country of origin's own agenda, overlooking local priorities and dynamics. These systems come with enabling technologies and products among other conditions. For example, to achieve LEED certification, prescribed technologies and products have to be employed. Therefore, manufacturer of "green materials", for example, will benefit from a global footprint through the validation and endorsement of their products by international governing institutions driving the agenda. Standardisation, therefore, becomes a strategy employed by large corporations – and governments – to ensure maximum profitability. This is exacerbated by the mental disposition of the "developing" world discussed earlier, that glorifies international solutions and products.

Since the inception of the concept of sustainable design, the Western world has introduced numerous regulations to guide sustainability. The view that introducing more stringent rules and regulations to the building industry in order to develop a sustainable built environment is widely held globally. However, this claim can be contested with regard to the extent of the impact these rules and regulations would have. An increase in regulations may not necessarily translate to a sustainable built environment. This is evident in the Western countries where these guides, frameworks, and assessment systems have been developed. Despite these tools having existed for decades now, there is little tangible evidence that suggests building environments have become more sustainable; on the contrary, evidence seems to point to a less sustainable built environment (Capeluto, 2022; Willis, 2000). However, the greater concern with the formation and adoption of sustainable design regulations in cities like Nairobi is the imposition and adoption of regulations that are termed as global standards, that have failed to bring about change even in the countries in which they were developed.

Furthermore, it is worth discussing the influence of the capitalist market as an integral part of the ideals of modernity and globalisation, in relation to sustainability. Evidently, economically, the colonial situation in Africa continues to exist largely through massive international (Western) organisations such as International Monetary Fund (IMF), the World Bank (WB), the North Atlantic Treaty Organisation (NATO), and more recently China, not to mention smaller international organisations that fund "development" projects across Africa. Therefore, capital remains a dominant axis that cements the imperialist hierarchical system. As they say, "he who pays the piper calls the tune." Another example with regard to sustainability is the capitalistic market-based solution referred to as "carbon emission (credits) trading" that privileges the Western world. The question once again is, who decides, and by what criteria are these emission allowances distributed? Who does it benefit? Certainly not the developing world. For transformative change to be achieved, countries like Kenya must de-link themselves from this systemic exploitative global system of domination and accumulation.

At its core, the call towards sustainability is fundamentally an ethical issue born out of a concern for the environment that supports human survival. However, unfortunately, on a global scale, there appears to be very little genuine concern for the environment or humanity for that matter, evidenced by the apparent reduction of this concept into a political issue, driven by a capitalistic global market. The profit-driven capitalist world continues to "steal" from the natural world. Kovel (2007), who in his book sub-title asks, "the end of capitalism or the end of the world?" is concerned with the possibility of maintaining the expanding capitalistic economy while re-establishing ethical responsibility to the planet and its inhabitants. The question as to whether capitalism and sustainability can co-exist remains unanswered.

In addition, the increasing commodification of knowledge contrary to the ethical precepts of sustainability cannot be overlooked. Often those who determine what counts as knowledge, lay claim to its ownership and therefore attribute value – monetary or otherwise – to it, so that "the rest" must pay to access and utilise that knowledge. In relation to sustainable design, this is again best exemplified by assessment systems and technological solutions mostly developed in the West that dictate what sustainable design should look like, while placing a premium price point to ascertain that buildings fit into that definition. The cost of certification based on these systems is considered unaffordable to the majority of developers in cities in Africa. Often, only large corporations seek certification, and not necessarily out of ethical concern for the environment but more as a badge for global (Western) recognition and marketing (Capitalist) strategy. This too, unfortunately, is representative of the global scenario. The "green" label has increasingly become a popular feature for buildings and products, often serving as a marketing tool to attract consumers. However, achieving certification such as LEED does not always equate to meaningful contributions to sustainability, particularly when such assessment systems are not tailored to the specific environmental, social, and economic context in which they are applied (Chelang'at & Lawrence, 2024). Similarly, the technological solutions advanced are geared towards economic benefit and maintaining current lifestyles, while deceptively being distinguished as "green" or "sustainable" (Kurnaz, 2021; Schoeman & Gunter, 2018).

Towards a situated sustainable design concept

> Decolonisation… does not mean and has not meant a total rejection of all theory or research or Western knowledge. Rather, it is about centring our concerns and world views and then coming to know and understand theory and research from our own perspective and for our own purpose.
>
> *(Tuhiwai, 1999, 39)*

Against the broad background of the perpetual epistemological impact of colonialism, Western modernity and globalisation, the foregoing discussion has challenged the hegemony embedded in the Western idea of universal knowledge and knowledge construction processes regarding the discourse of sustainability in countries like Kenya. Therefore, delinking from these imperialist forms of knowledge construction cannot be achieved using Western epistemologies. These countries must therefore deconstruct, reconstruct, and rearticulate their own narratives based on known contextual realities to dispel the false narratives that have been propagated by the Western world. Thus, epistemologically, it goes beyond impositions and detached observational constructs about particular contexts, to recognising contextual diversity and understanding the experiences of these contexts, towards specific goals and measures that are subjective to their context reference.

Consequently, the discourse of sustainable design is inextricable from historical, social, cultural, and political dispositions, and therefore cannot be understood without situating it. The concept of "situatedness" advanced by Haraway (1988), therefore, introduces a conceptual space that is aware and critical of epistemological, ontological, political, and even ethical positions from which existing sustainable design knowledge claims can be interrogated.

Situated knowledge challenges the universalist construction of knowledge, in other words it challenges what counts as knowledge, arguing fundamentally that knowing is context specific. In this case, the knowledge of sustainable design should be specific to the context of countries in Africa. It goes beyond simply taking an anti-colonial or anti-West stance to deliberate attempts towards self-determination by developing local epistemologies, methodologies, and approaches that consider local political, social, cultural, and material circumstances in Africa, and which influence and transform knowledge construction on sustainable design. Furthermore, it asserts the importance of understanding the manner in which local communities understand and theorise their own realities.

Generally, it is improbable to encounter an expert on Western cultural, developmental, or otherwise practices from Nairobi, for example. Conversely, it is acceptable, in fact expected, that the Western world should have "experts" on Kenyan (African) matters. It is more than likely, however, that Kenyans would know and understand their needs better than the so- called Western experts. As Irwin (cited in Tuhiwai, 1999, 38) argues, "we don't need anyone else developing the tools which will help us come to terms with who we are. We can and will do this work. Real power lies with those who design the tools – it always has. This power is ours." In the same vein, "developing" countries do not need the Western world developing tools that would dictate how their sustainable built environment should look within their contexts. This principle of self-determination gives local people the power to define priorities, determine issues that need to be highlighted, and the opportunity to discuss these issues amongst themselves as well as measure their progress.

It is imperative to note that while the argument is not opposed to Western knowledge and theory in its entirety, it is sensitive to its pertinence and the manner in which this knowledge and these theories are introduced, imposed, and applied within the local context. Beyond this, there is scepticism around the Western world's capacity, intentions, and methodologies to advance sustainable design solutions that would benefit local communities in Nairobi. It is not enough just to acknowledge multiplicity or plurality, as colonial power structures were also predicated on acknowledging these differences and using them to subjugate others. Therefore, it is crucial to critically interrogate the ontological and epistemological lenses through which this multiplicity is acknowledged (Fry & Willis, 2017). This then calls to question the use of Western epistemologies that often refuse to acknowledge the heterogeneity of knowledge, in the understanding of the multiplicity as evident in several sustainable frameworks, best practices and/or green building assessment systems that attempt to incorporate what are often termed "local components". More importantly, it emphasises the demand to recognise indigenous epistemologies towards the development of situated knowledge by deconstructing Euro-centric, geopolitical knowledge paths, and constructing pluriverse and interverse epistemologies (Mignolo & Walsh, 2018).

It has been argued (Dryzek, 1997; McLennan, 2004; Glassie, 1990; Oliver, 2000) that vernacular architecture was and is still sustainable, notwithstanding the term not necessarily being used or understood according to Western theory. Based on indigenous knowledge and reverence of their environment and their innate cultural disposition of a sense of community and inter-dependence, indigenous communities lived in harmony with nature, were efficient, self-sufficient, and even resilient; terms that are often used in Western theory to define sustainability. Perhaps, therefore, the development of situated knowledge on sustainable environments should be inclined towards interrogating and understanding the theories that vernacular communities within Africa were based upon, as opposed to imposing Western theory on local contexts.

Situating knowledge construction instigates a positionality shift from Western-oriented thinking to an alternative way of thinking that is centred on and bound by locally relevant realities and interests. It is a complex process that as a basis will require stakeholders to be cognisant of their colonial history, as well as current international power and pressure that influence knowledge construction. Similarly, it requires a level of awareness and appreciation of complex power relationships and interests among different actors. Therefore, there is a need to create local spaces for dialogue among different stakeholders that would allow different actors to negotiate meanings and approaches. Perhaps the first step towards situating sustainable design in cities in Africa would be to construct situated problems that sustainable design would solve. Furthermore, from a global perspective the different meanings of the concept of sustainable design are based on selective interpretations, depending on who is defining it. Often, groups with special interests have exploited the ambiguity in this concept to push their own agendas.

The process of constructing local knowledge requires a reflexive methodology that deconstructs and reconstructs knowledge. The process of deconstruction begins with a critical epistemological reflection and the recognition of how popular discourse – including its rationale and biases – sits within local conditions, and how local historical, socio-cultural, and political positionality shapes knowledge construction. This understanding sets the foundation for a continuous process of negotiation, reinterpretation, and reconstruction of knowledge cognisant of the evolving local situations (Suresh, 2002).

The chapter began by suggesting that perhaps the discourse of sustainability should be that of asking questions, the right questions. As part of the process of decolonising methodologies, asking these questions not only challenges the power, interests, and position of those who have placed (imposed) themselves as authorities in the construction and validation of knowledge, but more importantly, the idea of self-determination over local problems, priorities, procedures, and solutions is affirmed. The forgoing arguments begin to answer some of these questions, arguing that sustainable design problems and solutions should be defined and advanced by those living in a particular context, in this case Africa, ensuring they are relevant and beneficial to them. Without such contextual grounding, achieving a truly sustainable and inclusive world becomes unattainable. Approaches shaped by external methodologies, political agendas, and power imbalances, often overlook local realities, perpetuating inequalities and environmental challenges and ultimately undermining the potential of sustainable design.

Summary

This chapter attempted to theorise the positionality of the global South against that of the Western world and contribute to the ongoing discourse that challenges the hegemony embedded in Western Eurocentric knowledge forms and construction with reference to the concept of sustainable design. It establishes the correlation between knowledge, knowledge construction, and the idea of power, identity, and legitimacy. It also highlights Western imperialism, first as a consequence of its colonial past, closely followed by (and intertwined with) the Western construction of the concept of modernity, and finally, the culmination of hegemony as the concept of globalisation. In summation, modernity would be non-existent without coloniality. Whereas the main language of colonialism can be characterised as that of categorical subjugation, Western modernity on the other hand created a model and built narratives around that model as an aspiration through the imposition of Western norms, while that of globalisation was (is) the universalist language. Therefore, situating sustainability and by extension sustainable design would demand an awakening to the continued effects of coloniality, modernity, and globalisation.

The chapter also contests the notion that Western modernity provides a pathway for "the rest" of the world to follow (especially with regard to creating a sustainable world), by questioning and rejecting the hierarchical hegemonic structures that characterised the colonial system, and which privileged Western epistemology over local epistemologies. It further recognises the absurdity of the concept of universal global knowledge, arguing that if the process of knowledge construction is bound by the context in which it is being produced, then this "global" knowledge is in fact local knowledge, local to the West, despite claims of its universality. The West relies on the failure of countries like Kenya to distinguish their locality, thereby feeding its hegemony.

Informed by Mignolo's notion of "border thinking", Dussel's notion of "transmodernity", and Escobar's notion of the "pluriverse", it signals a departure from the singular Eurocentric forms of modernity and knowledge construction to the possibility of a network of alternative forms of local knowledge(s) that reconceptualise the power embedded in coloniality/modernity and knowledge construction. Decolonial thinking therefore provides a framework not only to decolonise knowledge, but more importantly to decolonise the process of knowledge construction. This involves an epistemological re-orientation from the Eurocentric imperial knowledge to knowledge construction that is centred on the African agenda as a deliberate attempt towards self-determination. This translates to the re-construction of concepts such as sustainability and sustainable design beyond the confines of Eurocentric epistemologies.

The strength of this argument is acknowledging that the concept of sustainable design does not and cannot bear a definitive, consistent, universal meaning or set of approaches, and therefore no single approach is, or should be, considered the "correct" approach. Instead, the construction of this concept should be cognisant of the range of possible approaches. Only in acknowledging this heterogeneity and remaining open to not only the physical but more importantly the socio-political contextual differences, can context-specific solutions that do not privilege some solutions over others be developed.

Finally, the core argument is that contrary to the abstract universalist and reductionist Western Eurocentric epistemologies, which do not recognise other epistemologies, decolonial methodologies should be adopted in order to re-conceptualise, re-construct, and re-articulate diverse contextualised alternative realities. This can be summarised in Escobar's notion of "other worlds" and "world otherwise"; "that is, worlds that are more just and sustainable and, at the same time, worlds that are defined through principles other than those of Eurocentric modernity" (Escobar, 2004, 220). This speaks to the intention of this chapter to not only continue the dialogue of developing a sustainable built environment (world otherwise), but to develop contextualised (other worlds) versions of that sustainable world.

TAKEAWAYS FROM CHAPTER 13

- The concept of sustainable design does not and cannot bear a definitive, consistent, universal meaning or set of approaches.
- Through the coloniality, modernity, and globalisation matrix, there exists a correlation between knowledge, knowledge construction and the idea of power, identity, and legitimacy.
- The discourse of sustainable design is inextricable from historical, social, cultural, and political dispositions, and therefore cannot be defined nor understood without situating it.
- Arguments should go beyond simply taking an anti-colonial or anti-West stance to deliberate attempts towards self-determination by developing local epistemologies, methodologies, and approaches that consider local political, social, cultural, and material circumstances.
- There is a need to create local spaces for dialogue among different stakeholders that would allow different actors to negotiate contextual meanings and approaches.

Bibliography

Achebe, C. *Things Fall Apart*. William Heinemann, 1958.

Appiah, K. *In my Father's House: Africa in the philosophy of culture*. Methuen, 1992.

Capeluto, I.G. The unsustainable direction of green building codes: A critical look at the future of green architecture. *Buildings*, 2022, 12, 773. https//doi.org/10.339/buildings12060773

Chatterjee, P. Talking About modernity in two languages. *A Possible India. Essays in Political Criticism*. 1998, pp. 263–285.

Chelang'at, F.N., & Lawrence, R. Always with the best intentions? Interrogating the use of sustainable building assessment systems in developing countries: Kenya. *Sustainability*, 2024, 16(9), 3868. https://doi.org/10.3390/su16093868

Comaroff, J., & Comaroff, J.L. Theory from the South: Or, how Euro-America is evolving towards Africa. *Anthropological Forum*, 2012, 22(2), 113–131. DOI: 10.1080/00664677.2012.694169

Cooper, F. *Colonialism in Question: Theory, knowledge, history*. University of California Press, 2005.

Dryzek, J. *The Politics of the Earth; Environmental Discourse*. Oxford University Press, 1997.

Du Bois, W.E.B. *The Souls of Black Folk*. C. McClurg & Company, 1903.

Dussel, E. Europe, Modernity, and Eurocentrism. *Nepantla*, 2000, 1(3), 465–478.

Escobar, A. Beyond the third world: Imperial globality, global coloniality and anti-globalisation social movements. *Third World Quarterly*, 2004, 25(1). After the Third World? pp. 207–230. Taylor & Francis Ltd. www.jstor.org/stable/3993785

Escobar, A. Worlds and knowledge otherwise. *Cultural Studies*, 2007, 21:2–3, 179–210, DIO: 10.1080/09502380601162506. https://dio.org/10.1080/09502380601162506

Etieyibo, E. *Africa Philosophy in the Eyes of the West*. Unisa Press, 2016.

Eze, M. (2010). *The Politics of History in Contemporary Africa*. Palgrave Macmillan.

Fry, T., Willis, A. Design and the Global South, Design Philosophy Papers, 2017, 15(1), 1–2, DOI: 10.1080/14487136.2017.1307537

Glassie, H. (1990). Architects, vernacular traditions, and society. *Traditional Dwellings and Settlements Review*, 1(2), 9–21. www.jstor.org/stable/23566248

Grosfoguel, R. Colonial difference, geopolitics of knowledge, and global coloniality in the modern/colonial capitalist world-system. *Review (Fernand Braudel Center)*, 2002, *25*(3), 203–224. www.jstor.org/stable/40241548

Grosfoguel, R. Decolonising post-colonial studies and paradigms of political – economy: transmodernity, decolonial thinking, and global coloniality. *TRANSMODERNITY: Journal of Peripheral Cultural Production of the Luso-Hispanic World*, 2011, 1(1). http://escholarship.org/uc/item/21k6t3fq

Haraway, D. Situated knowledges. *Feminism Studies*, 1988, 14(3): 575–599.

Kovel, J. *The Enemy of Nature: The End of Capitalism or the End of the World?*. Fernwood Publishing; Zed Books, 2007.

Kurnaz A. Green building certification systems as a greenwashing strategy in architecture. *Bartin University International Journal of Natural and Applied Sciences*, 2021, 31 July.

Maathai, W. *Unbowed: A memoir* (Vol. Anchor Books). Alfred A. Knopf, 2006.

Manuel, H. *African Modernism*. Park Books, 2015.

McLennan, J. The *Philosophy of Sustainable Design*. Ecotone Publishing Company, 2004.

Mignolo, W. D. Geopolitics of knowledge and the colonial difference. *South Atlantic Quarterly*, 2002, 103(1), 57–96.

Mignolo, W. D. Citizenship, Knowledge, and the Limits of Humanity. *American Literary History*, 2006, 18(2), 312–331. www.jstor.org/stable/3876708

Mignolo, W.D. Coloniality: The darker side of modernity. *Coloniality and Modernity/Rationality, Cultural Studies*, 2007, 21(2–3), 155–167.

Mignolo, W.D. Epistemic disobedience and decolonial option: A manifesto, *Subaltern Studies: An interdisciplinary Study of Media and Communication*, 2008, 2 February.

Mignolo, W.D., & Walsh C. *On Decoloniality. Concepts, Analytics, Praxis*. Duke University Press, 2018.

Mignolo, W.D., & Escobar A. (eds.). *Globalization and the Decolonial Option*. Routledge, 2010.

Moss, T. Infographics: *What is sub-saharan Africa's contribution to global CO2 emission?* Energy for growth hub, 2020. www.energyforgrowth.org/blog/infographic-what-is-sub-saharan- africas-contribution-to-global-co2-emissions/

Ngugi wa Thiongo. *Decolonising the Mind. The politics of language in African literature*. East African Educational Publishers Ltd, 1986.

Ngugi wa Thiong'o. *Something Torn and New: An African Renaissance*. Basic Books, 2009

Nhemachena, A., Malambo, & N., Kaundjua, M. The notion of the "field" and the practices of researching and writing Africa: Towards decolonial praxis. *Africology: The Journal of Pan African Studies*, 2016, 9(7).

Oliver, P. *Ethics and vernacular architecture:* In W. Fox, (ed.) (2000) *Ethics and the Built Environment*. Routledge, 2000. Available from: ProQuest Ebook Central. Accessed 3 October 2019.

Quijano, A. Coloniality of Power, Ethnocentrism, and Latin America, *Nepantla*, 2000, 1(3), 533–580.

Quijano, A. The return of the future and questions about knowledge. *Current Sociology*, 2002, vol. 50, SAGE Publications.

Quijano, A. 'Coloniality and Modernity/Rationality', Chapter 2, pp. 22–32, in *Globalization and the Decolonial Option*, ed. by Walter Mignolo & Árturo Escobar, Routledge, 2010.

Reiter B., (ed.). *Constructing the Pluriverse: The geopolitics of knowledge*. Duke University Press, 2018.

Scheub, H. *There was no Lighting: A conversation about race among South Africa's storytellers.* Parallel Press, 2010.

Schoeman, R.C., & Gunter, A.W. Greenwashing: Exploring the role of green building rating agencies on the building industry, the case of Johannesburg. *ResearchGate*, 2018.

Suresh, C. Reconstructing local knowledge. *Journal of Language, Identity and Education*, 2002, 1(4), 243–259. DOI:10.1207/S15327701JLIE0104_1

Tuhiwai, L. S. *Decolonising Methodologies: Research and Indigenous Peoples.* Zed Books Ltd, 1999.

UN-Habitat. *Sustainable building design for tropical climates. Principles and Applications for East Africa.* UN-Habitat, 2014.

UNEP. Building sector emissions hit record high, but low-carbon pandemic recovery can help transform sector – UN Report: UN Environment Programme, 2020.

Willis, A. The limits of "sustainable architecture". *Shaping the Sustainable Millennium,* Queensland University of Technology, 2000.

Wiredu, K. Formulating modern thoughts in African language: Some theoretical considerations, pp. 301–320. In V. Mudimbe, (ed.) *The Surreptitious Speech. Presence of Africa in the politics of Otherness, 1994-1987*, Chicago: University of Chicago Press, 1992.

14

BOUNDARIES AND BONDS IN A BOUNDLESS WORLD

The relationship of place and community to landscape

Maria Vidali

Introduction[1]

Tinos is a Cycladic island in the Aegean Sea that has been deluged by tourism every summer in recent years. Kampos is a village located approximately 5 miles from the main port of Tinos, in an area 1,000 feet above sea level. The Cycladic sea is not visible from the village and there are no hills protecting it against the cold, north wind. This is possibly the reason why the village remains 'introverted', built towards its 'interior' but also the reason why there are just a few front facades standing between its built boundaries and nature.

Over the island's long history, there is a communal perception of the value of land that hails from the period of Venetian occupation (1390–1715) (Sarafi, 2008, 21). The Venetians created a community there as part of a strategic plan to defend against the Ottomans and secure their dominance in the Aegean. At the time, the inhabitants of the island were described by travellers as being one of the most enterprising communities in Greece, in that they transformed most of the mountainsides' rocky and hostile terrain into terraced farmland that could be cultivated (Sarafi, 2008, 242).

Kampos is one of the island's oldest farming villages. The boundaries created by low stone walls and alleyways that define the farmland's boundaries are an integral part of village life. The landscape appears semi-artificial because of the countless rows of cultivation ridges and terraces. The exploration of the community life and the landscape surrounding it starts with the study of very formal narratives. These are contracts and testamentary wills that reflect the interpersonal relationships of the villagers in relation to sharing water and land.

Narratives of possession/ownership and the nature of boundaries

Contracts, *legata*[2] and testaments were the island's first written descriptions of farmlands and water. Moreover, these documents also provide an indication of the villagers' perception of the value of land and water, but also of the surrounding landscape at that time. Elements from each text reveal how the perception of the value of the land and water has

DOI: 10.4324/9781003293903-17
This chapter has been made available under a CC-BY-NC-ND 4.0 license.

changed in some cases, while in some others still remains the same, reflecting the villager's unchanged everyday habits and tasks.

Until 1960, both water and land were calculated in temporal terms and were recorded as such in all official contracts and testaments. Land was measured in the number of days it took two oxen to plough the arable parts of a field. Usually, it was described through the crops and trees that had been planted. This type of measurement stopped being used some years before Greece joined the European Union, though water is still measured in the same way even today, that is through the time it takes for it to flow into the villagers' cisterns or cultivated land.

In 1979, farmers and landowners had to measure their land in metres. Surface springs at the outskirts of the village provided water for the nearby gardens, while water was also collected in water tanks or reservoirs that were built at the gully by a running spring or stream. Water, even today, appears as a form of property only through its flow, except in cases of private wells and taps.

The stories of Kampos, as recounted by the villagers themselves, mention conflicts and agreements regarding the village's indefinable and disputed boundaries. Common water and land create a different type of bond for the villagers, not only with their land, but also their community and their village's landscape. In reference to property, Steinberg (1995, 9) writes about the law of property and about how it enters every aspect of daily life and, moreover, our understanding of place. He also claims that the complexity of nature can make property an extremely misunderstood, even unreal issue (Steinberg,1995, 5).

What lies between and among boundaries, as well as in the space that the boundary itself occupies, are village stories of conflicts, contradiction, disputes, love, collaboration and friendship. This contributes an additional dimension to 'place', something that gives meaning to and shapes the history of its landscape and experience.

Metaphor, narrative, fiction in architectural thinking: a methodology for understanding life and a changing landscape

Ricoeur talks about how words in a sentence can reveal a discourse in the world, allowing language to create images (Ricoeur, 2003, 213). Metaphors bearing a communal perception of place become a natural language of sharing a communal way of living connected with the natural and built environment. Stories connect language with mimetic action, habit, which again connects our bodily and mental experience with the environment, but also with place and space. Merleau-Ponty (1962, 177) writes that in language there are shared emotions and attitudes within every group and that learning their language is a way to share their world and become part of it. In this chapter, the village space is brought to life through excerpts of narratives created from the villagers' stories and metaphors about village life and its farming landscape.

The six excerpts that follow derive from fictional narratives based on the villagers' own metaphors of their reality, including the author's own personal experience in the village as a variation of truth. These fictional narratives were created out of their stories, their spaces, their life with animals, and the life they have together. However, the plot is imaginary so as to connect the various situations and events with each other, with the purpose of bringing to light another dimension of truth about this place. This methodology reveals the complexity of life with nature and its changes in Kampos through narratives. Excerpt 1 from the fictional narrative, *The Animal That I Lost:*[3]

> This early Monday morning, Gioura Island could not be seen on the horizon over the low hills that surround the village. This, for the farmers of Kampos, was a sign that heavy rain would soon arrive. An hour later, it started to pour. It always torrentially when the levantes[4] arrived, a restless and cold strong easterly wind that scared the villagers and made them rush to put their mules and donkeys in the safety of their stables. Any day now, Tassos was expecting one of his cows to give birth. Despite the heavy rain, he went to check on the cow every five hours, even though he was tired from his heavy work in the fields. He knew that calving time was approaching. He could read the signs on the cow's body and her behaviour. He wouldn't risk a minute of his animal's life by not being there, ready to help it give birth and remain alive.
>
> Tassos would spend the night in Livada, at 'the country house,' as they like to call it. This house is among farmlands, over the hill at the north of the village. It was built above the stables, close to the fields where their animals graze. This is where they spent every day and night in the summer, when the sun sets very late, waiting for the sunset, because it is always better to milk the cows after sunset to prevent the milk from turning bad. Ioanna, his wife, who helped him with the breeding and farming, had remained behind in the village. Her two younger daughters had arrived from Athens and she wanted to see them.

Excerpt 2[5]

> Thirty minutes after midnight, it was time to check on his cow again. He put his boots on and, carrying his jacket and a big torch, went down to the small stable to get the stick he used to walk over the rocky terrain and occasionally to prod the animals. Sometimes, the stick was more useful than the torch to help him walk at night. That night, the moon was a fine line in the sky and this made it hard to discern the dark outline of the terrain from the even darker skyline. The cow was in one of the bigger stables on the west side of the field. In a small enclosure made of stones and concrete, all the cats were lined up on the cement wall by the stable. There was an atmosphere of perseverance, supported instinctively by all the animals. Possibly, this would be the night. He went into the stable and switched on the light. He calmly approached the animal. The soft tissue around the tail head of the animal had sunk, allowing the pelvic bones to widen out. The time had come. Leaving his stick behind, he ran back to the house. He climbed up the steps and picked up the phone next to the kitchen door. He called Ioanna and shouted, 'Call Manolis, Marcos and Giannis, tell them to come. The cow is giving birth!!'
>
> Half an hour later, the light of five torches bobbing up and down and advancing slowly appeared at the edge of the field. Their lights broke the endless darkness between the sky and the earth that made it so difficult to distinguish anything. The *levantes* had dropped that night. When the three men and Ioanna arrived at the stable, the cow was on the ground on a thick layer of straw. The stable was humid and cold. Tassos felt somewhat relieved now and he observed the cow carefully, checking her every reaction. The three men stood around, waiting attentively. Ideally, the cow

FIGURE 14.1 The village grounds and the surrounding area, showing the routes taken in the fictional narrative, *The Animal That I Lost*.

should either have the strength to stand on her own legs, or the four men would need to lie on the floor to help her give birth.

Sooner than expected, Tassos realized that the time had come to pull out the newborn calf.

'Ioanna give me the rope.'

Kneeling on the floor, Tassos pushed his two hands deep into the cow's womb and tied the front legs of the yet unborn calf. This was something that only a few villagers and breeders could do. That knowledge was passed down by older breeders with many years of experience. Manolis took off his jacket; Marcos and Giannis rolled up their sleeves, and checked the space behind them. Breathing fast and acting very quickly, they kneeled on the wet concrete floor behind Tassos. Even though the cow was unable to stand up, the calf was big enough to be able to escape on its own from its mother's body. So, the four men would attempt to take out the calf by pulling on the rope that was tied to its legs. All the living beings in the fields, including the dogs, chickens and cats, seemed to be waiting patiently for something. Four pairs of male hands pulled on the rope in the warm air of the room with careful and precise movements and a steady and persistent effort guided by Tassos's voice and instructions.

'Eeeeooooop,' a loud cry of coordinated effort was heard.

'Check its head', Marcos sounded worried.

After a deep, endless 'moo' by the cow, the calf lay on the floor wet, with its big dark eyes wide open and its legs still tied.

Excerpt 3[6]

'Who exhumed the body of their deceased recently, Ioanna?' Tassos asked.

'I think Manolis exhumed his father a few months ago. It's been five years since his father's funeral and he had to put his bones in the ossuary. There was an empty spot next to my parents, so he made a nice new marble case with all the family's details inscribed on it.'

'What did you treat the villagers to in the memory of your father, this year?'

'Nothing,' Ioanna said with a lazy voice, 'we decided with the girls that since my father is not in a tomb any more, but in the ossuary, it is not necessary to offer something. There are so many treats offered every year... enough for me after five years.'

That day, Ioanna finished all her work cleaning the animals earlier and left Tassos alone to do the milking and then join her with their younger daughters later at the cemetery. She wanted to arrive home on time to get ready for the ceremony at the cemetery. It was around 5 o'clock when Ioanna, with three of her daughters, left their home. Walking in the main alleyway of the village, they met Eleni and her cousin, Foteini. Foteini was holding a huge platter of *xerotigana*.[7]

Excerpt 4[8]

Darkness had started to fall when the villagers began making their way back to the village from the cemetery. The commemoration would continue at the *leschi*,[9] the communal room by the side of the road separating the village from its farmlands. At the *leschi*, everyone sat down at tables set out in the empty room with its high ceiling. The villagers whose ancestors they were commemorating began to offer everyone sweets.

'*Sto sychorio tou*,[10] Foteini,' Ioanna mumbled checking her small handbag to see if she had brought a small plastic bag with her, so that she could accept all the sweets. Manolis offered *raki*[11] and Foteini offered *xerotigana* dipped generously in honey.

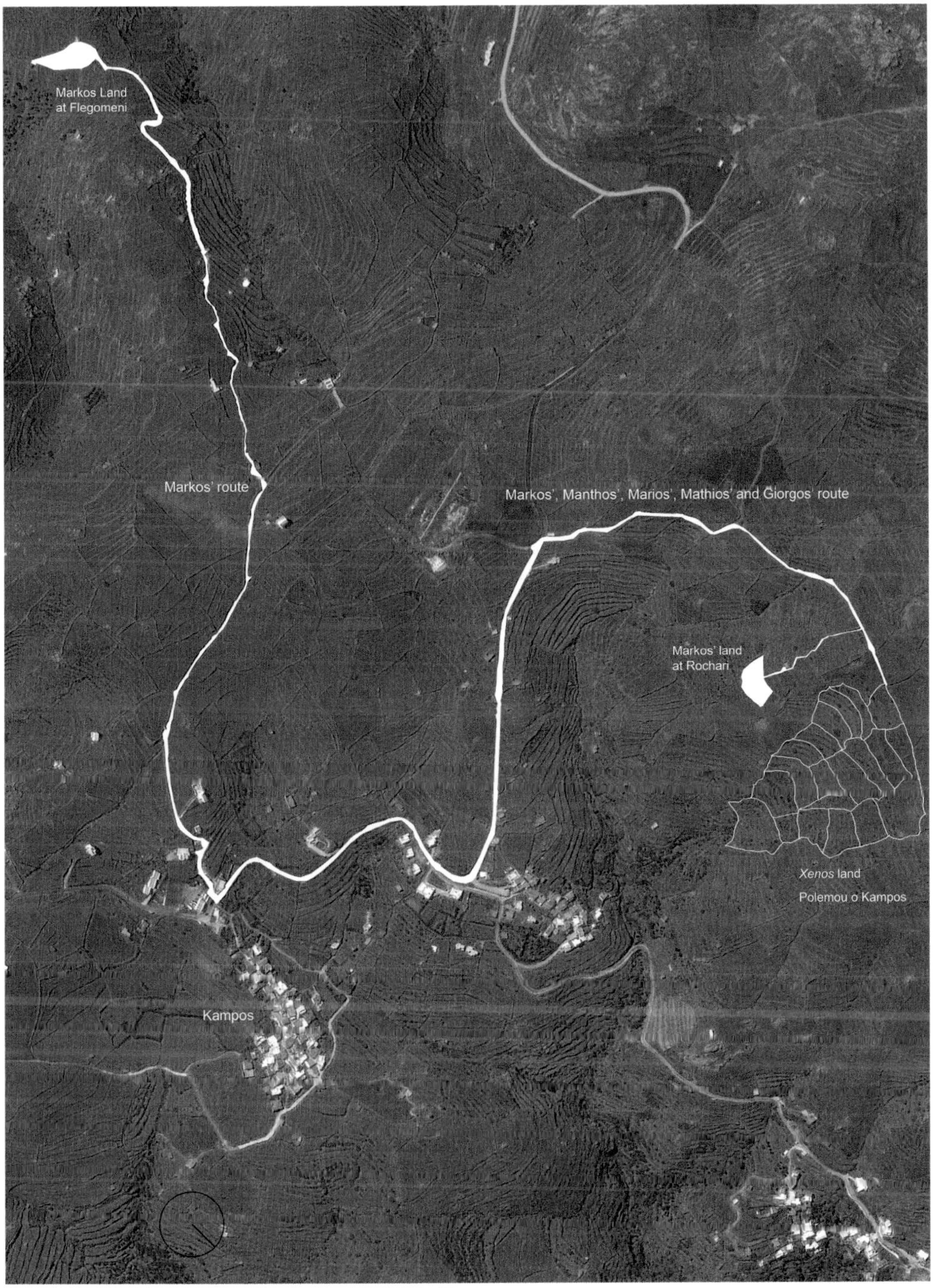

FIGURE 14.2 The village grounds and the surrounding area, showing the routes taken in the fictional narrative, *The animal that I lost*.

Villagers exchanged sweets and wishes for the forgiveness of their descendants and gossiped about both happy and sad things happening in the village.

Excerpt 5 from fictional narrative, 'The smell of honey was in the air'[12]

All the women got together on Wednesday at 7 o'clock at the *leschi,* a room provided by the church for communal use, located opposite to the entrance of the village, right across the main roadway. Everyone was on time, though some were absent. Our president from the village's official representatives was there, too, as he needed to note down all the ingredients for the recipes. However, nobody had the recipes, or the measurements of the recipes from last year's festival. They had been lost, so everyone had to figure out again the measurements for each sweet. Fragiska always managed to resolve such issues. She took her notebook and looked at it, saying: 'Let me know if something does not sound right. Fragiska, in other words myself, and my mother will prepare ten kilos of *melomakarona;* Maria will make three kilos, Antonia and Roza five; Georgia will prepare five kilos of *pasteli;* Maria and Ioanna will make six kilos of *halva*[13] with semolina; Anna and Poppi, three kilos of *pasteli;*[14] Margarita and Agni, *psarakia*;[15] Rosalie and Rita, *ravani*,[16] Katerina, two kilos of *pasteli*; Marina and Lambrini, *melokarido*;[17] and the pastry chefs of the village will help as well, they always offered to help in the preparations for this special day. So, do you want me to add or change anything before I give the list and the quantities for each ingredient to Nikos over the next few days?'

Excerpt 6[18]

It was 5:30 in the afternoon by the time we left home and walked through the alleyways and narrow streets. We realized that we were the last ones to arrive. The streets smelled of honey, and a hint of cinnamon floated in the air. When we arrived at the village's coffeehouse, the only place where the street widens up, allowing enough room to stand and look around, we met Popi with her daughters carrying *melomakarona.* The smell of nuts then overpowered the smell of the cinnamon that had started to dissipate in the breeze. By the time we arrived at the *plateia*,[19] most of the women were already there, decorating the tables to display the various types of sweets. A few of them were also on the tiny balcony of the small village office, trying to figure out where to put their dishes. They would start by offering *xerotigana,* followed by *pasteli* and then *melomakarona, melokarido, psarakia* and *halva*. They had to be careful with the portions to ensure that there would still be sweets left late in the evening, until after midnight.

It was around 6 o'clock when Fragiska and Eleni started heating up the oil to start frying the first *loukoumades*. The oldest couples of the village started arriving to make sure that they would find a seat and be able to attend the festival for most of the evening. After a few minutes, the band arrived. They sat next to the chairs, leaving enough room for a dancing area, and began tuning their instruments. Right next to them, five huge caldrons over a fire kept the honey syrup warm.

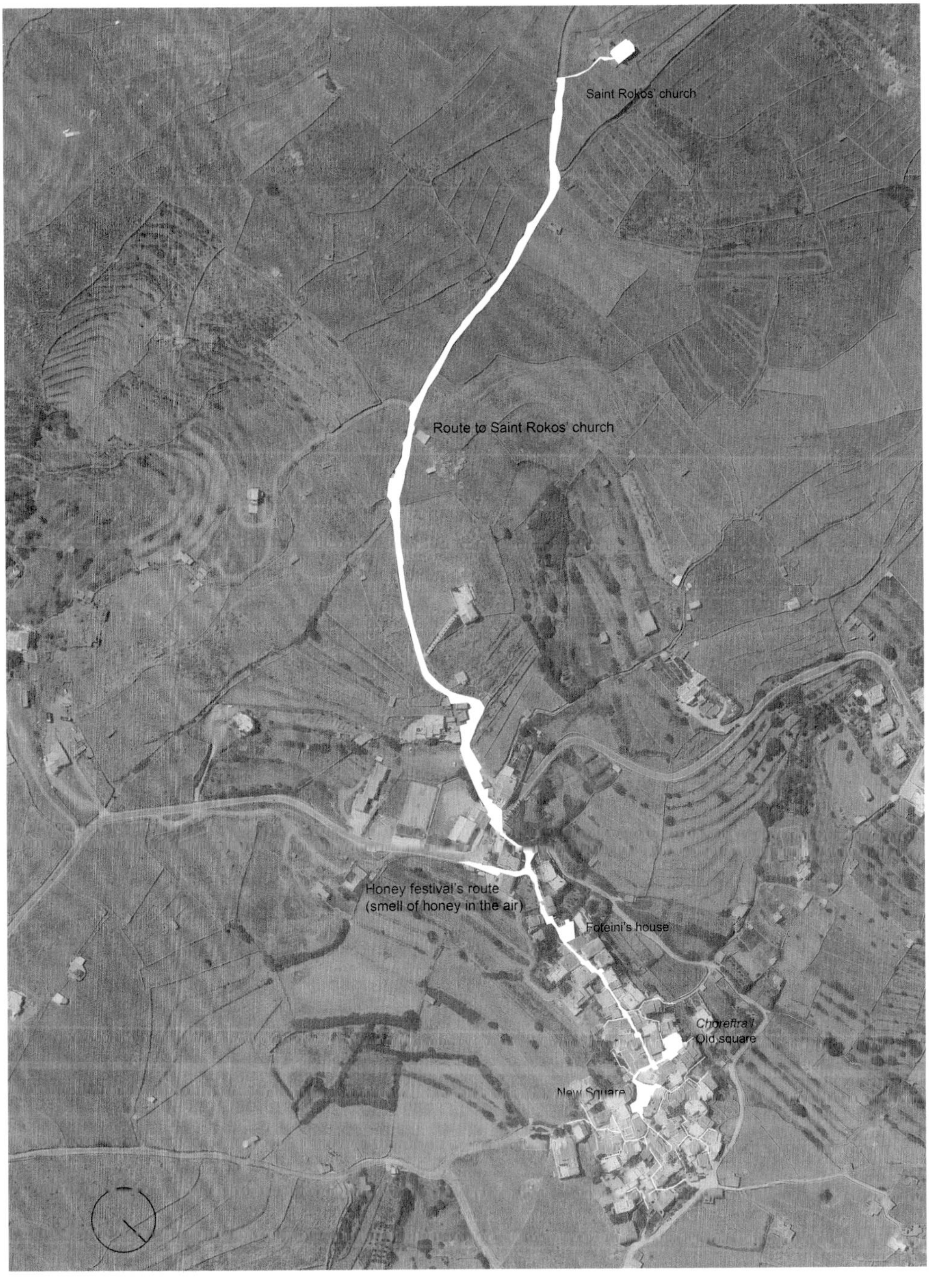

FIGURE 14.3 The village grounds showing the routes taken in the fictional narrative, *The smell of honey was in the air.*

Village life as a way of understanding place, architecture and climate emergency

Do we as architects understand enough about how people engage with the landscapes that we are called upon to serve? Are we close enough to nature, to communal life, to death as is often commemorated in the village of Kampos, so as to witness life together with humans, animals, nature? Do we allow room to be shaped by peoples' stories before we shape space? Could architecture break free from the obsessions of fashion and form and reignite the imagination of architects so that they may understand, reconnect and reconcile themselves with local cultures, beyond political and social preoccupations (Pérez-Gómez, 2016). Is this a way to help people reconnect with nature, inhabit the landscape and share a common understanding of the emergency facing the climate?

The farming landscape: A place of awareness and inhabitation

The villagers of Kampos inhabit a world in which they move daily or seasonally beyond the inner boundaries of the village core. Our capacity to move around within and use a space without having to consciously think about what we are doing creates, according to Merleau-Ponty, an understanding of the environment connected to one's body (Merleau-Ponty, 1962).

The breeders know that, during the winter, you cannot take the animals to a place exposed to the wind. It must always be sheltered and so they move towards a headland. When the summer arrives, they guide their animals uphill. They follow the seasons. They are aware of the water which needs to be transported over land, about the small cots

FIGURE 14.4 Kampos village cemetery, the day of the *Mera ton Psychon*.[21]

in the fields that can shield against the strong summer heat. They also know that their animals need to be protected against the summer insects, which penetrate their skin and causes them intense pain. These are just some of the many factors that breeders must always take into account. In winter, when the days are shorter, their work is less exhausting. During summer they often return home at midnight after having spent the evening milking.[20] Furthermore, over the last decade the winter snow that fell when sheep usually give birth, has led to a great loss of animals.

All this movement and transportation of the animals in the landscape over time has shaped villagers' habits and has made this landscape habitable, endowing it meaning and continuity. As Crossley (2001) claims, habits provide a set of meanings and preferences which affect our choices and the ways in which we inhabit the world. Clearly, what is common to indigenous oral peoples everywhere in the world is a common understanding and a sense of inhabiting of the landscape they live in (Abram, 2010). But Abram (2010) also observes that technologies do not allow us to connect with the flow of the earth's cycles.[21]

Paying attention to the flow of the wind over the land, noticing the first temperature change of the wind and predicting which of the seven winds have reached the island, affects their daily routine. This awareness relies on a very long tradition of breeding on the island, inhabiting the landscape and the place.

Moving from the house to the fields: Engaging with a landscape structure

Up until recently, the value of land in the village was not measured in terms of surface area, but in terms of how much of it could be farmed. The value of the land today is relative to its size and to how much can be built on it. The boundaries are marked by high, stiff, concrete walls that seal off the natural and social environment.

FIGURE 14.5 Cows are taken back to the fields from the village after being milked.

In the surrounding landscape of the village, just beyond the houses and further away from the centre where the vegetable gardens end and the larger fields begin, lie small stables and stalls, used mostly as cattle shelters. There is a constant movement of villagers and animals from the village to the cluster of gardens and orchards, and beyond, to the cultivated land and the fields. This is also because each family's property is made up of small lots spread over the surrounding area of the village. A farmer and breeder has to follow a specific route every day to reach them: 'Every farmer had plots from 4–12, that means 15 minutes-2 hours to move from one to another.'[22]

There are three professional breeders in the village of Kampos. These work with the animals, selling their milk, their meat and tending to their reproduction. These men take care of their animals on their own, milking by hand, guiding them and cleaning up after them.

The landscape is conceived and understood differently through farming and breeding. Working with nature on a daily basis, which for this island and village is an arid landscape that sees little rain, farmers develop a close bond with what they can never own: nature. Farmers work constantly on the earth: they follow an annual sowing schedule for every plant, they wait for the rain, they plan fallows for their land, share machinery, but also water and land. They enter into agreements with those who own wells so as to secure water and offer in return farming products. All this activity is based on the need to cultivate. Boundaries become permeable in order to obtain the products of nature such as vegetables and fruits, and to make wine and olive oil. This is probably why there is an understanding that everything must be created through a constant labour – even when farmers use machines on their land. The surface area of the land is never big enough for a proper farming machine and the small machines they use are never powerful enough to do all the work for them.

'From the earth and the sea comes everything. Bread, vinegar, olive oil, everything needs work to be made. Only God spoke and created, man needs to make,'[23] as one of the old farmers says. Although they live in an era when everything is available at the supermarket, work for their own vegetables, vinegar, wine and olive oil is carefully planned. In an effort to procure the products of nature, the village's boundaries disappear from the villagers' everyday routine, their house and family names extend to the fields. The property boundaries of water and land are merged for the sake of a good production.

The honey festival at the village square: a common world to be shared

The honey festival is organized through the participation of all the women of the village and the support of the island's municipality. For the festival to take place, the municipality and local women must work together. This festival came to life ten years ago when a village representative made a proposal to the municipality as food festivals where being scheduled around the island. Honey production was rich in Kampos and there were a few farmers who were good beekeepers. This made honey the basic ingredient of a communal event that brought centre-stage the daily life of the villagers and their village's 'image' through a series of events that they had never imagined before, given that tourism and local visitors were unknown in this area.

Before this festival, Kampos was known only as a farming village. Within the structure of this introverted village, a large festive event represents a different way for the

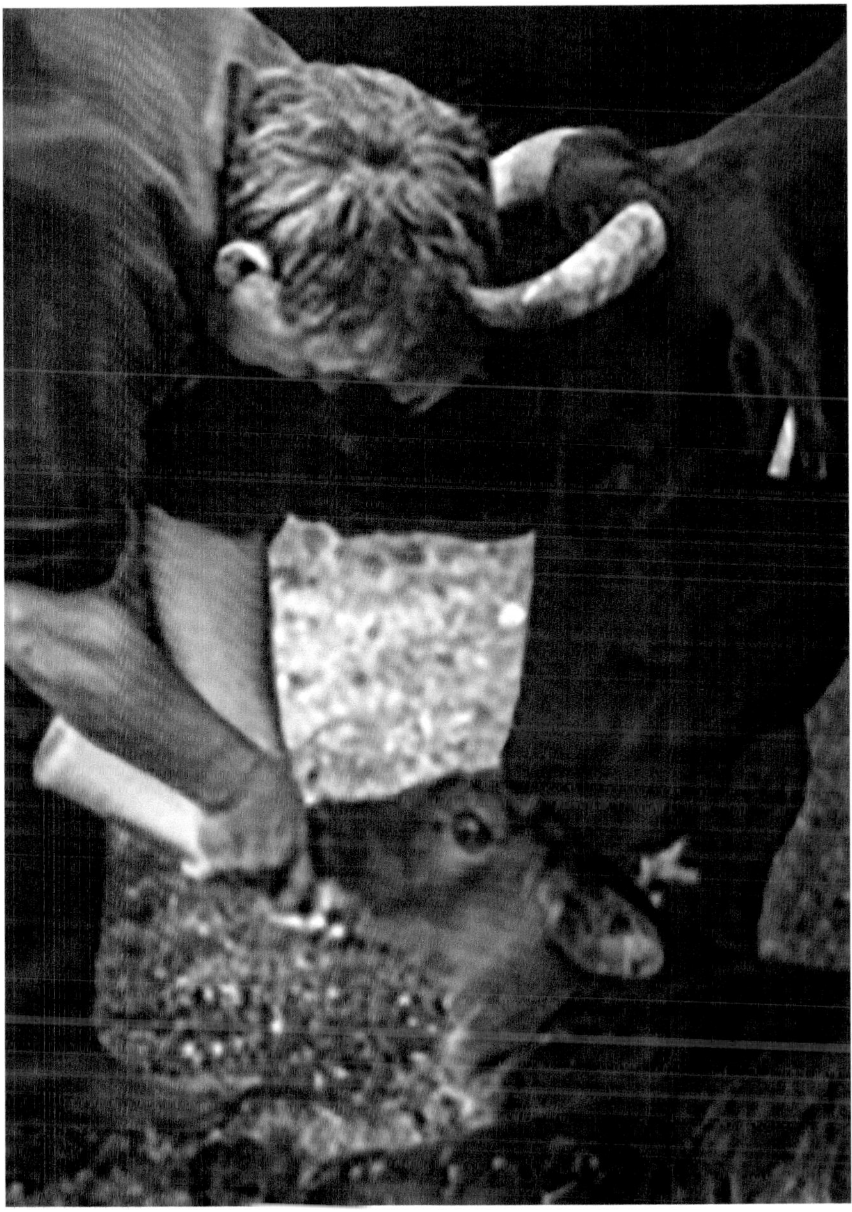

FIGURE 14.6 One of the breeders of the village looking after his animals.

FIGURES 14.7, 14.8 AND 14.9 Village women preparing sweets for the honey festival.

FIGURE 14.10 Procession towards St George's outlying chapel.

inhabitants to connect with each other. Yet simultaneously, the same event underlines the village's boundaries between 'us' and the 'strangers' visiting the village on the festival day. There is a relationship between inside and outside, which now involves the village and the outside world (Dubisch, 1993).

But honey also becomes an object of negotiation and mediation for the local festival and the identity of the village. The villagers and festival participants capture honey but the honey can also be said to capture them. Boundaries dissolve through honey. They manifest their bond with the village community and communal life spaces together by possessing and being possessed by their communal space and the honey festival.

Pérez-Gómez (2016) suggests that inhabitation derives from our bodies' habitual interaction with the environment. Furthermore, actions become habits which endow human beings with adaptability (Pérez-Gómez, 2016). Action and habits become traces which in turn preserve the histories of previous lives whose richness informs design, they are present testimonies that carve out a space for a regionalism with meaning (Pérez-Gómez, 2016). Ricoeur in *Architecture and Narrative* (Ricoeur, 1996) claims that a successful architect is able to turn his work into a reflection of the stories of people whose lives are connected to the places they live in. Perhaps Ricoeur can be seen to be responding to the crisis that architecture is facing today, its loss of novelty and its search for a highly technological, sustainable function that remains disconnected from humanity and environmental reality.

Summary

Life in Kampos appears as a structure of measures, relationships and distinctions that remain in a dialectical relationship with the perpetual continuity of earth, sky and seasons. Within the economy of possibilities, these possibilities are propitious for activities, largely understood through necessity, habit and custom.

Arendt (1998) claims that living together in the world means that our world of things lies in between those who have something in common, as a table between those who sit around it. This world that lies in between has the power both to relate and separate people. This applies to the stories that describe the village space, the surrounding landscape and daily life events. What if through some magic trick the table vanished and the villagers sitting opposite each other were no longer separated and realised that they are entirely unrelated to each other by anything tangible (Arendt, 1998). Arendt refers to the *koinon*, that which is common to all. The existence of a public realm and the world's subsequent transformation into a community of things, which brings people together and relates them to each other, depends entirely on permanence (Arendt, 1998). This is a permanence established through possession, participation, inhabitation that allow the villagers of Kampos to fully engage with the world they live in.

Language and narrative forms can become tools for understanding and revealing the truth of societies in relation to architecture and the environment. As the spatial nature/element of boundaries in stories, narratives and their interpretation, architecture can be used to achieve a deeper and better understanding of the different realities of a village, town, or city that are related to life and its complexities. It is another way for architects to bring forth the social and ethical function of architecture, related to humans and the environment, without responding to architecture and dwelling merely through form and the dominant architectural trends. This can lead to interpretations in architecture and design that are more intimate and connected to human living, closer to the actual complexity of life and the environment, to an awareness of seasonal changes, and to an understanding of the climate emergency as something that we all face and care to respond to.

KEY TAKEAWAYS FROM CHAPTER 14

- Language and narrative forms can become tools of understanding and revealing the truth of local societies in relation to architecture and the environment.
- It is very important to be freed from the obsessions of fashion and form, to reconnect and become attuned to local cultures (Pérez-Gómez, 2016, 191), to strive for an intimacy and understanding of the climate emergency as something that we commonly share and care for.
- The environment is not a 'machine to be fixed', it is a whole in which we humans are embedded.
- It is important to prioritise climate emergency as a communal issue under a communal awareness.
- If we can be clear about what we are aware of, have and love, we may be able to protect it.

Notes

1 Research from PhD thesis Maria Vidali, *Liminality, metaphor and place in the farming landscape of Tinos: The village of Kampos*, submitted to the University of Thessaly in partial fulfilment of the requirements of the degree of Doctor of Philosophy (University of Thessaly, School of Architecture, December 2017).

2 A type of covenant, which connected land and ownership with the church. This was the first attempt to create some form of land registry on the island, at the request of ruler, bishop Giustiniani (Foskolos, 1998, 15–16).

3 Excerpt from PhD thesis Maria Vidali, *Liminality, metaphor and place in the farming landscape of Tinos: The village of Kampos*, submitted to the University of Thessaly in partial fulfilment of the requirements of the degree of Doctor of Philosophy. (University of Thessaly, School of Architecture, December 2017), 288.
4 λεβάντες (ο) – a cold, strong easterly wind.
5 Excerpt from PhD thesis Maria Vidali, *Liminality, metaphor and place in the farming landscape of Tinos: The village of Kampos*, submitted to the University of Thessaly in partial fulfilment of the requirements of the degree of Doctor of Philosophy. (University of Thessaly, School of Architecture, December 2017), 288.
6 Excerpt from PhD thesis Maria Vidali, *Liminality, metaphor and place in the farming landscape of Tinos: The village of Kampos*, submitted to the University of Thessaly in partial fulfilment of the requirements of the degree of Doctor of Philosophy. (University of Thessaly, School of Architecture, December 2017), 294.
7 ξεροτήγανα (τα) – local sweet dredged in walnuts and honey, served at weddings or festivals.
8 Excerpt from PhD thesis Maria Vidali, *Liminality, metaphor and place in the farming landscape of Tinos: The village of Kampos*, submitted to the University of Thessaly in partial fulfilment of the requirements of the degree of Doctor of Philosophy. (University of Thessaly, School of Architecture, December 2017), 294.
9 λέσχη (η) – club, communal room for celebrations and communal events located at the edge of the village, by the roadway.
10 στο συγχώριο του (το) – a wish, for his/her forgiveness after death.
11 ρακί (το) – local spirit made of distilled grapes.
12 Excerpt from PhD thesis Maria Vidali, *Liminality, metaphor and place in the farming landscape of Tinos: The village of Kampos*, submitted to the University of Thessaly in partial fulfilment of the requirements of the degree of Doctor of Philosophy. (University of Thessaly, School of Architecture, December 2017), 295.
13 χαλβάς (ο) – sweet made of semolina.
14 παστέλι (το) – sesame seeds on lemon leaves served at weddings or festivals.
15 ψαράκια (τα) – (meaning 'little fish') fried local sweet containing walnuts and cinnamon served during Christmas or festivals.
16 ραβανί (το) – a cake with honey syrup.
17 μελομακάρονα (τα) – a sweet with honey syrup made during Christmas.
18 Excerpt from PhD thesis Maria Vidali, *Liminality, metaphor and place in the farming landscape of Tinos: The village of Kampos*, submitted to the University of Thessaly in partial fulfilment of the requirements of the degree of Doctor of Philosophy. (University of Thessaly, School of Architecture, December 2017), 180.
19 πλατεία (η) – Town Square
20 Interview with Mathios Filipoussis, Kampos, 13 September 2015, trans. From the Greek by Maria Vidali.
21 Μέρα των ψυχών (η) – All Souls Day or All Saints Day, a feast dedicated to the souls of the deceased.
22 Presentation of the agronomist Ioannis Aspromoungos, Tinos, September 2011, trans. from the Greek, Maria Vidali.
23 Interview with Marcos Filipoussis, Kampos, 3 January 2013. Trans. from the Greek, Maria Vidali.

Bibliography

Abram, D., *Becoming an Animal: An Earthly Cosmology*, (Pantheon Books, 2010).

Arendt, H., *The Human Condition*, (University of Chicago Press, 1998).

Crossley, N., *The Social Body. Habit, Identity and Desire*, (Sage Publications, 2001).

Dubisch, J., '"Foreign chickens" and other outsiders: gender and community in Greece', *American Ethnologist*, 20(2), pp. 272–287, (Wiley on behalf of the American Anthropological Association, 1993).

Foskolos, M. Τηνιακά Ανάλεκτα, Τόμος 3, Ανάτυπο, Το κτηματολόγιο των εκκλησιών της Τήνου και η καταγραφή των λεγάτων τους. ΑΚΤ, Κώδικας 4, (Αθήνα: Εκδόσεις Φιλιππότη, 1998).

Merleau-Ponty, M., *Phenomenology of Perception*, translated from French by Smith Colin, (Routledge & Kegan Paul; The Humanities Press, 1962).
Pérez-Gómez, A., *Attunement, Architectural Meaning after the Crisis of Modern Science*, (The MIT Press, 2016).
Ricoeur, P., 'Architecture and Narrative', in Derossi, P. (ed.), *Identity and Difference: Integration and Plurality in today's forms, Cultures between the Ephemeral and the Lasting. The Triennale in the city, the imageries of difference.* Translated by Huw Evans, 67–75. (Electa, 1996).
Ricoeur, P., *The Rule of Metaphor, The creation of meaning in language.* transl. by Czerny, R., McLaughlin, K. and Costello SJ. J., (Routledge Classics, 2003).
Sarafi, E.N., Τήνος, Χάρτες – Ενδυμασίες, (Εκδόσεις Καστανιώτη – Διάττων, Αθήνα, 2008).
Steinberg, T., *Slide Mountain or the Folly of Owning Nature*, (University California Press, 1995).

15

AN ARCHITECTURE OF PAYING ATTENTION

Scenarios for rehearsing climate futures otherwise

Renata Tyszczuk

Introduction: What on Earth?

What on Earth were we doing? Should we have been *paying attention*?

We live on a perilously disturbed planet. The Anthropocene is the catch-all scenario of humanity's role in destabilising Earth systems and shunting the Earth into a new geological epoch, one that has 'no analogue' in Earth history (Crutzen & Steffen, 2003). The Anthropocene is a planetary alarm, a cautionary tale and a call for action. It designates the climatic future as catastrophic, out of control and already here (Tyszczuk, 2018; 2022). The expansionary dynamics of global capitalist industries in tandem with fossil-fuelled rapid urbanisation and an insatiable demand for resources, has seen the extractive frontier constantly redefined, environments polluted and ecosystems disrupted. Existing international laws, treaties, protocols, rights, agreements and practices created to moderate such tendencies have in both their creation and application assumed best-case scenarios of relatively stable or potentially sustainable conditions, resources and geographies. Yet responding to a dynamic Earth and the intersecting and unstable conditions of environmental crises from climatic changes, ecological degradation, shifting coastlines, displacement of peoples, to species migrations and extinctions, requires finding new ways of navigating this climatic and geopolitical *terra incognita*. Moreover, the contribution to and experience of climatic disturbance and climate emergency is not equally distributed, in space or in time. Globally, economically and ethnically marginalised populations, including future generations, bear the brunt of rising waters, escalating temperatures and dangerous pandemics, and, as a consequence, suffer disproportionately. It's not that climate disruptions will end all life on Earth like the planet-destroying comet of *Don't Look Up* (2021). But rather that the planetary crisis has irrevocably altered and unsettled every aspect of inhabiting planet Earth: '*we are the asteroid*' (cf. Barnosky et al., 2011; Tyszczuk, 2016; 2018).

'These are the times of urgencies that need stories' (Haraway, 2016: 37). The present is a context of renewed fears of the world *as we know it* ending, along with concern over the lack of preparedness and appropriate governance in the face of interlinked crises.

DOI: 10.4324/9781003293903-18
This chapter has been made available under a CC-BY-NC-ND 4.0 license.

There is a widespread sense that this is a time of cultural, political and social reckoning for a 'planet of cities in planetary crisis' (Tyszczuk, 2014). There have been multiple declarations of climate emergency, and increasing emphasis that 'this is not a drill' (Farrell et al., 2019). Yet at the same time as announcing conditions of 'everything change' (Atwood, 2015; Klein, 2014), the planetary crisis induces a *crisis of agency*: what is to be done? Moreover, what kinds of politics does a 'state of emergency' serve to permit or erase? (Hulme, 2019; Anderson et al., 2020). In other words, it is not obvious what kinds of responses are needed in times of urgencies. Is this a time for 'climate improvisers'? (Hulme, 2017). The inevitable challenge to architecture, and to urbanising practices more generally, is not just to acknowledge this planetary predicament, the 'no analogue' condition of the Anthropocene – but 'to rethink the processes, products, practices as well as the profession of architecture through an exchange with the planetary dimensions of human existence' (Tyszczuk, 2014). Yet, 'If the world *as we know it* has ended how can we even begin to think about how we might respond and be responsible in a new world?' (Tyszczuk, 2012: 134). The challenge is both how to 'stay with the trouble' *and* 'change the story' (Haraway, 2016).

Isabel Stengers' *In Catastrophic Times* calls for 'the art of paying attention' (2015). She relates 'paying attention' to a temporal cosmopolitics of 'slowing down' and caring for earthly entanglements and unfinished endeavours (2018). The *Collective Scenarios* project has combined experimental, architectural inquiry, and civic and creative engagement around climate change.[1] It has explored scenarios as a mode of collaborative and transformative storytelling in times of urgencies, emergencies and catastrophe, that is also about *paying attention* to diverse and discordant experiences of living with climate changes and planetary disruptions. Scenarios were reconsidered in their potential as 'rehearsals' for speculative, improvised, and collaborative modes of acting on and thinking about uncertain climate futures in the *here and now*. The approach has been guided by speculative and pragmatist philosophy, and spatial and participatory practices. It has drawn inspiration from Stengers' mode of questioning as provocation: *what has happened to us?* (2011a: 14), when understood not as a lament, but rather as 'a lure for giving the possible a chance, for regenerating our imaginations, for experimenting with other ways of inhabiting our present, "for telling our stories in another way, in a way that *situates us otherwise* – not as defined by the past, but as able, perhaps, to inherit from it in another way"'(Stengers, 2011a: 14; cf. Savransky, 2018; my emphasis).

Through a series of explorations emerging from the question – *What on Earth?* – this chapter probes what an *architecture of paying attention* might entail. A preliminary assessment suggests that it is an architecture that embodies an experimental exchange between being attentive, or noticing – *pay attention*, and being alert – 'Pay Attention!' It indicates a provisional architecture that responds to the dilemmas of the present moment with probing and with urgency, but at the same time is wary of claims of emergency. *What if?* – It is an architecture whose approach is rooted in a scenario-making that is about improvising the future and not about forecasts, visions or predictions. *What now?* – It tests the potential of collective agency in rehearsing climate futures. *What works?* – It is an architecture that makes, or does, gets on with things, and yet, is also an architecture that speculates, is on the look-out, is vigilant and careful. *What matters?* – It is an architecture reimagined as both a creative practice and an infrastructure of care. *What for?* – And is supportive of emergent modes of Earth citizenship. *What next?* – It involves a weaving together of imagining and practically enacting *how things could be otherwise*. In other words, it is about *rehearsing* the future by attending to the present. Moreover, it is not an accident that the research process – exploring both the thinking and doing inherent in *scenarios* – has been one of rehearsal and improvisation.

Scenarios in times of urgencies: *What if?*

What if? scenarios are a common method of getting a better grip on the future, particularly when the future is understood to be in crisis. Resorting to scenarios has become an indispensable means to hypothesise upon, imagine and construct futures in this troubled time, provisionally called the Anthropocene (Tyszczuk, 2018; 2019a). As 'anticipatory practices', scenarios are enrolled in various ways of calculating, imagining and performing futures, in often disputed modes of 'pre-emption, prefiguration, and preparedness' (Anderson, 2010). They can serve to 'bring imagined futures into the present' (Andersson, 2018). Scenarios thus contribute to processes through which the present is transformed, intervened in and ultimately governed or managed as a prerequisite of an optimised future (Adams et al. 2009; Anderson, 2010; Granjou et al., 2017). Scenario techniques are used in risk management and financial forecasting, in speculative extractive industries, in climate modelling, including the science-policy interface of climate science and United Nations Framework Convention for Climate Change (UNFCCC) processes. Scenarios are deployed in contingency and adaptation planning for smart or resilient urban futures, pandemic preparedness and securitisation; in speculative design practices that try to anticipate post-growth or eco-future lifestyles, and in the catastrophic (for the most part) imaginings of climate fiction. Scenarios (especially the worst-case ones) have also informed the rush of 'wake-up' and 'panic' calls of 'climate emergency' that demand immediate climate justice and climate action. This troubled time is also 'scenario time': replete with multiple, proliferating *what ifs* (Tyszczuk, 2022).

Scenarios shore up a myriad architectural projects for future inhabitation of planet Earth responding to the climate crisis, in tandem with either aspirational targets of limiting average global warming to 1.5°C, or the threat of exceeding critical thresholds. The wide-ranging temporally agile visions include amphibious buildings in flooded cities, regenerative post-disaster landscapes, reactivation of sites in adaptive re-use, re-purposing of obsolete fossil fuel infrastructures, and re-rooting of urban resilience in collective eco-practices. Indeed, there has been an expansion of disciplinary scope in the focus on designing scenarios rather than temporally impoverished objects, buildings or territories. Taking account of, for example, resource procurement, food security and logistics, labour processes, social value systems and climate justice, even species modifications, future rituals, and off-world construction within the remit of an architectural project. Future planetary scale visions, informed by speculative design research tell fantastic stories about possible Earths in the context of climatic disruptions, geopolitical manoeuvres and terraforming experiments (e.g., Ghosn et al., 2021). More down-to-earth scenarios about changing times make an account of the economic value of urban-ecological transition through citizen involvement (Petcou and Petrescu, 2019).

Paying attention to the history of scenarios in the *Collective Scenarios* project has been part of a broader effort to understand the authoritative status of scenarios within formal climate change research and policy processes, their ubiquity in narratives that underpin our economic and political systems, their role in pre-emptive, algorithmically determined techniques of managing future risks as well as their grip on the cultural imaginary of climatic and urban futures (Tyszczuk, 2019ab; 2021). In a world increasingly depicted in terms of catastrophe and limits, scenarios enrol all kinds of creative, calculative and speculative knowledges to envisage future possibilities 'ahead of time' (Amoore, 2013). Yet, looking back to the origins of the term *scenarios* in sixteenth-century Italian street theatre, reveals that scenarios were not the visions or stories *per se*, but were rather the prompts or provocations to storytelling within a framework for improvisation (Tyszczuk, 2019a; 2021). The *scenario* indicated the skeletal synopsis or rough outline of an *unscripted* play, which would then be fleshed out by the actors *all'improvviso* during performances. In other words, it was the improvised action and how it negotiated everyday concerns that mattered. Moreover, traditions of improvisation are connected to the idea of *phronesis* or practical wisdom, the ability to respond to unforeseen situations where there is no precedent, indeed, *no script for action* (Tyszczuk, 2011). Therefore, rather than simply privileging spontaneity or endorsing an immediacy of experience in creating something unplanned, collective improvisations rely on the skills and capacities of all those taking part and are about *paying attention* in conditions of uncertainty.

In a messy, disrupted world, rather than focusing on scenarios as finished stories or ideal visions summoning the end point of projects, or even as methods for generating a *plurality of narratives* that summon more or less, 'possible', 'probable', 'plausible', 'preferable' or 'prepared for' futures, could scenarios be understood instead as the speculative infrastructure for collaborative exercises in *improvising the future*? Approaching scenario-making as a collective, experimental and ethical practice proposes to reclaim the territory of scenarios and speculation from pre-emptive techniques and practices of 'future-proofing', which either aim to capture the climatic future in advance, or 'stop the future happening' (Wilkie et al., 2017). The aim has been to explore the potential of scenarios as a way of *rehearsing* climate futures that are uncertain and unknowable. This approach to scenarios as rehearsed improvisations does not promise predictable or disciplined outcomes, nor is it expected or intended to drive all participants towards a shared imagination of the future. Rather than simply performing, anticipating or visualising climate futures, it encourages, among other things, approaching uncertain and diverse knowledges of the future with care, and honing particular qualities of *paying attention to the future in the present.*

Rehearsing climate futures: *What now?*

What now? The context of interlinked crises and multiple possible responses helps to emphasise why the radical uncertainties inherent in climate change futures make civic improvisation, rehearsal and experiment all the more important. The Covid-19 pandemic has been posited as a rehearsal for the far greater challenges, not only of impending climate and ecological catastrophe, but also the far-reaching disruptions accompanying system transformation (Latour, 2020). Indeed, 'rehearsal' invites a sense of the ongoing testing and experimentation that will be required for future ways of inhabiting the Earth (Tyszczuk, 2022). The rehearsal spaces of climate change deliberation and action can take many forms and span more formal decision-making and radically open participatory forms of 'democratic conversation', including collaborative scenario making (Rickards et al., 2014; Cairns et al., 2013), policy-relevant 'storylistening' (Dillon and Craig,

2021), and citizens' assemblies that foster societal resilience (e.g. Climate Assembly UK, 2020). These pre-figurative political practices seek to work out a future in conversations, through consensus or dissent, but they nevertheless tend to be (more, or less) directed and curated.

The *Collective Scenarios* project has drawn on insights from a series of experimental projects in both the design studio and in public engagement on climate change, where a scenario mode, conceptually understood as 'rehearsal', was tested (Smith et al., 2017; Tyszczuk & Smith, 2019; Tyszczuk & Udall, 2018; Tyszczuk, 2019a; 2021). The aim was to explore improvisational modes of scenario making to engage with the possibilities and limits of civic interventions and social imaginaries with respect to climate futures, at the same time avoiding the logics of pre-emption or persuasion. The interest has been in opening up the possibilities for experience, but where the outcomes are unpredictable, involving a kind of speculative practice around climate change that is 'crucial to political agency, democratic pluralism and innovation' and where 'political agency is only disclosed *collectively*' (Diprose, 2017: 41; 44, emphasis in original). The collaborative scenario workshops in the various projects involved improvised experiments into rehearsing climate futures, which at the same time cultivated more speculative modes of inquiry. These activities, together understood as *speculative improvisations*, worked across different media and included architectural propositions and urban projects, sound installation, interactive documentary, and immersive theatre (Tyszczuk, 2021; 2022). The stories, projects, performances, installations, mappings and prototypes generated in collective scenario-making allowed for transformative, social, intellectual, adaptive and affective engagements with alternative futures in relation to climate change. Through engaging with disrupted climates, errant materialities, climate and energy justice, future generations and intractable and divisive climate politics, they also drew attention to the impossibility of realising either a stable climate or a perfectly worked-out society.

Exploring scenarios practice as 'rehearsal' was not about creating a guide or method for rehearsing the future, in the sense of securing the 'best possible future' (Adams et al., 2009), nor about indicating the relevant people, practices, materials and tools needed for ingeniously devised, rescued or restored future worlds. Instead, the scenario making was considered to be both speculative and transformative and, 'like prototyping, a way of working out what to do next' (Smith & Tyszczuk, 2018). It was about testing how a more improvised and speculative undertaking might challenge preconceptions, accommodate differences, and at the same time respond to widely diverging experiences of climate change. An engagement with scenarios *as* rehearsal cautions against the scripting or counter-scripting operations of speculative design, scenario-making and prototyping, which are driven by experts, designers or researchers, generating scenarios that corroborate expectations, build consensus or involve pre-emptive curatorial and authorial processes. Instead, scenarios are considered as an anticipatory infrastructure that can make room for working with a multiplicity of (sometimes contradictory and often unexpected) perspectives, generating the possibility for multiple *improvised what ifs* – and thus accommodating cultural shifts and futures that are as yet unknown and uncertain.

Speculative improvisations: *What works?*

What works? Or, how does it work? Reconceptualising scenarios as *speculative improvisations* (Tyszczuk, 2021) redirects the focus from the narrative, visual or descriptive content of scenarios and instead draws on the dramaturgical dimensions or the *activity* of speculative, collaborative and improvised storytelling inherent in scenarios. This challenges current scenario-thinking within a climate problem-solutions frame, positing instead a more collective and inclusive improvisation of *'what ifs'*, relevant to the reconsideration and reconfiguration of societal futures in the context of climate change. This mode of speculation is, 'related to a way of thinking which challenges business-as-usual explanatory frameworks... not a contemplative activity but one of the look-out, of resisting reassuring appearances' (Savransky & Stengers, 2018).

To speculate is to establish a rapport with futures yet to come, whereby thinking in the present and futures become responsive to each other (Savransky, 2017). It is this *rapport* with possible futures in the here and now that is revealed in *speculative improvisations* – an entangling of speculation and improvisation inherent in conditions of rehearsal. This approach to scenarios draws on ideas of improvisation that go beyond any association with ad hoc spontaneity, adaptive capacities or responsiveness in performance, and explores the relation of the improvisatory to the speculative, to sense-making and to *paying attention* in conditions of precarity and uncertainty. Improvisation is understood as the shared practice of actors or improvisers rather than a director's instruction. It requires paying attention to others – and thus a sensitivity and responsiveness to the situation, as well as the possibility of endless reconfigurations (Ingold & Hallam, 2007). Improvisation involves a collective, social and generative process, or 'collaborative emergence', that relies on shared structures and experience as well as careful listening (Sawyer, 2000). It is an intuitive form of social practice that resists commodification (Svendsen, 2017).

Scenarios, when considered as making room for anticipatory, improvised rehearsal of climate futures, invite *paying attention* to the habitable thresholds of a dynamic and altered Earth. The broader resonances of *faire attention*, 'to pay attention' – in Stenger's original French – are important in this regard, 'where "*faire*" means "to make" or "to do,"' and *faites attention* means something like "look out" or "be careful"' (Stengers, 2015: 62, n1.) This approach to rehearsing or improvising climate futures invites both a *making with*, and a *looking out for*, the world. It leaves room for the troubled, unscripted, surprising, marginalised, unanticipated, contested and provisional. Speculative improvisations take place in the diegetic and mimetic space of civic rehearsal, with full acknowledgement however, that this is neither a space nor a spatial practice that offers solutions and reassurances about the future. Instead, rehearsal is understood as a situated noticing and as a way of practising the skills, capacities, responses and relations that a range of conflicting possibilities and alternative futures might require in times of urgency and precarity. In other words, this is about *improvising for the unforeseen*.

Care of the possible: *What matters?*

What matters? Surely, it is about working out *what matters* in times of crisis. Stengers understands 'paying attention' as an art: 'If there is an art, and not just a capacity, this is because it is a matter of learning and cultivating, that is to say, making ourselves pay attention [*faire attention*]' (2015: 62). This aligns with an understanding of improvisation as 'the exercise of care, judgment and dexterity' [...] where the practitioner has continually to make adjustment to keep on course, in response to a sensitive monitoring of the conditions of the task as it unfolds' (Ingold and Hallam, 2007: 13). Scenario-making in modes of improvisation can thus be attentive to 'matters of care' (Puig de la Bellacasa, 2017), where care is understood as a relational and affective activity with the capacities to create new worlds. Moreover, paying attention implies risk-taking and experimentation and thinking through the consequences of actions, or, 'care of the possible' (Stengers, 2011b). Care is understood here as a profound engagement with the world, simultaneously 'a vital affective state, an ethical obligation and a practical labor' (Puig de la Bellacasa, 2012: 197; 2010). Puig de la Bellacasa modifies Fisher and Tronto's widely cited definition of care (1990), such that care is, 'everything that is done (rather than everything that "we" do) to maintain, continue, and repair "the world" so that all (rather than "we") can live in it as well as possible' (2017: 5; 161).

The troubled world is where human, and more than human and non-human distributed agencies (future generations, other lifeforms, materials, surroundings, technologies and infrastructures) are implicated in a web of interdependencies, a 'thick mesh of relational obligation' (Puig de la Bellacasa, 2017: 20). 'Care for' is thus inevitably a contested and compromised practice, grounded in all of the mundane and 'inescapable troubles of interdependent existences', with no guarantee of a 'smooth harmonious world' (Puig de la Bellacasa, 2012: 197–199). Moreover, 'matters of care' are more extensively binding than 'matters of concern' (Latour, 2004) in so far as they require active involvement in the changing conditions of planetary co-existence. Care, 'makes of ethics a hands-on, ongoing process of recreation of "as well as possible" relations and therefore one that requires a speculative opening about what a possible involves' (Puig de la Bellacasa, 2017: 6). This invites a re-imagination of architecture as both a speculative and provisional practice that actively explores the terms of its own production and practice, that is, with what *makes things possible*, and an infrastructure for 'care of the possible'.

Intra-active climate citizenship: *What for?*

What for? Or, what's it all about? Scenarios, when re-configured as rehearsed improvisation are about *paying attention* to the creative, adaptive and affective aspects of practising alternative futures with others. This has been approached as a way of 'staying with the trouble', which, 'requires learning to be truly present, not as a vanishing pivot between awful or Edenic pasts and apocalyptic or salvific futures, but as mortal critters entwined in myriad unfinished configurations of places, times, matters, meanings' (Haraway, 2016). The dominant use of scenario techniques has tended to focus on envisioning trajectories to and from climate-changed futures or consensus building around options, rather than imagining what it would *be like* and *feel like* to inhabit and negotiate a world in transformation. More than simply visions of future civic life, engaging with scenarios in structured improvisations can encourage the invention and trial of provisional words, actions, objects, practices, sensibilities, moods, gestures, laws, protocols and institutions which, through rehearsal can make imagined worlds both tangible *and* changeable. This creative, shape-shifting dimension of improvisation, and the capacity to imagine and enact mutable practices that are tentative or even at odds with prior experience – *things never said or done before* – can help explore modes of citizenship or *who we might be* as future citizens in a transformed and disrupted world (Svendsen, 2019; Tyszczuk 2019a; 2022). Indeed, rehearsed improvisations can help reveal the kinds of skills and capacities needed to respond to the 'problematic togetherness' of the world (Stengers, 2010), among them, agility, collaborative agency and the exercise of care through relational alteration.

This approach to scenarios as rehearsal draws attention to the possibilities of developing 'an *intra-active citizenship* with respect to questions of individual and collective agency around responses to climate futures in the present' (Tyszczuk, 2022). 'Intra-action' is a way of understanding agency not as an individual property but as a dynamism of relations, emergence and transformation (Barad, 2007: 141; 2010). Working with scenarios as speculative and rehearsed improvisations with respect to climate futures and relations of co-existence opens up modes of 'climate intra-action' and the potential of re-asserting 'climate action as a series of unpredictable, experimental, collaboratively enacted reconfigurations of worldly relationships' (Verlie et al., 2020). An intra-active climate citizenship is thus understood as a material-spatial-temporal practice – a mutually constitutive working out, a working out what next, and a reworking of the relations and unexpected entanglements between Earth citizens, infrastructures, entities, species, societies, phenomena and everything that this entails. It is about unsettling expertise and dissolving boundaries that worked to place environment as a resource 'out there' and recognising the world's iterative and uncertain intra-activity. It is also humbling in so far as it is about citizen-improvisers recognising the limits of human agency in either futile attempts to control the climate or in apprehending what's next.

An architecture of paying attention: *What next?*

What next? Or, where do we go from here? An *architecture of paying attention* indicates both an affectively and ethically engaged practice of intra-active climate citizenship and an infrastructure for *rehearsing the future otherwise*. It is suggested here as a practice and a supportive structure that also works to situate research, teaching and practice – however modest their impacts – as practical acts of care and risk-taking that can draw others into a sense of curiosity and concern for a changing world. 'We know how to pay attention to all sorts of things... And yet we can also say that once it is a matter of what one calls "development" or "growth," the injunction is above all to not pay attention' (Stengers, 2015). To counter this, Stengers calls for a hesitation – an interfering idiocy or a slowing down which acknowledges times of catastrophe and upheaval that interrupt modern ideas of 'the project', of acceleration, progress and knowledge: 'paying attention means slowing down and accepting that intrusive interstices open up even in the midst of an urgency' (Stengers, 2018). The question is how *paying attention* might help to remake not just architecture but established ways of thinking and being in the world: what might it mean to take care, to be curious about, and accountable for provisional worlds in the making? This is also an invitation to re-imagine what architecture, and urban spatial practices and ecologies might yet become, when reconfigured as infrastructures of collaborative citizenship or *infrastructures of care*, for honing and supporting the capacities for inhabiting uncertain, turbulent worlds. Moreover, as Stengers' says, 'We need practitioners who have learned what embracing the messiness of the world demands' (Savransky & Stengers, 2018: 145). This suggests being attentive *and* responsive to what the world throws at us. Architects when considered as climate researchers and climate practitioners – in effect climate improvisers, require being skilled in both a making with, and a looking out for, the world.

An *architecture of paying attention*, both speculative and improvised, is involved with re-working daily practices of living, recalibrating the scope of possible interventions, to unsettling and re-thinking the broader frameworks of value and dwelling and accountability that render current approaches as *the way things* are, rather than asking *how they could be otherwise*. This provisional series of investigations has probed what an *architecture of paying attention,* as both practice and as infrastructure might do. It is an architecture which, *What on Earth?* – engages with conditions of climate crisis; *What if?* – unsettles dominant approaches to future scenarios of climate changed worlds; *What now?* – tests the potential of rehearsing alternative climate futures; *What works?* – improvises across material, social, spatial and temporal dimensions; *What matters?* – considers care in making things possible; *What for?*– explores shifting, relational modes of Earth citizenship; and (not so) finally, *What next*? – attends to climate futures, as a collective endeavour, as a matter of urgency, in trial and error, by taking the time it takes, and not giving up. It is a provisional practice agile enough to cope with the unruly temporalities of climatic disruptions, and does so with care and humility: '[I]t is a matter of learning *from now* on' (Stengers, 2015: 57). It is a matter of *paying attention to the future in the present*.

Acknowledgements

The *Collective Scenarios* project (2019–2022) is supported by a Leverhulme Trust Major Research Fellowship [MRF-2018-015]. The *Culture and Climate Change: Scenarios* project (2014–2019) was funded by the Jerwood Charitable Foundation, The Ashden Trust, The University of Sheffield, The Grantham Centre for Sustainable Futures at the University of Sheffield and The Open University's Open Space Research Centre. The *Stories of Change: exploring energy and community past present and future* project (2014–2018) was supported by the Arts and Humanities Research Council (AHRC) Connected Communities programme [Award Number AH/L008173/1]. I gratefully acknowledge the contributions of all the participants and collaborators in these different projects testing scenarios for rehearsing the future otherwise.

KEY TAKEAWAYS FROM CHAPTER 15

- Architectural scenarios are opportunities for rehearsal rather than forecasting. Using scenarios in this way promotes adaptability and supports architects to experiment and respond to the uncertainties of climate futures without being bound to rigid predictions.
- An architecture of 'paying attention' prioritises careful, ethically engaged practices, designing with awareness of the interconnected impacts of climate change and fostering structures that support care, resilience and proactive citizenship.
- Architects have the capacity to create spaces and frameworks that support intra-active climate citizenship through which collective agency can be exercised. Through collaborative improvisation, architects can contribute to resilient futures.
- Instead of aiming for perfect, stable or fixed solutions, architects should embrace uncertainty and allow space for trial, error and iteration.

Note

1 The *Collective Scenarios* project brings together a series of research, teaching and public engagement initiatives on climate scenarios based at the University of Sheffield, UK, most recently funded by the Leverhulme Trust. The project has explored how speculative, improvisational and reflexive approaches can refresh climate scenarios and allow for more open and energetic engagements in climate futures.

Bibliography

Adams, V., Murphy M. & Clarke A.E. (2009) 'Anticipation: technoscience, life, affect, temporality', *Subjectivity*, 28(1), 246–265.

Amoore, L. (2013) *The Politics of Possibility: Risk and security beyond probability*. Duke University Press.

Anderson, B., Grove, K., Richards, L. & Kearns, M. (2020) 'Slow emergencies: temporality and the racialized biopolitics of emergency governance', *Progress in Human Geography*, 44(4), 621–639.

Anderson B. (2010) 'Preemption, precaution, preparedness: anticipatory action and future geographies', *Progress in Human Geography*, 34, 777–798.

Andersson, J. (2018) *The Future of the World: Futurology, futurists, and the struggle for the post-Cold War imagination*. Oxford University Press.

Atwood, M. (2015) 'It's not climate change – it's everything change'. *Matter*. https://medium.com/matter/it-s-not-climate-change-it-s-everything-change-8fd9aa671804

Barad, K. (2010) 'Quantum entanglements and hauntological relations of inheritance: dis/continuities, spacetime enfoldings, and justice-to-come', *Derrida Today*, 3(2), 240–268.

Barad, K. (2007) *Meeting the Universe Halfway: Quantum physics and the entanglement of matter and meaning*. Duke University Press.

Barnosky, Anthony D. et al. (2011) 'Has the Earth's sixth mass extinction already arrived?', *Nature*, 471, 51–57.

Cairns, G., Ahmed, I., Mullett, J. & Wright, G. (2013) Scenario method and stakeholder engagement: Critical reflections on a climate change scenarios case study. *Technological Forecasting and Social Change*, 80(1), 1–10.

Climate Assembly UK (2020) The path to net zero: Climate Assembly UK Full Report. www.climateassembly.uk/report/read/final-report.pdf

Crutzen, P.J. & Steffen W. (2003) How long have we been in the Anthropocene era? An editorial comment. *Climatic Change*, 61, 251–257.

Dillon, S. & Craig, C. (2021) *Storylistening: Narrative evidence and public reasoning*. Routledge.

Farrell, C., Green, A., Knights, S. & Skeaping, W. (2019) *This is not a Drill: An Extinction rebellion handbook*. Penguin.

Diprose R. (2017) Speculative research, temporality and politics. In Wilkie et al. (Eds.) (2017). *Speculative Research: The lure of possible futures*. Routledge. pp. 39–49.

Ghosn R. et al. (2021) *The Planet after Geoengineering*. Actar Publishers.

Granjou, C., Walker J. & Salazar J.F. (2017) The politics of anticipation: on knowing and governing environmental futures. *Futures*, 92, 5–11.

Haraway, D. (2016) *Staying with the Trouble: Making Kin in the Chthulucene*. Duke University Press.

Hulme, M. (2019) Climate emergency politics is dangerous. *Issues in Science and Technology*, 36(1), 23–25.

Hulme, M. (2017) *Weathered: Cultures of climate*. Sage.

Ingold, T. & Hallam, E. (2007) 'Creativity and Cultural Improvisation: An Introduction,' in *Creativity and Cultural Improvisation*, ed. Hallam and Ingold. Berg, 1–24.

Klein, N. (2014) *This Changes Everything: Capitalism vs. the climate*. Allen Lane.

Latour, B. (2020) 'Is this a dress rehearsal?' *Critical Inquiry*, 26 March 2020, https://critinq.wordpress.com/2020/03/26/is-this-a-dress-rehearsal/
Latour B. (2004) 'Why has critique run out of steam? From matters of fact to matters of concern', *Critical Inquiry*, 30, 225–248.
Petcou, C. and Petrescu, D. (2019) 'R-Urban and its three-year scenario of how to involve everyone in "saving the planet"' in Tyszczuk et al., *Culture and Climate Change: Scenarios*. Shed, pp. 96–98.
Puig de la Bellacasa, M. (2010) Ethical doings in nature cultures. *Ethics, Place and Environment*, 13, 151–169.
Puig de la Bellacasa, M. (2012) 'Nothing comes without its world': Thinking with care. *The Sociological Review*, 60, 197–216.
Puig de la Bellacasa, M. (2017) *Matters of Care Speculative Ethics in More Than Human Worlds*. University of Minnesota Press.
Rickards, L., Ison, R., Fünfgeld, H. & Wiseman, J. (2014) Opening and closing the future: climate change, adaptation, and scenario planning. *Environment and Planning C: Government and Policy*, 32(4): 587–602.
Savransky, M. (2018) 'Isabelle Stengers and the dramatization of philosophy', *SubStance*, 47(1), 3–16.
Savransky, M. (2017) The wager of an unfinished present. In Wilkie et al. (Eds.) *Speculative Research: The lure of possible futures*. Routledge, pp. 25–37.
Savransky, M. & Stengers, I. (2018) 'Relearning the art of paying attention: A conversation'. *Substance*, 47(1), 130–145.
Sawyer, K. (2000). Improvisational Cultures: Collaborative emergence and creativity in improvisation. *Mind Culture and Activity*, 7(3), 180–185.
Smith, J. et al. (2017) 'Gathering around stories: interdisciplinary experiments in support of energy transitions', *Energy Research and Social Science*, 34.
Smith, J. & Tyszczuk, R. (Eds.) (2018). *Energetic. Exploring the Past, Present and Future of Energy*, AHRC Stories of Change project book. Shed.
Stengers, I. (2010) *Cosmopolitics*. Minneapolis: University of Minnesota Press.
Stengers, I. (2011a) *Thinking with Whitehead: A free and wild creation of concepts*, trans. M. Chase. Harvard University Press.
Stengers, I. (2011b) 'The Care of the Possible': Isabelle Stengers interviewed by Erik Bordeleau in *Scapegoat*, Issue 01.
Stengers, I. (2015) *In Catastrophic Times: Resisting the Coming Barbarism*, trans. Andre Goffey. Open Humanities Press.
Stengers in Savransky, M. & Stengers, I. (2018). 'Relearning the art of paying attention: A conversation', *Substance*, 47(1), 130–145.
Svendsen, Z. (2019) Epic not tragic: We know not what we may be. In Tyszczuk et al. (Eds.) (2019) *Culture and Climate Change: Scenarios*. Shed, pp. 49–56.
Svendsen, Z. (2017) The dramaturgy of spontaneity. Improvising the social in theater. In G. Born, E. Lewis & W. Straw (Eds.) *Improvisation and Social Aesthetics*. Duke University Press.
Tyszczuk R. (2011) 'On constructing for the unforeseen'. In Butler R, Margolies, E., Smith, J. & Tyszczuk, R. (Eds.), *Culture and Climate Change: Recordings*. Shed, pp. 23–27.
Tyszczuk R. (2012) 'Future Worlds: to-ing and fro-ing' in Tyszczuk R., Smith J., Clark N. & Butcher M., (Eds.) (2012) *ATLAS: geography architecture and change in an interdependent world*. Black Dog Publishing.
Tyszczuk R. (2014a) 'Architecture of the Anthropocene. the crisis of agency', *Scroope*, 23, *Cambridge Architecture Journal*, July 2014, 66–73.
Tyszczuk, R. (2016) '*Anthropocenophobia*: the stone falls on the city', *Harvard Design Magazine*, 42, 38–42.
Tyszczuk, R. (2018) *Provisional Cities: Cautionary tales for the anthropocene*. Routledge.

Tyszczuk, R. (2019a) 'A Brief History of Scenarios' in Tyszczuk R., Smith J. & Butler, R. *Culture and Climate Change: Scenarios*. Shed, pp. 15– 23.
Tyszczuk, R. (2019b) 'Scenarios of interactive citizenship' in S. M. Figueiredo, S. Krishnamurthy & T. Schroeder, *Architecture and the Smart City* (Routledge), pp. 139–153.
Tyszczuk, R. (2021) 'Collective scenarios: speculative improvisations for the Anthropocene', *Futures*, 134.
Tyszczuk, R. (2022) 'Times of urgencies: Speculative improvisations for the Anthropocene' in K. Facer, J. Siebers and B. Smith (Eds.) *Time as Method. Working with time in qualitative research: case studies, theory and practice*. Routledge, pp. 50– 66.
Tyszczuk, R. & Smith, J. (2019) 'Culture and climate change, experiments and improvisations – an afterword' in G. Feola, H. Geoghegan & A. Arnall, (Eds.) *Climate and Culture: Multidisciplinary perspectives on a warming world*. Cambridge University Press, pp. 309–326.
Tyszczuk R., Smith J. and Butler, R. (Eds.) (2019) *Culture and Climate Change: Scenarios*. Shed.
Tyszczuk, R. & Udall, J. (2018) 'Future Works: Stories of Energy, Industry and Resilience', in I. Bauman, R. Lawrence & D. Petrescu & K. Trogal (Eds.) *Architecture and Resilience: Interdisciplinary dialogues*. Routledge, pp. 32–48.
Verlie, B. and CCR 15 (2020) 'From action to intra-action? Agency, identity and 'goals' in a relational approach to climate change education', *Environmental Education Research*.
Wilkie A., Savransky M. & Rosengarten M. (Eds.) (2017) *Speculative Research: The lure of possible futures*. Routledge.

INDEX

Note: **Bold** page numbers refer to tables and *italic* page numbers refer to figures.

For Product Safety Concerns and Information please contact our EU representative GPSR@taylorandfrancis.com Taylor & Francis Verlag GmbH, Kaufingerstraße 24, 80331 München, Germany

Batch number: 10722104

Printed by Printforce, the Netherlands